W9-CYY-688

INFRARED
FIBER
OPTICS

Edited by

Jasbinder S. Sanghera, Ph.D.
Ishwar D. Aggarwal, Ph.D.

Naval Research Laboratory
Washington, D.C.

INFRARED FIBER OPTICS

CRC Press
Boca Raton Boston London New York Washington, D.C.

Acquiring Editor:	Felicia Shapiro
Project Editor:	Andrea Demby
Marketing Manager:	Jane Stark
Cover design:	Dawn Boyd
PrePress:	Kevin Luong
Manufacturing:	Carol Royal

Library of Congress Cataloging-in-Publication Data

Infrared fiber optics / edited by J.S. Sanghera and Ishwar D. Aggarwal.
    p. cm.
    Includes bibliographical references and index.
    ISBN 0-8493-2489-0 (alk. paper)
    1. Fiber optics. 2. Infrared technology. 3. Electrooptics.
I. Sanghera, J. S. (Jasbinder S.). II. Aggarwal, Ishwar D., 1945–
.

TA1800.I528   1998
621.36′92--dc21
                                                97-39324
                                                   CIP

Introduction

In 1966 Kao and Hockham[1] described a new concept for a transmission medium. They suggested the possibility of information transmission by optical fibers. In 1970, scientists at Corning, Inc.,[2] fabricated silica optical fibers with a loss of 20 dB/km. This relatively low attenuation (at that time) encouraged scientists from around the world that perhaps optical communications could become reality. A concerted effort followed, and by the mid 1980s AT&T[3] and Sumitomo[4] reported losses of 0.157 and 0.154 dB/km at 1.55 μm, respectively, which were very close to the theoretically predicted losses. And so, optical telecommunications was born.

As we approach the end of the millennium, silica-based fiber optics is a mature technology with major impacts in telecommunications, laser power transmission, sensors for medicine, industry, and the military, as well as other optical and electro-optical systems. While silica-based fibers exhibit excellent optical properties out to about 2 μm, other materials are required for transmission to longer wavelengths in the infrared (IR). These materials can be glassy, single crystalline, or polycrystalline. Examples of such materials include fluoride and chalcogenide glasses, single-crystalline sapphire, and polycrystalline silver halides. Some plastic optical fibers are also currently being developed primarily for visible applications and are therefore outside the scope of this book. The figure below shows the transmission spectrum of several bulk materials representative of these different classes of optical materials that can be utilized in fiber form for IR transmission. Depending upon composition, these materials can transmit to beyond 20 μm. Consequently, optical fibers made

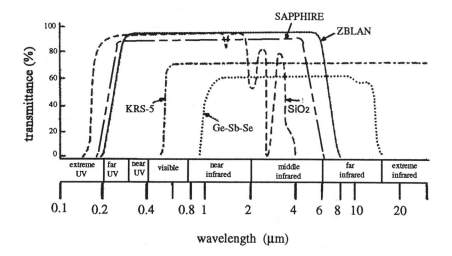

from these materials enable numerous practical applications in the IR. For example, IR transmitting fibers can be used in medical applications such as for laser surgery and in industrial applications such as metal cutting and machining using high power IR laser sources (e.g., Er:YAG, CO, and CO_2 lasers). More recently, there is considerable interest in using IR transmitting fibers in fiber-optic chemical sensor systems for environmental pollution-monitoring using absorption, evanescent, or diffuse reflectance spectroscopy since practically all molecular species possess characteristic vibrational bands in the IR.

Aside from chemical sensors, IR fibers can be used for magnetic field, current and acoustic sensing, thermal pyrometry, IR imaging, IR countermeasures, and laser threat-warning systems. While low-loss silica fibers are highly developed for transmission lines in telecommunication applications, the IR transmitting glasses and fibers because of their low phonon energy are becoming excellent candidates for rare earth doping to produce IR sources that could be made into optical amplifiers and lasers. Actually, fiber amplifiers and lasers are currently being developed at the telecommunication wavelengths of 1.3 and 1.5 μm using these glasses. This may be a future major area of development for these glasses considering that optical communication is a huge industry and that optical amplifiers are playing a key role in this industry.

The materials described above are used to fabricate solid-core fibers; however, there is another class of fibers based on hollow waveguides which has been investigated primarily for CO_2 laser power transmission. These waveguides possess a hollow core and are based on metallic or glassy tubes with or without internal dielectric coatings.

Since IR transmitting fibers have become increasingly important in recent years because of optical communication, sensors, spectroscopy, and numerous other applications, we felt it necessary to review the current status of the different types of IR transmitting fibers being developed around the world. With this in mind, it is most appropriate to give an overview of optical transmission theory for the benefit of those unfamiliar with fiber optics or of those who just need refreshing. Therefore, Chapter 1 describes the fundamentals of fiber optics with particular emphasis on propagation theory in solid-core fibers and hollow waveguides, as well as dispersion. Absorption and scattering mechanisms responsible for attenuation in a fiber-optic material are addressed along with predictions of the loss minimum. Macroscopic bending losses in optical fibers are also described since they play an important role in system design.

The optical, mechanical, and other physical properties of a fiber are highly dependent upon the material properties as well as the techniques used to make the fibers. Consequently, the subsequent chapters describe the IR transmitting fiber fabrication processes and properties of the various types of fibers, including appropriate applications. Chapters 2, 3, and 4 discuss glass-based optical fibers made from silica, fluoride, and chalcogenide based–glasses, respectively. Silica-based fibers are prevalent today, primarily in telecommunications. Chapter 2 introduces the reader to telecommunication fiber design and specifications. The chapter also focuses on the composition, fabrication, and performance of telecommunication fibers. More recently, the reliability and longevity of fiber systems have become major issues, and this is also addressed. This chapter also reviews several classes of specialty silica fibers, especially highlighting the erbium-doped fiber-optical amplifiers at

1.5 μm, fibers containing UV-written gratings, as well as the new Raman amplifiers for 1.3 μm telecommunications.

Chapter 3 deals with heavy metal fluoride–based fibers such as the fluorozirconates. These glasses were discovered quite by chance in the mid 1970s. Their uniqueness lies in the fact that they possess lower theoretical losses than silica, provide transmission to longer wavelengths, and can accommodate large quantities of rare earth dopants. This chapter discusses the different glass compositions, their fabrication processes, and their general properties. Owing to their relatively lower glass stability compared with silicate glasses, novel fiberization techniques have been developed to inhibit crystallization. These, along with the current status of both the optical and mechanical properties of the fibers, are reviewed. Some applications utilizing heavy metal fluoride glass fibers are also presented. These include both low-loss and medium-loss applications. Examples of the former include telecommunications, while the latter consist of chemical sensors, IR laser power delivery, and, perhaps the most significant of all applications, fluoride fiber–based lasers and amplifiers.

Chapter 4 deals with chalcogenide-based glass fibers. These are based on the chalcogen elements sulfur, selenium, and tellurium, and the addition of other elements leads to stable glass formation. Chalcogenide fibers are technologically important, since, depending upon composition, they transmit between approximately 1 and 12 μm. This chapter gives examples of the various glass-forming compositions, their synthesis, and purification routes, as well as the techniques used to prepare fibers from these materials. Since these glasses possess high vapor pressures at elevated temperatures, they require different fabrication and fiber-drawing procedures compared with silica and fluoride glasses. The theoretically predicted attenuation is discussed, and experimentally measured losses are presented.

Chapters 5 and 6 discuss crystalline fibers made from single-crystal and polycrystalline materials, respectively. Unlike the glass fibers discussed in Chapters 2, 3, and 4, these fibers require alternative fabrication procedures. Chapter 5 describes the fabrication and properties of single-crystal sapphire fibers, since these have received the most attention to date. These fibers possess excellent thermal and mechanical properties. Although these fibers do not transmit much beyond 3 μm in the IR, they are suitable for numerous applications where temperature stability becomes important, such as chemical sensing in high-temperature environments or delivery of high-power Er:YAG laser energy for medical applications. Chapter 6 discusses fabrication and properties of polycrystalline halide fibers. These fibers transmit to about 15 μm and therefore are ideal for chemical sensor applications such as detection of chlorinated hydrocarbons whose main absorption bands lie beyond 10 μm. Furthermore, they have been used for delivery of high-power CO_2 laser energy for medical applications.

Chapter 7 describes the properties of hollow waveguide fibers made from metal, glass, and crystalline components. While metal tubes have been around for a long time, recent developments have led to some significant improvements in transmission and attenuation. These types of fibers possess no solid end face, and so reflection losses are not a problem. Consequently, these types of fibers are well suited for high-laser-power delivery, especially utilizing Er:YAG, CO, and CO_2 lasers for a plethora of applications.

Chapter 8 describes some state-of-the-art commercial chemical-sensing applications using IR transmitting fibers, especially chalcogenide fibers since these are not only chemically and mechanically durable, but possess optical losses low enough for commercial use. Commercially available reflectance and evanescent spectroscopic probes are described, along with examples of applications using these probes in the field.

In summary, this book deals with most optical materials that transmit in the near, middle, and far IR wavelength region and can be fabricated into optical fibers. It contains a comprehensive description of all pertinent materials, such as silica, fluoride, and chalcogenide glasses, single and polycrystalline materials, as well as glass and metal hollow waveguides. The fabrication techniques for bulk materials and fibers, their properties and representative applications are also described. It is hoped that this book will be very useful for researchers working in the field of IR fiber optics and for those who are considering entering into it.

Finally, we would like to thank all the authors for their excellent contributions and their patience during the production of this book.

REFERENCES

1. C. K . Kao and G. A. Hockham, *Proc. IEE,* 133, 1158 (1966).
2. F. P. Kapron, D. B. Keck, and R. D. Maurer, *Appl. Phys. Lett.,* 17, 423 (1970).
3. R. Csencsits, P. J. Lemaire, W. A. Reed, D. S. Shenk, and K. L. Walker, *OFC'84,* TU13, 54 (1984).
4. H. Yokota, H. Kanamori, Y. Ishiguro, G. Tanaka, S. Tanaka, H. Takada, M. Watanabe, S. Suzuki, K. Yano, M. Hoshikawa, and H. Shimba, *OFC'86,* PD3-1, 11 (1986).

The Editors

Jasbinder S. Sanghera, Ph.D., received his doctorate in Materials Science from the Imperial College of Science, Technology, and Medicine, University of London in 1985. His thesis topic was based on the research, development, and characterization of semiconducting oxide glasses and glass–ceramics. Following this he was a postdoctoral scholar at the University of California, Los Angeles, where he investigated the fabrication and structural, physical, mechanical, and optical properties of infrared transmitting glasses such as oxide, halide, chalcogenide, and chalcohalide glasses. During this time, he also assisted in editing the *Journal of Non-Crystalline Solids*.

In 1988, he went to work at the Naval Research Laboratory (NRL) in Washington, D.C. His initial work was based on the research and development of low-loss and stable fluoride glasses for making ultra low-loss fibers and windows. Over the last five years, his work has involved the research and development of infrared transmitting glasses and fibers for the 1 to 12 μm region, with particular emphasis on chalcogenide, chalcohalide, and heavy metal oxide glasses for numerous Department of Defense applications. More recently, his interests have expanded to include rare earth doping for active applications such as sources and amplifiers in the infrared. He is currently Section Head of Infrared Materials in the Optical Sciences Division at NRL.

Dr. Sanghera has published over 10 papers in various technical journals, published a chapter in a book, been awarded several patents, and presented over 50 papers in conferences and technical meetings. Furthermore, he is a member of several technical societies and associations including the Materials Research Society, American Ceramic Society, and the International Society of Optical Engineers.

Ishwar D. Aggarwal, Ph.D., received his doctoral degree in Materials Science from the Catholic University of America, Washington, D.C., in 1974. He subsequently worked at Corning, Inc., as Senior Materials Scientist, investigating new glass compositions for optical fibers. He then joined Galileo Electro Optics Corporation, Sturbridge, MA. As Associate Director for Research and Development, he worked on communication fibers and coherent fiber optics for medical and other applications. In 1978, he started working at Valtec Corporation (subsidiary of U.S. Philips Corporation). As Vice President for Research, Development, and Engineering at Valtec, he was responsible for developmental work on fibers, cables, connectors, and installation of fiber-optic cables. Dr. Aggarwal also worked at Lasertron, Inc., in Burlington, MA as Vice President, responsible for manufacturing and quality engineering. At Lasertron, he directed engineering work on semiconductor lasers, detectors, transmitters, and receivers for communications applications.

In 1986, Dr. Aggarwal joined the Optical Sciences Division at the Naval Research Laboratory. He is currently head of the Infrared Materials and Fiber Optic Chemical Sensors Section. He is responsible for directing research in infrared glasses and fibers for various defense applications and for chemical sensors for hazardous and nuclear waste detection, as well as for other industrial and defense applications.

Dr. Aggarwal has published over 75 papers in various technical journals, edited a book on fiber optics, and presented over 50 papers in conferences and technical meetings. He is a member of several technical societies and associations, including the Optical Society of America and the American Ceramic Society.

Contributors

Ishwar D. Aggarwal
Naval Research Laboratory
Washington, D.C.

Allan J. Bruce
Lucent Technologies
Murray Hill, New Jersey

Lynda E. Busse
Naval Research Laboratory
Washington, D.C.

Leonid Butvina
General Physics Institute
Russian Academy fo Science
Moscow, Russia

Robert S. F. Chang
Department of Physics
University of South Florida
Tampa, Florida

Nicholas Djeu
Department of Physics
University of South Florida
Tampa, Florida

Mark A. Druy
Sensiv, Inc.
Waltham, Massachusetts

Christopher Gregory
Sarnoff Corporation
Princeton, New Jersey

Paul Klocek
Raytheon TI Systems
Dallas, Texas

Malcolm E. Lines
Lucent Technologies
Murray Hill, New Jersey

Junji Nishii
Osaka National Research Institute
Optical Materials Division
Osaka, Japan

Charles F. Rapp
Owens Corning Fiberglass
Granville, Ohio

Jasbinder S. Sanghera
Naval Research Laboratory
Washington, D.C.

Toshiharu Yamashita
Non-Oxide Glass Company Ltd.
Hoya Corporation
Tokyo, Japan

Dedication

To my wife, Kulvinder, for her understanding, patience, and support for this book. To all my children, Jas, Steven, Sean, and Symren, for just being themselves.

J.S.S.

To my wife, Shail, and my children, Puneet and Anjali, for their support and understanding.

I.D.A.

Table of Contents

1 Optical Transmission Theory

Malcolm E. Lines and Paul Klocek

CONTENTS

1.1 FUNDAMENTALS OF FIBER OPTICS

1.1.1 PROPAGATION

1.1.1.1 Reflection and Refraction

In optics, and particularly infrared (IR) optics, we are dealing with electromagnetic radiation with wavelength λ very much larger than atomic spacings d. As a consequence, radiation propagating in solids can be treated classically, described in terms of macroscopic fields that are averages of rapidly varying atomic fields over volumes V for which $d^3 \ll V \ll \lambda^3$. The electric and magnetic fields $\mathbf{E}$ and $\mathbf{H}$ so defined then not only determine the nature of the propagating radiation, but also act on the medium as stimuli to induce macroscopic responses like polarization $\mathbf{P}$, magnetization $\mathbf{M}$, and current density $\mathbf{J}$ in the forms:

$$\mathbf{P} = \varepsilon_0(\varepsilon - 1) \cdot \mathbf{E} = \varepsilon_0 \boldsymbol{\chi} \cdot \mathbf{E}$$

$$\mathbf{M} = \mu_0(\mu - 1) \cdot \mathbf{H} = \mu_0 \boldsymbol{\chi}_m \cdot \mathbf{H} \tag{1.1}$$

$$\mathbf{J} = \boldsymbol{\sigma} \cdot \mathbf{E}$$

defining such material compliances as dielectric constant ε, polarizability χ, permeability μ, magnetic polarizability χ_m, and conductivity σ (where ε_0 and μ_0 are, respectively, the permittivity and permeability of empty space). Such compliances are in general tensor quantities,[1] but reduce to scalars for macroscopically isotropic materials.

In vacuo the velocity of "light" is $c = (\varepsilon_0\mu_0)^{-1/2} \approx 3 \times 10^8$ ms^{-1} and is frequency independent. On entering a transparent dielectric medium, the light frequency f is unchanged, but the light velocity $v = c/(\varepsilon\mu)^{1/2}$ and wavelength $\lambda = f/v$ are changed. The ratio c/v, which is a frequency-dependent characteristic of the material, then defines the "refractive index" n. It follows that $n = \sqrt{\varepsilon}$ for nonmagnetic materials ($\mu \approx 1$) which (with few exceptions) dominate the field of fiber optics.

Classically, light waves display a transverse electromagnetic wave motion involving oscillating electric and magnetic fields perpendicular both to each other and to the direction in which the wave travels. Light radiating from a point is representable by a train of spherical wave fronts. However, when the light wavelength λ is much smaller than the object it encounters, a plane wave representation often suffices. This is the "ray," or geometric optical, representation with the ray drawn in the direction of energy flow which is perpendicular to the wave front.

Mathematically, a light plane wave traveling in direction z with its $\mathbf{E}$ vector parallel to direction x through a homogeneous isotropic transparent medium can be expressed as

$$E_x(z,t) = E_{ox} \exp[i(\omega t - kz)]$$

$$H_y(z,t) = H_{oy} \exp[i(\omega t - kz)], \tag{1.2}$$

where $\omega = 2\pi f$, $k = 2\pi/\lambda$, and $H_{0y} = (\varepsilon\varepsilon_0/\mu_0)^{1/2} E_{0x}$. The frequency ω controls the time dependence of the wave and the material wave vector $\mathbf{k}$ the position dependence. The wave is said to be linearly or plane polarized with $E \parallel x$, see Figure 1.1. A more general state of polarization can be defined by combining with it an orthogonally polarized wave ($E \parallel y$) of different phase, e.g., $\propto \exp[i(\omega t - kz + \delta)]$. For general phase δ the resultant ray is elliptically polarized with $\mathbf{E}$ vector (and $\mathbf{H} \perp \mathbf{E}$) rotating and changing in magnitude as the wave progresses. In particular, for $\delta = \pm \pi/2$ we have circularly polarized light with clockwise or anticlockwise senses of rotation. However, light coming from ordinary sources is normally unpolarized, the $\mathbf{E}$ vector changing direction rapidly and randomly although remaining perpendicular to the direction of propagation.

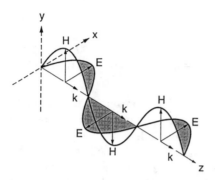

FIGURE 1.1 The plane-polarized light wave of Equation 1.2.

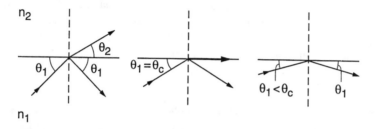

FIGURE 1.2 Representation of the critical angle and total internal reflection at an $n_1 > n_2$ interface.

The concepts of reflection and refraction at a boundary are most simply conveyed in terms of plane waves traveling in isotropic dielectrics. At the boundary, reflected light and transmitted light remain in the plane of incidence but suffer changes of direction (Figure 1.2). The laws governing the behavior are derived from Maxwell's equations for electromagnetic radiation.[2] If the media (n_1 and n_2) are transparent, then the transmitted, or refracted, ray is bent according to Snell's law

$$n_1 \cos\theta_1 = n_2 \cos\theta_2. \tag{1.3}$$

As the angle of incidence θ_1 in the optically denser material ($n_1 > n_2$) becomes smaller, the refracted angle θ_2 decreases and goes to zero when $\theta_1 = \theta_c = \cos^{-1}(n_2/n_1)$. When $\theta_1 < \theta_c$, no refraction is possible and total internal reflection results. Light intensity ratios in the two media are in general a function of light polarization conditions and are governed by Fresnel's reflection and refraction formulae.[2] In particular, unpolarized light can become partially (or even totally) plane polarized upon reflection. In the regime of total internal reflection, light also experiences an angle-dependent phase change upon reflection. The phase shifts of wave components parallel and perpendicular to the plane of incidence are then, respectively,

$$\delta_{\parallel} = 2\ \tan^{-1}\left[n\left(n^2 \cos^2 \theta_1 - 1\right)^{1/2} \Big/ \sin \theta_1\right]$$

$$\delta_{\perp} = 2\ \tan^{-1}\left[\left(n^2 \cos^2 \theta_1 - 1\right)^{1/2} \Big/ n \sin \theta_1\right],$$

(1.4)

where $n = n_1/n_2$, values ranging from zero at $\theta_1 = \theta_c$, to 180° at grazing incidence $\theta_1 \rightarrow 0$. This phase change is important in understanding mode selection and propagation in dielectric waveguides.

When a medium has a nonzero conductivity, its refractive index can be considered complex: $\nu = n - iK$, where K is called the extinction coefficient and is related to conductivity by $2nK\varepsilon_0\omega = \sigma$. This is a convenient way of including both ε and σ in the reflection/refraction formalism.[2] Reflectivity as a function of angle of incidence from a transparent ($\sigma = 0$) to an absorbing ($\sigma \neq 0$) medium is qualitatively similar to that involving two transparent media, except that the magnitude of the reflectivity is generally higher though never any larger than unity. For the $K \gg 1$ case of reflection from a metal, the reflectivity is large at all angles, even for normal incidence, but can reach totality only in the theoretical limit of infinite conductivity.

1.1.1.2 Dielectric Fibers

A dielectric fiber is a cylindrical waveguide which, via internal reflection, guides radiation along its length. The simplest configuration is that of a circular cylinder (radius a_1, refractive index n_1), called the core, surrounded by a cladding of radius $a_2 > a_1$ and index $n_2 < n_1$, both materials being transparent dielectrics. The cladding serves many purposes, including reducing losses at the core–clad interface, adding to fiber strength, and protecting the core from contaminants. In addition, fibers are usually protected further by an outer jacket, often of plastic composition.

The details of electromagnetic propagation in this environment is sought through a solution of Maxwell's equations subject to the relevant boundary conditions at the interface.[3-6] Although this "modal" method of solution is essential for quantitative study, the ray-tracing approximation can aid in a pictorial representation (Figure 1.3), actually becoming fairly quantitative when a_1/λ is large. In general, it is found that a fiber can support a number of "trapped" modes, each defined by its electromagnetic (**E** and **H**) field pattern normal to the fiber axis z. To obtain maximum transmission bandwidth through such fibers, precisely controlled graded-index profile cores are usually prepared. On the other hand, the easiest situation to envisage and to describe in detail is that of a "step-index" fiber for which n_1 is constant for radius $r < a_1$ and constant again at a lower value $n_2 = n_1(1 - \Delta)$ for $r > a_1$. In practice, the relative index difference $\Delta = (n_1 - n_2)/n_1$ is a few percent, at most.

Ray transmission for a step-index fiber is depicted in Figure 1.3a, where, for simplicity, we confine our attention to "meridional rays" containing the core axis. Skew rays, not confined to a single plane, can also propagate (in a helical-like path) but will not be discussed here. Using a ray picture in combination with Snell's law, we see that the fiber can trap and channel rays only for a limited solid angle of incidence defined by the angle θ_0 for which

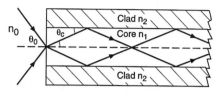

(a) Multimode Step-Index Fiber

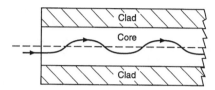

(b) Graded-Index Fiber

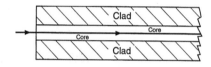

(c) Single-Mode Fiber

FIGURE 1.3 Schematic of light-ray propagation in step-index (a), graded-index (b), and single-mode (c) optic fibers.

$$n_0 \sin \theta_0 = n_1 \sin \theta_c = \left(n_1^2 - n_2^2\right)^{1/2} = n_1(2\Delta)^{1/2}, \tag{1.5}$$

where $n_0 \approx 1$ is the index of air. The quantity $n_1(2\Delta)^{1/2}$ is called the numerical aperture (NA) of the fiber.

At first sight the ray theory appears to allow any ray with $\theta < \theta_c$ to propagate along the fiber. However, when the phase changes under reflection of Equation 1.4 are taken into account, it can be shown that only a discrete number of modes remain. This number, in a step-index fiber, is controlled by a "V-number" parameter

$$V = \left(2\pi a/\lambda\right)n_1(2\Delta)^{1/2}, \tag{1.6}$$

in which λ is now, and henceforth, the vacuum wavelength. They are, in general, hybrid modes (designated HE_{pq} and EH_{pq}) with both H_z and E_z nonzero, where z is the fiber axis, although some modes do have $E_z = 0$ (TE modes) or $H_z = 0$ (TM modes). The lowest-order mode (traditionally designated HE_{11}, see Figure 1.4) can propagate for any nonzero value of V and, for $V < 2.405$, is the only guided mode. For larger V values, other "higher-order" modes cut-on sequentially until, for large V, the total number of guided modes (including polarization) approaches $V^2/2$.

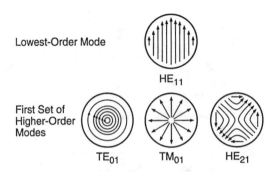

FIGURE 1.4 Cross-sectional projections of the transverse electric field vectors for the four lowest-order modes in a step-index fiber.

The order of the mode correlates not only with V, but also with the degree of complexity of the transverse EH-field pattern (Figure 1.4) and with the incidence angle θ of mode reflection, the lowest-order mode propagating most closely parallel to the axis (Figure 1.3c, $\theta \approx 0$, although a ray picture is inadequate for this case). It is also important to recognize that the guided mode fields are not completely confined to the core but extend (with exponentially decreasing amplitude) into the cladding. As a rule, the higher-order modes penetrate farthest into the cladding, but, for each mode, this degree of penetration is a function of V, increasing as V decreases toward cutoff.

One of the more important effects of this clad penetration arises in the propagation around fiber bends. Deviations from an ideal right-cylindrical core geometry can never be completely avoided in practice. Although all such imperfections induce energy losses of one form or another, macroscopic bending can become a very significant source of loss if the radius of curvature decreases below a certain threshold that is a function of mode penetration into the clad. Since any bound core mode has an evanescent tail in the cladding, this tail moves along the fiber with the field in the core. When a fiber is bent, the field tail on the outside of the bend therefore has to move faster than the energy in the core until, near a critical bend radius, a significant fraction of the tail energy is required to move faster than the speed of light in order to keep up. As this is not possible, the field tail energy involved radiates away. Since the higher-order modes in general penetrate farthest into the clad, it is they that radiate out of the fiber first. The fraction of modes lost in this fashion for a multimode step-index fiber has been calculated[7] to be

$$f = \left(1/2\Delta\right)\left[\left(2a/R\right)+\left(3/2n_2kR\right)^{2/3}\right], \tag{1.7}$$

where R is the bending radius and $k = 2\pi/\lambda$.

Radiation bending losses of a different character can be induced by small-scale bending fluctuations caused by nonuniformities arising in manufacture and particularly in cabling. These "microbends" often have rms (root-mean-square) amplitudes of only a few nanometers (and a mean repeat distance of a few hundred micrometers), but induce repetitive coupling in energy between modes and, particularly, between guided and nonguided modes, the latter inducing energy losses from the fiber. In general, both types of bending loss sources are a function of the degree of confinement of modes to the core and are therefore smallest for single-mode operation and, in this case, for operation close to the cutoff condition of $V \approx 2.4$, for which mode confinement to the core is maximum.

1.1.1.3   Hollow Fibers

By far the most technologically advanced fiber transmission systems to date involve fibers with cores and clads based on cationic dopings of fused silica. Although not a perfectly transparent dielectric, the absorption loss for silica is extremely small in the near IR (out to $\lambda \approx 1.6$ μm) and total attenuations ~ 1 dB/km are not uncommon in communications applications with fibers of this kind. At longer wavelengths, however, silica becomes highly absorbing. Although, in theory, other dielectrics (such as chalcogenide and heavy halide glasses) can maintain low attenuation out to $\lambda \approx 10$ μm, technical problems concerning purification, mechanical viability, and glass stability have, as yet, severely limited their application.

Difficulties of this kind have led to an interest in the study of wave propagation in hollow waveguides for longer-wavelength radiation transmission.[8,9] Using a core of air, efficient propagation of radiation can be achieved if a sufficient degree of internal reflection can be induced at the (air) core to clad boundary. Two types of clad suggest themselves in this context, namely, metals and dielectrics for which anomalous dispersion makes $n < 1$ at the wavelength of operation. The simplest theoretical analysis results for the academic limit of a perfectly conducting clad ($n = K \rightarrow \infty$). in this case, regardless of the cross-sectional shape of the hollow guide, all modes can be classified as TM or TE, no penetration of the clad can exist, and the boundary condition is almost trivially simple (specifically, that the tangential component of $\mathbf{E}$ should vanish at the core/clad wall). Total internal reflection takes place at all incident angles, making the NA equal to unity, and there are no unguided modes. It follows that the resulting guide is lossless. Unfortunately, no real metal comes close to this ideal.

The simplest description of hollow-fiber wave propagation for a real metal clad is for a rectangular core of, say, cross section $a \times b$. Ray or modal methods of solution show that the allowed modes are TE_{pq} and TM_{pq} with $\mathbf{E}$ and $\mathbf{H}$ vectors, respectively, perpendicular to the guide axis (Figure 1.5a). The integers p and q for this case denote the number of half wavelengths that the mode has in its a axis and b axis cross-sectional field pattern, respectively. Each member of each family of TE and TM modes can propagate if its frequency exceeds its particular cutoff value, larger m and n tending toward higher cutoff frequencies. However, unlike the case for the dielectric fiber of the last section, no mode can propagate below the lowest cutoff frequency.[10]

The lowest-frequency mode is the TE_{10} or TE_{01} mode depending upon whether $b > a$ or $a > b$. The TE_{10} mode for $b \gg a$ is shown in Figure 1.5b. Losses are incurred, for smooth walls, via the penetration of the mode fields into the lossy metal guide

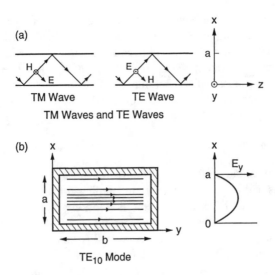

FIGURE 1.5   (a) TM and TE mode propagation in an $a \times b$ ($b \gg a$) rectangular waveguide and (b) The lowest-order (TE$_{10}$) mode and its electric field pattern for this same guide.

walls. The TE$_{p0}$ modes, when $b \gg a$, have the lowest losses (TE$_{10}$ lowest of all) since, for them, **E** must be small (**E** $\rightarrow 0$ as $\sigma \rightarrow \infty$) all along the long wall (see Figure 1.5b) and field penetration is consequently small. These modes travel at an angle $\theta_p = p\lambda/2a$ with respect to the fiber (z) axis.[11] This angle is smallest for TE$_{10}$, which is called the fundamental mode.

The analogous problem for hollow metal guides with cylindrical geometry has also been analyzed. The lowest-loss mode is TE$_{10}$[12] (see Figure 1.4) for which **E**, being parallel to the walls, is again small at the core/clad boundary. Initially, however, severe problems for cylindrical metal guides were encountered in the form of high bending losses.[12,13] The cause, and its solution, will be discussed later. As a result, however, the focus of attention for cylindrical guides broadened to include hollow dielectric guides for which the clad (usually a glass) is in its anomalous dispersion wavelength regime with $n < 1$.

Most oxide glasses undergo a lattice-vibration-induced (complex) refractive index dispersion near $\lambda = 10$ μm of the qualitative form shown in Figure 1.6.[14,15] With $n_{core} = 1$ and $n_{clad} < 1$ one anticipates that efficient internal reflection at the air/clad interface should be achievable in a manner paralleling that for a conventional dielectric fiber. However, since a significantly nonzero value of K is necessary (via the Kramers–Kronig equations[2]) in order to induce values of $n < 1$, a retention of a nonzero value of K in the theory of mode propagation is now essential. Although the mode symmetries obtained for the allowed propagating modes are essentially still those of their dielectric core counterparts (e.g., Figure 1.4), perfect internal reflection can never be achieved for any angle of core/clad reflection and the guide is necessarily lossy.

All hollow guides are therefore subject to significant energy losses induced by the nonperfect reflectivity of the core/clad interface. These losses are essentially

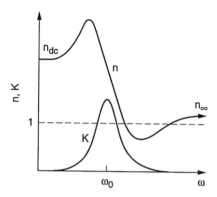

FIGURE 1.6 Schematic of the complex refractive components n and K as functions of angular frequency ω near a strong lattice vibrational absorption centered at $\omega = \omega_0$.

absent in dielectric core fibers for which K is close to zero in the operating frequency range. Dielectric fiber losses (as will be discussed in Section 1.2 below) are primarily of material, rather than waveguide, origin and are consequently (since core and clad are of very similar material composition in these fibers) essentially independent of the dimensions of the core. Hollow guides, on the other hand (as will also be detailed in Section 1.2), have losses that decrease rapidly as a function of increasing core width or radius. Hollow guides therefore normally operate in a grossly multimode regime with transverse core dimensions (≈ 1 mm) being much larger than the normal operating wavelength ($\lambda \lesssim 10\ \mu m$).

Hollow guide modes propagate as TE or TM modes in any guides rectangular cross sections or in guides with circular cross section in the limit of a perfectly conducting clad. In other cases EH and HE hybrid modes can form, the first letter denoting which field is closest to the transverse plane. For hollow cylindrical guides the eigenmodes retain the symmetry restrictions dictated by the circular cross section, but their detailed intensity patterns are dependent on the values of n and K, as well as on the ratio a/λ, where a is core radius.[16] Although operating in a grossly multimode regime, efforts are usually made to excite and propagate only a single low-loss mode, usually one of low order for which the reflection angle (and therefore the number of reflections per unit length) is small. Unfortunately (see Section 1.2.6), guide bending induces a mixing of modes in general and specifically induces high bending loss by coupling this low-loss mode with higher-order, high-loss modes.

1.1.2 DISPERSION

1.1.2.1 Intermodal

An optical signal becomes increasingly distorted as it travels along a fiber or guide as a consequence of two dominant mechanisms, referred to as intermodal and intramodal dispersion, respectively. These distortions can be explained by focusing on the group velocity, or speed, at which energy in a particular mode travels along the fiber length.

Intermodal dispersion is present whenever more than one mode is excited in a guide. It can contribute to distortion both in dielectric and hollow core guides. Since different modes zigzag along the guide (Figures 1.3a and 1.5a) with different reflection angles at the core/clad boundary, those with steeper angles of propagation (typically the higher-order modes) progress more slowly along the guide than those with less steep angles. The energy pulse broadening resulting from this intermodal source is measured as the difference in travel time between the longest (L_1) and shortest (L_2) paths and is simply obtained by ray tracing in the form:[5]

$$\tau_m = n_1 (L_1 - L_2)/c, \tag{1.8}$$

where the core refractive index $n_1 = 1$ for hollow guides. The factor $L_1 - L_2$ can be more formally cast in terms of λ, mode numbers, fiber length, and core and clad indexes n_1 and n_2 via a theoretical calculation of reflection angles for any geometry of interest.

This particular distortion mechanism can in principle be eliminated entirely by single-mode operation, and modern-day communications systems operating over long distances do actually achieve this in silica-based dielectric core fibers by operating below the cutoff $V = 2.4$, for which conditions only the HE_{11} mode can propagate. Single-mode operation is much more difficult to accomplish in hollow fibers because of the need to operate in a grossly multimode wavelength regime (to reduce loss) coupled with the potential for mode interaction discussed above.

1.1.2.2 Intramodal

Even in a single-mode operational configuration, an optical energy pulse is still subject to distortion as it travels along the fiber or guide. This single mode, or intramodal, dispersion occurs because, via its Fourier transform, a single mode pulse is composed of a distribution of propagating waves of wavelength λ centered about a spectral mean value. As the signal propagates along the guide, each spectral component undergoes a time delay t_g per length L of guide given by[5]

$$t_g = (L/c)(d\beta/dk) = -(\lambda^2 L/2\pi c)(d\beta/d\lambda), \tag{1.9}$$

where $k = 2\pi/\lambda$ is the free-space wave vector and β is the propagation constant which measures the component of $n_1 k$ along the guide length. Thus, $\beta < n_1 k$ by definition, and, for a step-index fiber, $n_2 k < \beta < n_1 k$ can be shown to define the range for which modes are guided.

If the spectral width is not too wide, say, $\pm \Delta\lambda/2$ about $\lambda = \lambda_0$, then the time delay difference (or "group" delay) over a guide length L is, using Equation 1.9,

$$\tau_g = \left(\frac{dt_g}{d\lambda}\right)\Delta\lambda = \left(\frac{-L\Delta\lambda}{2\pi c}\right)\left[2\lambda\frac{d\beta}{d\lambda} + \lambda^2\frac{d^2\beta}{d\lambda^2}\right], \tag{1.10}$$

evaluated at $\lambda = \lambda_0$. The factor

$$D = \frac{1}{L}\left(\frac{dt_g}{d\lambda}\right) = \left(\frac{-1}{2\pi c}\right)\left[2\lambda\frac{d\beta}{d\lambda} + \lambda^2\frac{d^2\beta}{d\lambda^2}\right], \qquad (1.11)$$

is then defined as the intramodal dispersion and is usually measured in units of ns/nm·km.

In general, D contains contributions from both the core material itself (involving $dn_1/d\lambda$) and from the guiding aspects of internal reflection (involving $d\theta/d\lambda$). The former is referred to as *material dispersion* and the latter as *waveguide dispersion,* as if the two were independent. In reality, at least for dielectric core guides, the two are intricately related. However, for simplicity, it is often assumed that each can be calculated separately and the results added to give the total intramodal dispersion.

1.1.2.3 Material Dispersion

Material dispersion is calculated by neglecting the zigzagging of rays along the fiber completely. This is accomplished by setting $\beta = 2\pi n_1/\lambda$, assuming the rays to propagate in a straight line path parallel to the fiber length. Substituting into Equation 1.11 now provides a relationship $D = M$ for material dispersion in the form[17]

$$M = -(\lambda/c)(d^2n/d\lambda^2), \qquad (1.12)$$

where we have dropped the "core" subscript from refractive index. Material dispersion, therefore, arises whenever the refractive index n of the core composition is a nonlinear function of λ. This is the case for all dielectrics, but not, of course, for hollow guides for which $n = 1$. It follows that hollow guides possess no material dispersion at any wavelength.

Within their wavelength regimes of highest optical transparency (called the *optic window*), the refractive index is real ($K \approx 0$) and for most dielectrics can be fairly accurately expressed by a two-term Sellmeier formula of the form:[18]

$$n^2 - 1 = E_d E_o/(E_o^2 - \hbar^2\omega^2) - E_\ell^2/\hbar^2\omega^2, \qquad (1.13)$$

in which ω is the frequency of the radiation. The first term on the right-hand side is the contribution from electronic excitations (with E_o denoting a mean electronic energy gap and E_d the corresponding oscillator strength) and the second from lattice vibrations (with E_ℓ being a measure of the lattice oscillator strength). Methods for efficiently deducing these E parameters from experimental refractive index data have been discussed by one of us in the literature (see, for example, Reference 25). Differentiation of Equation 1.13 with respect to $\lambda = 2\pi c/\omega$ and insertion into Equation 1.12 now gives[19]

$$M \approx -15,400(E_d/E_o^3)/n\lambda^3 + 2170E_\ell^2 \lambda/n, \quad \text{ps/nm·km}, \qquad (1.14)$$

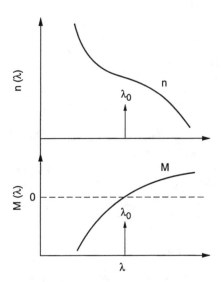

FIGURE 1.7 Refractive index n and material dispersion M as functions of wavelength λ near the zero material dispersion wavelength λ_0.

if λ is in units of μm and E_o, E_d, and E_ℓ are in eV. This equation admits a solution $M = 0$ for the particular wavelength

$$\lambda = \lambda_0 = 1.63\left(E_d/E_o^3 E_\ell^2\right)^{1/4} \quad \mu m, \tag{1.15}$$

called the "zero material dispersion wavelength." Thus, for every dielectric there is a wavelength $\lambda = \lambda_0$ for which the curvature of n as a function of λ, and therefore the material dispersion, goes to zero. This wavelength is often close to, but will not normally coincide with, the wavelength λ_{min} of lowest material attenuation. For example, for pure silica, $\lambda_0 = 1.27$ μm while $\lambda_{min} = 1.55$ μm. The qualitative situation for dielectrics in general is made clear by the schematics shown in Figure 1.7 for n and M as functions of λ.

Using Equation 1.15, together with broadly applicable trends in electronic and phonon oscillator strengths, Nassau[20] has estimated λ_0 for a wide range of metal halides and oxides, as shown in Figure 1.8, spanning a range from $\lambda_0 = 1.2$ μm for BeO to $\lambda_0 \approx 8.5$ μm for TlBr.

1.1.2.4 Waveguide Dispersion

Waveguide dispersion is calculated by assuming a zero value for material dispersion and then investigating the dependence of the reflection angle θ on λ for any particular mode. The angle θ affects the intramodal dispersion D of Equation 1.11 via its influence on the propagation constant β. In general, the angle θ is different for different modes, but, even for a specific mode, θ is still a function of wavelength; this variation produces a λ-dependent time delay τ_{wg} along a fiber length L as λ

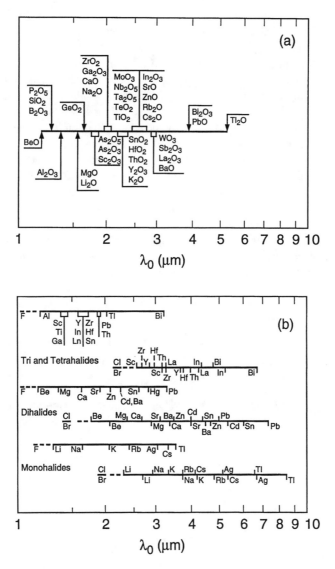

FIGURE 1.8 Zero material dispersion wavelength λ_0 for (a) oxides and (b) halides. (Adapted from Nassau, K., *Bell Syst. Tech. J.*, 60, 327, 1981.)

spans the spectral width $\Delta\lambda$ of the mode resonance. The resulting ratio $D_{wg} = \tau_{wg}/L\Delta\lambda$ is then a measure of waveguide dispersion.

Waveguide dispersion is usually neglected in a multimode context, since it is small compared with intermodal dispersion. Nevertheless, it is often important for single-mode propagation since it is not, in general, small compared with material dispersion M. This is particularly true near a zero material dispersion wavelength (for which $M \to 0$). Waveguide dispersion is naturally a function of waveguide geometry and, in the context of communications fibers, of the shape of the refractive index envelope in the core and clad. It is, however, well documented for the simple step-index cylindrical geometry and is usually plotted as a function of V number for each mode.[21]

For all except the lowest (HE_{11}) mode, which has no cutoff, waveguide dispersion is largest for each mode at frequencies just above cutoff. For the HE_{11} mode, which is the propagating mode of monomode communications fibers, D_{wg} peaks near $V = 1.3$ and is of negative sign, such that, when combined with material dispersion (Figure 1.7), it moves the zero-dispersion wavelength (which is not the wavelength for which the total intramodal dispersion $D_{wg} + M$ goes to zero) to larger values. Long-distance (e.g., transoceanic) communications fibers are designed to make use of this effect and to shift λ_0 from its material value of $\lambda_0 \approx 1.27$ μm to a value much closer to $\lambda_{min} \approx 1.55$ μm. Referred to as "dispersion-shifted" fibers they therefore have minimum pulse distortion and minimum attenuation at the same operational wavelength, specifically, ≈ 1.5 μm.

Waveguide dispersion may also be calculated in a straightforward manner both for rectangular and cylindrical hollow waveguides. However, little attention is generally paid to intramodal dispersion in hollow-core guides because their intrinsically higher losses appear to exclude their use for communication or any other purpose involving transmission over distances long enough for intramodal mode delay limitations to be of concern.

1.2 ATTENUATION OF LIGHT IN FIBER-OPTIC MATERIALS

Attenuation, or power loss, as light travels a distance L through any material, can be measured in terms of the fractional power $[P(0) - P(L)]/P(0)$ lost in transit by writing

$$P(0) - P(L) = P(0)\left[1 - e^{-\alpha L}\right], \tag{1.16}$$

defining a "loss coefficient"

$$\alpha = L^{-1} \ln_e[P(0)/P(L)], \tag{1.17}$$

with the dimensions of reciprocal length (usually cm^{-1}). Frequently, however, and particularly in the context of attenuation in optical fibers, the loss coefficient is quoted in units of dB/km, where

$$\frac{dB}{km} \equiv \frac{10 \log_{10}[P(0)/P(L)]}{L(km)}. \tag{1.18}$$

The conversion factor is $1 \ cm^{-1} = 4.343 \times 10^5$ dB/km.

Let us first consider the sources of attenuation in dielectric core fibers. Loss mechanisms fall into two categories, one involving the absorption of radiation, with

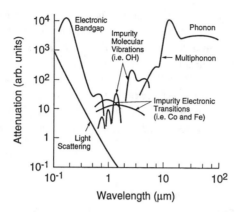

FIGURE 1.9   Attenuation contributions as a function of wavelength for fused silica.

light energy being converted to heat, the other concerned with scattering, in which radiation is diverted (with or without change of frequency) from its primary propagational path. Both of these processes would still be present even in the unattainable ideal limit of propagation through pure and structurally perfect materials (intrinsic losses) although, in more realistic circumstances, each is always embellished by extrinsic contributions from impurities and a variety of structural imperfections. It is, of course, a primary objective in the preparation of efficient dielectric fiber transmission systems to eliminate extrinsic contributions to as high a degree as possible, particularly those that contribute losses close to the planned propagation frequency.

The frequency regime of maximum intrinsic transparency in crystals and glasses is referred to as the *optic window*. It is bounded on the long-wavelength side by absorption from polar modes of lattice vibration (the multiphonon edge) and on the short-wavelength side by absorptions from electronic band gap (valence to conduction band) excitations. Within the window, attenuation is usually dominated by a combination of processes, including intrinsic scattering from frozen or propagating structural modes and extrinsic absorptions from the electronic transitions of impurity transition metal or rare earth ions and from the vibrational fundamentals and overtones of impurity molecular complexes. A schematic of the optic window regime for fused silica is shown in Figure 1.9.

1.2.1   ABSORPTION

1.2.1.1   The Multiphonon Edge

Absorption in the far IR is dominated by the one-phonon lattice vibrational bands (the Reststrahl). However, in addition to these bands, other features are present that can be explained by processes in which a photon is absorbed by the lattice via interactions involving two or more phonons. Such effects are induced either by the anharmonicity of the fundamental phonon modes or by any nonlinearity in the relationship between induced polarization and stimulating electric field. For example, two-phonon lines can appear at the sum and difference frequencies of the phonons involved, with the sum frequency (corresponding to the creation of two phonons) appearing on the short-wavelength (or optic window) side of the Reststrahl. The complete set of such "sum resonances," involving, respectively, 2,3,4,5 ... phonons, tails off into the optic window from the top of the Reststrahl and makes up the multiphonon edge.

Although some structure is usually observed on this edge, reflecting selection rules and density of states characteristics, there is a persistent exponential decrease in intensity with increasing energy (i.e., decreasing λ) which can be expressed in the form

$$\alpha_{multiphonon} = Ae^{-a/\lambda}. \tag{1.19}$$

The exponent parameter a is proportional to ω_0^{-1}, where ω_0 is some weighted mean frequency of the Reststrahl band. Qualitatively, one expects ω_0 to scale with the frequency of the optically strongest mode (in which cations and anions vibrate against each other) and therefore to be proportional to $(F/\mu)^{1/2}$, where F is an electrostatic force and μ the relevant reduced mass. We therefore anticipate that

$$a \propto (\mu/F)^{1/2}, \tag{1.20}$$

and, consequently, that the multiphonon edge should move out to longer wavelengths as μ increases (heavier atomic masses) and as F decreases (weaker bond strengths). This, indeed is the general behavior observed, as indicated schematically in Figure 1.10.[22] For single component dielectrics, a more quantitative estimate for the multiphonon exponent has been given[23] in the form

$$a = 12\left[\mu V_M / s Z_A^{1/2} Z_C^{1/2}\right]^{1/2} \quad \mu m, \tag{1.21}$$

where μ is in atomic mass units, V_M is the molar volume (in cm³), s is the number of atoms per molecular unit, and Z_A (Z_C) is the magnitude of the formal anionic (cationic) valency. The prefactor 12 should be reduced to 10 for "open network"

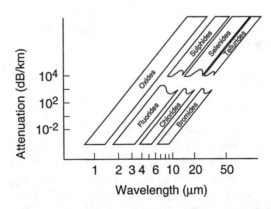

FIGURE 1.10   Absorption plots for various material classes in the multiphonon edge regime.

structures that contain two-coordinated anions (e.g., SiO_2, BeF_2). As an example, we obtain for fused silica $a = 10[(7.5 \times 27)/(3 \times \sqrt{2} \times \sqrt{4})]^{1/2} \approx 49$ μm.

1.2.1.2   The Electronic Edge

Attenuation on the high-frequency side of the optic window is dominated by band gap absorption which is primarily the result of one-electron excitations between the filled valence and empty conduction bands. In the simplest band scheme, it might naively be expected to "cutoff" sharply at a wavelength corresponding to the energy gap between the top of the valence band and the bottom of the conduction band. In real crystals, however, there are several complications. First, the excited electron and the vacated "hole" in the valence band are created at the same site and interact via an attractive electrostatic force. The consequence of this attraction is most pronounced near the band edge where bound electron–hole excitations can form and produce resonant absorptions below the conduction band.[24] They are more pronounced for wide–band gap materials.

In addition to these electron–hole interactions, further embellishments of the simple one-electron band picture are produced by deviations of the lattice from exact periodicity. Such deviations are present even in the intrinsic limit, either via thermally excited lattice vibrations (in crystals) or by the essential nonperiodicity inherent in glasses even at absolute zero.

In crystals the resulting perturbations induce joint electron–phonon excitations in which only the total (electron plus phonon) wave vector **k** need be conserved. The electron wave vector alone is then no longer conserved and electrons in different k regions of the valence and conduction bands can now be excited in a manner forbidden in the perfectly periodic system.[24] The resulting absorption tail below the excitonic resonances is referred to as the *Urbach tail* and is temperature dependent via the thermal population of phonons. This crystalline Urbach tail is particularly pronounced (at room temperature) for the common situation where the energy maximum of the valence band and energy minimum of the conduction band occur at different k values.

In glasses, the lattice periodicity is naturally disrupted in a much more essential fashion, which is not so markedly affected by temperature.[25-27] The perturbation induced by the glass disorder is now so severe that near the valence and conduction band edges the "band" description of electron actually cease to propagate as band electrons and become trapped in potential wells with a distribution of depths.[27] The density of the localized states so produced extends into the band gap from both the conduction and valence band ends to produce an Urbach tail that extends well into the optic window below the excitonic resonances.

The precise mathematical form of the Urbach tail in these various contexts is extremely complicated and still the subject of much discussion in the literature. In practice, however, many crystals and most glasses are observed to have Urbach tails at room temperature which approximate the simple exponential form:

$$\alpha_{\text{Urbach}} = Ca^{c/\lambda}, \tag{1.22}$$

in which C and c are λ-independent material parameters.

In wide–band gap materials (with a formal gap $\gtrsim$ SeV), the Urbach tail is sufficiently "steep" that its contribution to attenuation is usually not significant in the wavelength region of total minimum loss. These include most fluoride glasses and many oxides. However, for smaller–band gap materials, the Urbach tail may remain a significant component of optical attenuation at minimum loss. Such is particularly the case in many amorphous semiconductors (including most chalcogenide glasses), where the steep Urbach tail goes over to a much less steep "weak absorption tail" at longer wavelengths.[25,26] It is believed that this weak tail arises from electronic transitions induced by defect states inside the gap. By *defect,* in an amorphous context, we refer to a disruption of perfection in the topological pattern of bonding or coordination implied, for example, by the concept of a continuous random network (CRN).[27] Thus, in As_2S_3 glass it might refer to the existence of As–As and S–S bonds in addition to the As–S bond which alone determines the structure in a CRN model. While proper annealing can control this weak tail to a degree, the tail is not necessarily of entirely extrinsic origin since it seems likely, in many cases, that the perfect CRN topology is not energetically the most stable glassy state, and that defects are therefore present as an essential component.[27]

1.2.1.3 Extrinsic Contributions

Technology has unfortunately by no means yet developed to the point where we can gloss over the attenuation induced by material impurities and imperfections. In fact, for dielectrics other than fused silica, extrinsic absorption (and to a lesser degree scattering) still completely dominates the losses deep inside the optic window for all materials. More specifically, point defects, dislocations, grain boundaries, and impurity atoms can all create electronically active states within the band gap that enable significant low energy absorption to occur in what would, in their absence, be a highly transparent wavelength regime.

Point defects and structural imperfections are more important in crystals than in glasses, although, as set out above, topological defects can also exist in glasses. Impurity atoms and molecules are a problem in both crystals and glasses. In particular, rare earth (RE) and transition metal (TM) ions are of great concern because of the ability of f and d electrons to absorb efficiently at wavelengths inside the optic window. The RE resonances are numerous, but each is fairly well localized in energy even in a glass because the f electrons are well shielded from the bonding environment, which determines the nature of the electron distribution for the outer electrons. Specific RE resonances can therefore often be avoided by a judicious choice of transmission wavelength. The same is not true for TM d electrons, which can give rise to widely extended absorption bands in the near IR when embedded in the disordered atomic environment of a glass. Impurity absorption is usually quoted in units of dB/km/ppm (where ppm denotes parts per million).[28] A range of impurity absorption values as a function of wavelength are shown in Figure 1.11.[29]

Impurities, in their bonding with the primary material anion, can also give rise to impurity-driven vibrational modes which can absorb either at the fundamental or overtone frequencies. The most problematic impurities in the 1 to 10 µm range of primary interest for IR fibers and guides are light atoms and, in particular, hydrogen.

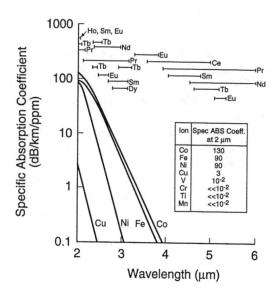

FIGURE 1.11 Typical wavelength bands for RE and TM impurities. (Adapted from Sparks, M., Report R3nr 18354, Science Research Center, Dec. 1983.)

For example, the hydroxyl ion (OH) in silica gives rise to an intense localized absorption at 2.72 μm and to an important overtone absorption at 1.38 μm (which is close to the intrinsic minimum loss wavelength λ_{min} for pure silica). Other common impurity vibrations include those resulting from H_2O and from O_2. Vibrational impurities are important not only as dissolved species but also as macroscopic heterogeneities and as surface impurities.

It is also important to recognize that in semiconductors the assumption of a filled valence band and an empty conduction band never holds precisely. Either via thermal activation (in narrow–band gap semiconductors) or more usually via dopant impurities (in extrinsic semiconductors) carriers always exist in the form of electrons, holes, or both.[30] Since these carriers are mobile to a significant degree, they can be accelerated by the electric field component of the electromagnetic radiation and subsequently scattered by phonons or impurities in a manner that produces an intraband contribution to absorption. Its intensity is naturally a function of carrier concentration, but its energy range covers the complete optic window. In the simplest model of itinerant electron response (the Drude model)[1] the absorption varies with wavelength as $\lambda.^2$ More-precise theoretical treatments are complicated and can produce λ exponents varying anywhere between 1.5 and 4 depending on the specific character of the dominant microscopic mechanism involved.

1.2.1.4   Hollow Waveguide Attenuation

Thus far we have discussed attenuation processes for dielectric materials. The results, in the context of dielectric fibers (for which core and clad materials are normally of comparable composition) are qualitatively applicable for all core radii and all modes of propagation. In hollow fibers, the difference in composition between the (air) core and clad is always large. A consequence is that the dimension and shape of the core, and the symmetry of the mode of propagation, figure explicitly in the calculation of hollow guide attenuation. In hollow waveguides the attenuation obviously arises exclusively from the clad. In the academic limit of a hollow metal waveguide with a perfectly conducting clad and a perfectly smooth and specular air/clad interface, the guide would be lossless. Unfortunately, no real air/clad interface comes close to this ideal.

Considering first a hollow metal straight guide of circular cross section, the lowest loss modes are those of TE_{oq} symmetry (for which the **E** vector is always parallel to the walls and decreases to small values at the walls). By expressing the complex refractive index of the metal in the form $v = n - iK$ (where n and K are both frequency dependent but, for good metals at longer wavelengths, ≈ 10 μm, have values $n \approx 10^1$, $K \approx 10^2$, these TE_{oq} modes are found to have losses that decrease with λ/a (where a is the core radius) and, when $\lambda/a \ll 1$, to approach the form:[9]

$$\alpha_{oq}^{TE} = \left(\lambda^2/a^3\right)\left(u_{oq}/2\pi\right)^2 \operatorname{Re}(1/v). \tag{1.23}$$

In this equation Re stands for "real part of" and u_{oq} is a mode geometric factor with value 3.8, 7.0, 10.1, 13.3, ... for $q = 1,2,3,4$.... In the limit of very high conductivity $(n - K \gg 1)$, this loss equation can be reexpressed as

$$\alpha_{oq}^{TE} = \left(\lambda^2/a^3\right)\left(u_{oq}/2\pi\right)^2\left(\omega\varepsilon_0/2\sigma\right)^{1/2}, \tag{1.24}$$

where σ is electrical conductivity. Low losses for a hollow straight cylindrical metal guide are therefore most favored by propagation in the TE_{01} mode (Figure 1.4) and by using a high-conductivity metal and a core of large diameter. Diameters of order 1 mm are usually preferred as a compromise between loss and guide flexibility and convenience. The TM and HE mode losses are typically a factor $n^2 + K^2$ (i.e., a few orders of magnitude) larger than their TE counterparts. Any coupling of a propagating TE_{0q} mode with TM or HE modes is therefore disastrous. Unfortunately, just such a coupling is necessarily induced by guide bending, leading to severe bending losses.

An additional difficulty concerns the fact that TE_{0q} modes in general, and TE_{01} in particular, have a mode geometry that makes them difficult to excite with commercially available lasers. Strongest laser coupling is achieved by exciting the HE_{11}-symmetry mode which, however, is normally highly attenuated, even in a straight metal guide. This problem has been attacked (together with the related bending problem) by coating the internal metallic guide surface with a thin (≈ 0.5 μm) layer, or layers, of dielectric.[31] By modifying the surface impedance in this way, the losses of the HE and TM modes can be greatly reduced. In this manner straight guide losses as low as 0.1 dB/m have been obtained with powers as high as 1 kW (CW) using an HE_{11} mode at 10.6 μm in a (Ge-coated Ag) 1.5 mm diameter guide.[32]

Hollow metallic waveguides have also been made with rectangular geometry $a \times b$ $(b \gg a)$. The lowest-loss modes for this configuration are of TE_{p0} symmetry (with **E** vector parallel to the longer wall) with straight guide losses:[11]

$$\alpha_{oq}^{TE} = p^2\lambda^2 \left[\frac{Re(1/v)}{a^3} + \frac{Re(v)}{b^3} \right]. \tag{1.25}$$

For highly conducting metal walls this can again be cast in terms of electrical conductivity σ:

$$\alpha_{p0}^{TE} = p^2\lambda^2 \left[a^{-3} \left(\varepsilon_0 \omega/2\sigma \right)^{1/2} + b^{-3} \left(\sigma/2\varepsilon_0 \omega \right)^{1/2} \right], \tag{1.26}$$

from which we see that loss is minimized by keeping $b \gg a$ and by propagating in the TE_{10} mode (of Figure 1.5b). It is also apparent from the final term that losses can be reduced by replacing the shorter wall with a less-conductive material (e.g., a dielectric). Straight line losses as low as 0.2 dB/m have been achieved at 10.6 μm in a 0.5×10 mm aluminum guide.[11] Bending losses are in general less severe than for a cylindrical guide, although clearly a rectangular cross section with $b \gg a$ is a less-versatile geometry for guide flexibility.

Hollow dielectric cylindrical waveguides with a clad of dielectric constant $v = n - iK$ exhibit straight guide losses[33]

$$\alpha_{pq} = \left(\lambda^2/a^3 \right) \left(u_{pq}/2\pi \right)^2 Re[f(v)], \tag{1.27}$$

where

$$f(v) = \begin{cases} \left(v^2 - 1 \right)^{-1/2}; & TE_{0q} \text{ modes} \\ v^2 \left(v^2 - 1 \right)^{-1/2}; & TM_{0q} \text{ modes} \\ (1/2) \left(v^2 + 1 \right) \left(v^2 - 1 \right)^{-1/2}; & EH_{pq} \text{ modes, } p \neq 0. \end{cases} \tag{1.28}$$

These equations are valid for any v and, therefore, also cover the hollow metal case of Equations 1.23 for which $|v| \gg 1$.

By using these equations, with $n < 1$ and realistic values of $K \lesssim 1$, calculations[15] show that the resulting attenuation is smallest for the fundamental HE_{11} mode, although the attenuation difference among TE, TM, and HE modes is much smaller for these hollow dielectric guides than was the case for hollow metal guides. Theoretical loss values for a core radius $a = 0.5$ mm and HE_{11} mode propagation are shown in Figure 1.12 for $\lambda = 10$ μm and for values of n and K covering those of real materials at this wavelength.[34] The loss, for a given $n < 1$, is smallest for small K and reaches a maximum near $K = 0.5$. Above this K value the loss decreases again as the electric field becomes increasingly short-circuited by the conductive K. It is also of interest to note from Figure 1.12 that, as a function of wavelength, the minimum loss does not generally occur at the minimum value of n.

1.2.2 SCATTERING

When an atom is subjected to a propagating electromagnetic wave, an electric dipole is created that radiates in all directions with characteristic dipole intensity distribution.

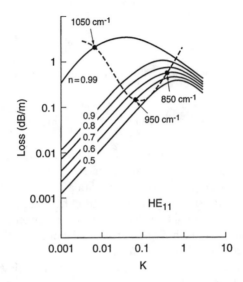

FIGURE 1.12 HE_{11} mode loss as a function of n and K for a 1 mm diameter hollow dielectric waveguide. The dashed line is the loss profile for a germania-based guide as a function of wavenumber (Adapted from Hidaka, T. et al., *J. Appl. Phys.*, 53, 5484, 1982.)

In condensed media the distance d between atoms is such that $d \ll \lambda$, so that the scattered wavelets from neighboring scatterers are essentially in phase and the resultant can be calculated by adding amplitudes. In this manner the scattered wavelets for a macroscopically homogeneous medium are found to cancel in all directions except forward, the forward scattering then combining with the incident radiation to define a wave velocity c/n. Hence, a perfectly homogeneous medium does not scatter light energy away from the direction of propagation.

Even an intrinsically pure and nonconducting real solid is never homogeneous in this sense. At room temperature many diffusive and propagating excitations exist that can scatter light energy away from the incident direction. Their effect can be defined in terms of the variation of dielectric constant $\varepsilon(r)$ from a constant value as r ranges through the macroscopic medium. For an unpolarized incident beam, the resulting attenuation coefficient when scattering is summed over all angles takes the form:[1,35]

$$\alpha_{scatt} = \left(8\pi^3/3\lambda^4\right) < \int \Delta\varepsilon(r)\,\Delta\varepsilon(0)d^3\,r>, \qquad (1.29)$$

where De(r) is the dielectric fluctuation of e at the point r from its mean value, d^3r runs over spatial extent of the fluctuation, and <...> is an ensemble average.

Equation 1.29 is actually valid only for fluctuations of spatial extent that are small with respect to λ (known as the Rayleigh scattering regime). In this case, the angular distribution of scattering is symmetric with respect to a plane normal to the direction of propagation (specifically varying as $1 + \cos^2\theta$, where θ is the angle between scattered and incident radiation) and the scattering is independent of the detailed shape of the scattering volume. If the scattering regions approach or exceed λ in size, we refer to Mie scattering[36] for which details of the shape of the scattering regions are relevant. In particular, for sufficiently elongated fluctuations along the direction of fiber length, this can lead to a forward bias in the scattering and a deviation from the λ^{-4} Rayleigh wavelength dependence.

If the scattering loss of Equation 1.29 is recast in the fashion

$$\alpha_{scatt} = \left(8\pi^3/3\lambda^4\right) <(\Delta\varepsilon)^2> v \qquad (1.30)$$

as the product of a mean square ε fluctuation and an associated correlation volume v, then the detailed contributions of the various ε-perturbing mechanisms Δx_i can be included by expanding

$$\Delta\varepsilon = \sum_i \left(\partial\varepsilon/\partial x_i\right) \Delta x_i, \qquad (1.31)$$

and substituting in Equation 1.30.

In crystals, the primary variables Δx_i contributing are the temperature ΔT and volume ΔV. Together, they induce a loss of the form $\alpha_{scatt} = B/\lambda^4$, where[1,37]

$$B = 5 \times 10^{-5} \, n^8 p^2 \, TK_T \; (\text{dB/km})(\mu m)^4, \qquad (1.32)$$

where p is the elasto-optic coefficient, n is the refractive index, K_T is isothermal compressibility (in units $10^{-11}/\text{Pa}$), and temperature T is in Kelvin. The dominant contribution arises from volume fluctuations induced by acoustic phonons (Brillouin scattering). This component is inelastic, the scattered frequency being shifted from the incident frequency by an acoustic phonon frequency. There are sometimes also inelastic losses induced by optic phonons (Raman scattering), although these are subject to crystalline symmetry restrictions and may be forbidden in some structures.

Equation 1.32 is still valid for a glass melt above its glass temperature T_g, although K_T is then much larger because of the diffusive structural configurational contributions to ΔV allowed by the fluid state. At T_g the latter freeze out,[38] while the phonon contributions remain in equilibrium down to room temperature. In these circumstances the scattering loss coefficient B for a glass becomes[38]

$$B_{glass} = 5 \times 10^{-5} \, n^8 p^2 \left[T_g \, \Delta K_T\left(T_g\right) + TK_S(T)\right] \; (\text{dB/km})(\mu m)^4, \qquad (1.33)$$

where $\Delta K_T(T_g)$ is the fluid-to-glass change in K_T that takes place at T_g and K_S is adiabatic compressibility. The first term is the elastic scattering (no frequency shift) from frozen-in volume (or density) fluctuations, while the second is the Brillouin term. Inelastic Raman scattering is also always present in glasses although its contribution to loss cannot be case in terms of thermodynamic variables.

For multicomponent glasses, additional scattering can result from fluctuations in ε induced by local compositional variations.[39] In particular, for a two-component glass with relative concentrations c and $1 - c$, the resulting concentration scattering in a perfect mixing approximation adds to the B of Equation 1.33 an amount:[40]

$$B_{conc} = 2.4 V n^2 (dn/dc)^2 c(1-c) \ (dB/km)(\mu m)^4, \tag{1.34}$$

where V is the molar volume (in cm^3). However, the result must be treated with caution since perfect mixing is rarely achieved even if the compositional components in the glass structure are currently identified. For example, calcium aluminate glasses[37] with the formal composition $cCaO \cdot (1 - c) Al_2O_3$ exhibit essentially no concentrational scattering at all, since Al enters the glass as the network-forming complex $(AlO_2)^-$, mimicking SiO_2, and requires Ca^{2+} in its immediate vicinity for charge balance. It follows that there are virtually no Ca-rich or Al-rich macroscopic regions and that ε is essentially uniform in this context. In other cases, for which phase separation is imminent,[41] B_{conc} may increase to quasi-divergent magnitudes.[40] These various scattering losses are often the dominant contributions to intrinsic attenuation within the optic window for dielectric core fibers.

Additional scattering losses can arise if the core/clad interface of a dielectric (or hollow) guide is not geometrically perfect. For example, if a cylindrical core possesses radial fluctuations or a nonzero ellipticity varying in a magnitude or in principal axis direction, then induced dipoles at the interface induce a forward-biased scattering, the degree of bias depending on the ratio of the mean axial correlation length of the fluctuation in question to λ.[42,43] In the context of metal guides, surface roughness is also a serious problem since it induces diffuse scattering and interferes with specular reflection. The quality of the metal surface is therefore of primary importance if losses are to approach their intrinsic levels in hollow fibers.[44]

All the above mechanisms give rise to losses involving the incoherent scattering of radiation. However, the scattering cross section from interactions with lattice vibrational modes also contains a quantum component representing coherent scattering.[45] This scattering term, referred to as stimulated Brillouin (Raman) scattering if resulting from interactions with acoustic (optic) phonons, results in a nonlinear coherent transfer of energy from the incident "carrier" wave to energy-shifted sideband modes traveling either parallel or antiparallel to the carrier. The process is proportional to the sideband intensity and, therefore, as the latter increases, eventually becomes catastrophic and removes all power from the carrier. It therefore defines the limits of the power-handling capability of any dielectric fiber.[46] For a hollow fiber, on the other hand, the mechanism is essentially absent, enabling hollow guides, in general, to carry much higher power radiation than their dielectric counterparts.

1.2.3 PREDICTION OF LOSS MINIMA

The theoretical loss minimum α_{min} of an IR optical material as a function of λ can be determined by assessing the intrinsic attenuation processes that define the optic window. These are usually assumed to be the multiphonon edge, the Rayleigh scattering, and the Urbach edge in the form:

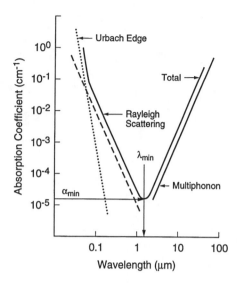

FIGURE 1.13   The optic window V plot.

$$\alpha = Ae^{-a/\lambda} + B/\lambda^4 + Ce^{c/\lambda}. \tag{1.35}$$

The schematic of such an "instrinsic window," sometimes known as a V plot, is shown in Figure 1.13. This schematic indicates that α_{min} and its corresponding wavelength λ_{min} are determined dominantly by only the multiphonon edge ($Ae^{-a/\lambda}$) and Rayleigh (B/λ^4) terms and such does appear to be the case for most halide and oxide glasses. It has been shown[47] that this leads to the approximate findings:

$$\lambda_{min} \approx 0.03a \; (\mu m), \tag{1.36}$$

$$\alpha_{min} \approx 1.4 \times 10^6 \; \left(B/a^4\right), \; (dB/km) \tag{1.37}$$

with a minimum loss dominated (≈ 85 to 90%) by Rayleigh scattering.

Relating a and B to valency and mass demonstrates[47] that the lowest-loss glasses should be sought among materials of lowest valency and highest mass. One is immediately led to the heavy halide glasses for which intrinsic losses as low as $\approx 10^{-3}$ dB/km are predicted for some compositions.[23] For the halide glasses, values of λ_{min} vary from ≈ 2 μm for fluorides to as high as ≈ 9 μm for iodides. Analogous predictions can be made for intrinsic chalcogenide (i.e., sulfide and selenide) glasses. Here, however, there is much less assurance that the Urbach edge (or more precisely its intrinsic weak absorption tail extension) can be neglected, and no really reliable intrinsic loss values have yet been given for these glasses. In crystals, where the intrinsic Rayleigh scattering is much less than in glasses, a (possibly intrinsic) weak absorption edge may influence α_{min} in an even wider range of materials.[48]

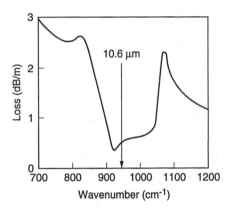

FIGURE 1.14 Theoretical loss spectrum for an HE_{11} mode in a 1 mm diameter hollow GeO_2 guide. (Adapted from Worrell, C. A. and Skarda, V., *J. Phys.*, D22, 535, 1989.)

In hollow waveguides, the absolute value of α_{min} is controlled by the cross-sectional dimension of the hollow core and is therefore not wholly a material property. Nevertheless, for a given mode and guide dimension, the loss as a function of λ is still an important characteristic. For hollow dielectric guides, λ_{min} is close to the wavelength for which $n < 1$ reaches its minimum value. This occurs just below the wavelength of the major IR-active phonon resonance band and, for oxide glasses, is typically between 8 and 11 μm. Halide and chalcogenide glasses have λ_{min} values a little larger. To date, focus on finding efficient guides for the 10.6 μm radiation of the CO_2 laser has limited most studies to compositions based on the network glass formers SiO_2 and GeO_2.[14,49] The theoretical loss curve for a GeO_2 hollow fiber is shown in Figure 1.14.[50]

For hollow metal waveguides, the λ^2 factor in Equations 1.23 to 1.26 points to a somewhat featureless variation of loss, with losses decreasing with decreasing λ in a monotonic fashion. Quantitative predictions, however, must also incorporate the implicit λ dependence of the factor $Re(1/v) = n/(n^2 + K^2)$. Although n and K for metals are very λ dependent in the IR spectral range, the ratio $n/(n^2 + K^2)$ is less so, usually increasing with decreasing λ to reduce the explicit λ exponent.[44,50] In dielectrically coated cylindrical hollow metal guides, the loss is a sensitive function of layer thickness,[51] and the dielectric film adds a structure to the otherwise rather featureless λ dependence described above. This structure (Figure 1.15) can be tailored by choice of dielectric layer thickness to induce a minimum loss at any convenient mid-IR wavelength. Dielectric coating has also been used to advantage in hollow rectangular guides, particularly to produce a low-loss guide with a square cross section in order to enhance flexibility.[52]

1.2.4 MACROSCOPIC BENDING LOSSES

Practical usage of any fiber or guide requires bending and, hence, demands both flexibility and low bending losses. The primary mechanism responsible for bending loss in dielectric fibers has been discussed in section 1.1.1.2. For this case, losses

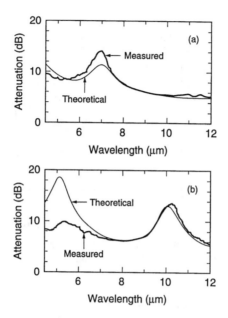

FIGURE 1.15 Attenuation spectra for germanium-coated nickel cylindrical waveguides with 1.5 mm core diameter, 1 m length, and coating thicknesses (a) 0.4 μm and (b) 0.61 μm; multimode propagation. (Adapted from Hongo, A. et al., *IEEE J. Quantum Electron.*, 26, 1510, 1990.)

vary exponentially with bending radius R, specifically, as e^{-R/R_c}, where R_c is a critical bending radius. For R significantly greater than R_c, the bending losses are negligible. This critical radius R_c is a function of mode-field diameter and decreases with the degree of confinement of mode to the core. For typical silica-based communications fibers, R_c varies from about 10 cm for multimode fibers to as low as 2 cm for monomode fibers. Thus, particularly in monomode operation, it is not difficult to ensure that bending losses do not limit the performance of communications fibers.

In hollow metal waveguides, the circular cross-sectional geometry is obviously most convenient from a practical standpoint since it bends equally in all directions. In straight guide operation (see Equation 1.23) the lowest loss is achieved by adopting multimode core dimensions ($a \gg \lambda$) but propagating a single, low-order TE mode, since high-order modes (and particularly TM modes) are heavily attenuated. Unfortunately, guide bending inevitably induces mode coupling to all other modes, the coupling to TM modes providing a channel for potentially large bending losses. It follows that bending losses are a very serious source of energy loss in hollow metal guides. The situation is particularly acute for the lowest-loss straight guide mode TE_{01} since it has long been known that this mode is closely degenerate with (i.e., travels at the same speed as) the lossy TM_{11} mode for good conductors. So severe is this loss that the bend radius R_0 (at which bend losses begin to exceed straight guide losses) is typically many meters for this case.[12]

The situation can be dramatically improved, not only in terms of removing the TE_{01} and TM_{11} degeneracy, but also in reducing TM attenuation in general by lining

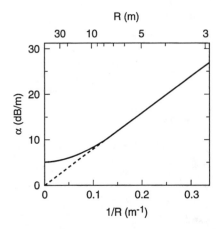

FIGURE 1.16   Attenuation α of the HE_{11} mode for a 1 mm diameter hollow Ni waveguide at $\lambda = 10.6\ \mu m$ as a function of bend radius R. (Adapted from Migaya, M. et al., *Proc. SPIE,* 484, 117, 1984.)

the metal wall with a thin dielectric coating of appropriate thickness.[53] The reduced TM attenuation also opens up the possibility[54] of operating in the lowest-order TM mode (TM_{11}, or HE_{11} to be more precise), which has the advantage of coupling strongly with the output of commercial lasers since it is linearly polarized.

For all these cases there are qualitatively two bending loss regimes.[54] In the first, with $R \geq 4a^3/\lambda^2$, the bending losses are proportional to $a^3/R^2\lambda^2$. For smaller bending radii the electric field patterns begin to deviate significantly from their straight guide forms and to concentrate more and more near the outer edge of the guide. In a ray picture, the radiation goes over to a propagation in which it reflects only from the curved outer surface as it travels around the bend (whispering gallery modes).[55] In this regime, bending loss becomes independent of guide radius a and varies with bend radius as $1/R$. Practical bending radii (e.g., $R < 1$ m) are normally well inside this whispering gallery regime (see, for example, Figure 1.16). The basic theory seems also to be valid for hollow dielectric guides and has been experimentally tested in several cases.[56,57] A probe of the effects of surface roughness has also been presented.[58]

Bending losses in a hollow metal guide with rectangular cross section (Figure 1.5b) are, in general, less severe than for an equivalent core-area cylindrical metal guide because, in the lowest-loss TE_{10} mode (see Equation 1.26), TM-like losses from the sidewalls can be minimized by the geometry ($b \gg a$). In addition, the TE_{10} mode is linearly polarized and is readily excited by conventional laser launch geometry. On the other hand, a guide with a ribbonlike cross section is obviously less adaptable for flexing in two dimensions. Bends in the plane normal to the longer wall are readily accomplished, and, for them, the light propagates in a whispering gallery configuration for most practical bend radii, specifically, when $R \lesssim 2a^3/\lambda^2$ for the TE_{10} mode. In the whispering gallery configuration the *total* guide loss (assuming specular reflection) takes the extremely simple form:[11]

$$\alpha = (2/R)\ \mathrm{Re}(1/v). \qquad (1.38)$$

In a bend of total angle ϕ (i.e., bend arc length ϕR), this can be recast as an attenuation $2\text{Re}(1/v)$ per radian of bend angle, which is independent of λ, R, and waveguide dimension, and amounts to only a few hundredths of decibel of loss per 90° bend for a normal metal at $\lambda = 10.6 \ \mu\text{m}$.

For full flexibility, the rectangular guide must also undergo twists about the longitudinal guide axis. This twist configuration induces losses primarily via TM-symmetry effects at the sidewalls. By defining a twist angle ϕ_T per unit guide length, this loss can be expressed as[11]

$$\alpha_T = 4a\phi_T^2 \ \text{Re}(v), \qquad (1.39)$$

and can be reduced by dielectrically coating the sidewalls.[52] An alternative bending technique for rectangular guides avoids this twist by confining the radiation to the whispering gallery mode in a flexible helix.[59] However, this geometry has an inherent loss per turn which limits the number of turns that can be used.[60]

REFERENCES

1. M. E. Lines, in *Handbook of Infrared Optical Materials*, P. Klocek, Ed., Marcel Dekker, New York (1991).
2. See any introductory optics text, e.g., F. A. Jenkins and H. E. White, *Fundamentals of Optics*, McGraw-Hill, New York (1976).
3. D. Marcuse, *Theory of Dielectric Optical Waveguides*, Academic Press, New York (1974).
4. D. Gloge, *Rep. Prog. Phys.*, 42, 1777 (1979).
5. S. E. Miller and A. G. Chynoweth, Eds., *Optical Fiber Telecommunications*, Academic Press, New York (1979).
6. A. W. Snyder and J. D. Love, *Optical Waveguide Theory*, Chapman and Hall, New York (1984).
7. D. Gloge, *Appl. Opt.*, 11, 2506 (1972).
8. E. A. J. Marcatili and R. A. Schmeltzer, *Bell Syst. Tech. J.*, July, 1783 (1964).
9. E. Garmire, T. McMahon, and M. Bass, *Appl. Opt.*, 15, 145 (1976).
10. P. Diament, *Wave Transmission and Fiber Optics*, Macmillan, New York (1990).
11. E. Garmire, T. McMahon, and M. Bass, *IEEE J. Quantum Electron.*, QE-16, 23 (1980).
12. M. E. Marhic and E. Garmire, *Appl. Phys. Lett.*, 38, 743 (1981).
13. M. E. Marhic, *SPIE*, 320, 79 (1982).
14. T. Hidaka, Y. Mitsuhashi, and J. Shimada, *JEE*, Nov., 72 (1983).
15. T. Hidaka, *J. Appl. Phys.*, 53, 93 (1982).
16. Y. Kato and M. Miyagi, *IEEE Trans. Microwave Theory Tech.*, 40, 679 (1992).
17. There is some confusion in the literature whether or not to include the negative sign in Equation 1.12 in the definition of material dispersion.
18. M. DiDomenico, *Appl. Opt.*, 11, 652 (1972).
19. S. H. Wemple, *Appl. Opt.*, 18, 31 (1979).
20. K. Nassau, *Bell Syst. Tech. J.*, 60, 327 (1981).
21. G. Keiser, *Optical Fiber Communications*, McGraw-Hill, New York (1983).
22. J. R. Gannon, *J. Non-Cryst. Solids*, 42, 239 (1980).
23. M. E. Lines, *J. Non-Cryst. Solids*, 103, 265 (1987); unpublished comments.

24. W. A. Harrison, *Electronic Structure and the Properties of Solids,* W. H. Freeman, San Francisco (1980).
25. P. Klocek and L. Colombo, *J. Non-Cryst. Solids,* 93, 1 (1987).
26. D. L. Wood and J. Tauc, *Phys. Rev.,* B5, 3144 (1972).
27. N. E. Cusack, *The Physics of Structurally Disordered Matter,* Adam Hilger, Bristol, U.K. (1987).
28. T. Izawa and S. Sudo, *Optical Fibers: Materials and Fabrication,* KTK Scientific Publ., Tokyo (1987).
29. M. Sparks, Report R3nr 18354, Science Research Center (Dec. 1983).
30. F. Bassani, G. Iadonisi, and B. Preziosi, *Rep. Prog. Phys.,* 37, 1099 (1974).
31. M. Miyagi and S. Kawakami, *J. Lightwave Tech.,* 2, 116 (1984).
32. M. Miyagi, A. Hongo, and Y. Matsuura, *Proc. SPIE,* 1228, 26 (1990).
33. E. Garmire, *Appl. Opt.,* 15, 3037 (1976).
34. T. Hidaka, K. Kumada, J. Shimada, and T. Morikawa, *J. Appl. Phys.,* 53, 5484 (1982).
35. B. Chu, *Laser Light Scattering,* Academic Press, New York (1974).
36. L. P. Bayvel and A. R. Jones, *Electromagnetic Scattering and Its Applications,* Applied Science Publ., London (1981).
37. M. E. Lines, J. B. MacChesney, K. B. Lyons, A. J. Bruce, A. E. Miller, and K. Nassau, *J. Non-Cryst. Solids,* 107, 251 (1989).
38. N. I. Laberge, V. V. Vasilescu, C. J. Montrose, and P. B. Macedo, *J. Am. Ceram. Soc.,* 56, 506 (1973).
39. J. Schroeder, in *Treatise in Materials Science and Technology,* Vol. 12, M. Tomozawa and R. H. Doremus, Eds., Academic Press, New York (1977), p. 157.
40. M. E. Lines, *J. Non-Cryst. Solids,* 171, 209 (1994).
41. O. V. Mazurin and E. A. Porai-Koshits, *Phase Separation in Glass,* North Holland, Amsterdam (1984).
42. D. Marcuse, *Appl. Opt.,* 23, 1082 (1984).
43. E. G. Rawson, *Appl. Opt.,* 13, 2370 (1974).
44. Y. Matsuura, M. Saito, M. Miyagi, and A. Hongo, *J. Opt. Soc. Am.,* A6, 423 (1989).
45. R. H. Stolen, in *Optical Fiber Telecommunications,* S. E. Miller and A. G. Chynoweth, Eds., Academic Press, New York (1989), 125.
46. P. Klocek and M. Sparks, *Opt. Eng.,* 24, 1098 (1985).
47. K. Nassau and M. E. Lines, *Proc. SPIE,* 484, 7 (1984).
48. H. Mori and T. Izawa, *J. Appl. Phys.,* 51, 2270 (1980).
49. N. Nagano, M. Saito, M. Miyagi, N. Baba, and N. Sawanobori, *Appl. Phys. Lett.,* 58, 1807 (1991).
50. C. A. Worrell and V. Skarda, *J. Phys.,* D22, 535 (1989).
51. A. Hongo, K. Morosawa, T. Shiota, Y. Matsuura, and M. Miyagi, *IEEE J. Quantum Electron,* 26, 1510 (1990).
52. H. Machida Y. Matsuura, H. Ishikawa, and M. Miyagi, *Appl. Opt.,* 36, 7617 (1992).
53. M. E. Marhic, *Appl. Opt.,* 20, 3436 (1981).
54. M. Migayi, K. Harada, Y. Aizawa, and S. Kawakami, *Proc. SPIE,* 484, 117 (1984).
55. H. Krammer, *Appl. Opt.,* 16, 2163 (1977).
56. C. A. Worrell, *Electron. Lett.,* 25, 570 (1989); *Proc. SPIE,* 843, 80 (1987).
57. S. J. Wilson, R. M. Jenkins, and R. W. J. Devereaux, *IEEE J. Quantum Electron.,* QE23, 52 (1987).
58. O. B. Danilov, M. I. Zintchenko, Y. A. Rubinov, and E. N. Sosnov, *J. Opt. Soc. Am.,* B7, 1785 (1990).
59. M. Marhic, L. I. Kwan, and M. Epstein, *Appl. Phys. Lett.,* 33, 874 (1978).
60. M. E. Marhic, *J. Opt. Soc. Am.,* 69, 1218 (1979).

2 Silica Glass-Based Fibers

Allan J. Bruce

CONTENTS

2.1 INTRODUCTION

Silica-based glasses can be optically transparent from the near-ultraviolet to the mid-infrared range of the electromagnetic spectrum. Optical fibers made from these glasses are widely used in the near-infrared (NIR) at wavelengths close to the zero material dispersion (1310 nm) and minimum loss (1550 nm) wavelengths of silica (Figure 2.1). Such fibers provide the backbone of modern optical telecommunication networks.

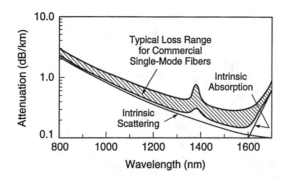

FIGURE 2.1   NIR attenuation of high-silica, single-mode telecommunication fibers.

A great deal has been written on all aspects of optical fiber telecommunications, and several texts are referenced for further reading.[1-6] It is a key technology in the "information age," enabling the construction of extremely high capacity networks for analog and digital transmission of data, including voice and video. These systems were first conceived in the 1960s and became practical in the late 1970s with advances in laser and detector technology and the development of fibers with optical transmission as high as 96% (0.2 dB) per kilometer.

In the operation of a basic system, data are encoded through the modulation of a NIR light source (e.g., a GaInAsP semiconductor laser). This can be accomplished by electrical modulation of the laser or with in-line devices such as electro-optic lithium niobate waveguide modulators. The modulated signal is launched into a fiber and is ultimately transmitted to a photodetector receiver (e.g., a PIN photodiode) where an electrical signal is reconstituted for decoding. By using in-line optical fiber amplifiers (OFAs) to maintain adequate signal intensity, the span between transmitter and detector can be intercontinental in range.

Existing networks include (1) intra- and interoffice networks for dedicated or secure communications, (2) local area networks (LANs), with fiber links up to tens of kilometers in length and many branching points for local distribution, (3) terrestrial trunks for intercity and transcontinental transmission, and (4) undersea links, for intercontinental traffic. Because of preexisting infrastructure, and cost competitiveness, many of the final linkages are still made by electrical transmission over coaxial cables or copper wire. Copper LANs are steadily being replaced by fiber, and cost-effective options for installing fiber (and/or wireless links) all the way to the curb, to the home, or to the desktop are still being evaluated to provide individual users with full access to optical communication systems. The projected lifetime of many telecommunication systems is between 15 and 25 years. This motivates the incorporation of excess data handling capacity and backup components and makes long-term optical and mechanical reliability of fibers critical.

The technology for producing telecommunication fibers and cables has been developed since the early 1970s. It is now highly advanced and similarly practiced worldwide. As of 1996, the total length of telecommunication fiber deployed worldwide was close to 100 million km. Current demand and production capacity is in the range of tens of millions of kilometers per year and growing. "Standard" fibers are

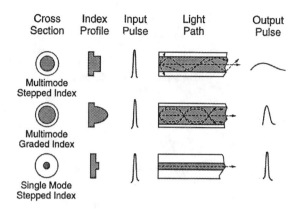

| Cross Section | Index Profile | Input Pulse | Light Path | Output Pulse |

FIGURE 2.2 Modal dispersion of transmitted signals in multimode and single-mode fibers.

virtually a commodity item, and their specifications are recommended by both national organizations and the International Telecommunications Union/Telecommunications Standardization Sector (ITU/TSS). These specifications are periodically revised to keep pace with advances in systems technology and are voluntarily adhered to by all major fiber manufacturers.

It is the intent in this chapter, with a brief digression to cover multicomponent glass fibers, to focus on the design, composition, fabrication, performance, and reliability of high-silica-content, single-mode telecommunication fibers. Specialty fibers, including erbium-doped fibers for OFAs and fibers with photoinduced gratings, which have made a significant impact on the design and operation of telecommunication systems, will also be considered. Fibers for nontelecommunications applications, including imaging, power delivery, spectroscopy, and sensors, will not be dealt with explicitly. The materials and operating principles for such fibers are similar to those for telecommunications, but their design and performance requirements are often less demanding.

2.2 DESIGN AND SPECIFICATIONS OF TELECOMMUNICATION FIBERS

The conventional core-clad structure of fibers and the optical principles that govern the propagation of light therein have been described in Chapter 1. The design of single-mode telecommunication fibers is primarily determined by the required optical performance including bandwidth and/or data transmission rate. A particular concern is the time-based dispersion of transmitted light pulses in a fiber. This limits the maximum transmission rate and span length that can be used for coherent digital transmission. As described in Chapter 1, there is intermodal (for multimode, see Figure 2.2) and intramodal dispersion. The latter comprises material dispersion and waveguide dispersion contributions. Materials dispersion derives from variations in refractive index of the core glass as a function of wavelength, and waveguide dispersion derives from incomplete mode confinement of light by the core-clad interface and is also a function of wavelength. The net dispersion of a fiber is the

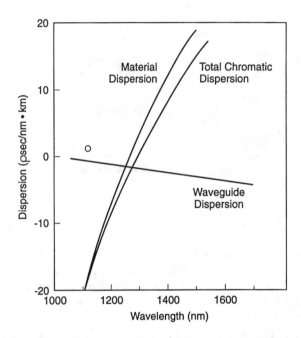

FIGURE 2.3 Total chromatic dispersion of a 1300 nm optimized single-mode fiber and underlying contributions from material and waveguide dispersion.

sum of the individual dispersion contributions (Figure 2.3). While the materials dispersion is essentially fixed, the magnitude and wavelength dependence of the waveguide dispersion can be strongly influenced by the refractive index profile of the core and immediate cladding in single-mode fibers. It can also be opposite in sign to the materials dispersion, which is the key to minimizing chromatic dispersion for different operational wavelengths (Figure 2.4).[7]

Polarization mode dispersion (PMD) is also of concern for high-bandwidth (e.g., cable TV) and polarization-sensitive applications. It arises because single-mode fibers in reality support two degenerate orthogonally polarized modes. Deviations from circular symmetry in the fiber core can introduce birefringence and introduce phase velocity differences between these modes, which can impact the coherence of transmitted signals. This phenomenon may be enhanced by external perturbations, such as pressure, temperature, and bending. PMD is minimized by reducing the magnitude or constancy of orientation of the birefringence of the fiber core. Also, as will be discussed in Section 2.6.3, the birefringence may be deliberately enhanced to fabricate polarization-maintaining fibers.

Several types of single-mode fiber are widely used. The ITU/TSS has published recommendations for their specification and performance which are summarized in Tables 2.1 and 2.2. Typically, performance and physical characteristics are recommended, as opposed to glass compositions and index profiles. In addition to the more common nondispersion-shifted (G.652) and dispersion-shifted (G.653) fibers, 1550 nm loss-minimized (G.654, e.g., Sumitomo Z-fiber) and nonzero dispersion (G.65x, e.g., Lucent TrueWave fiber) fibers have been evaluated. The latter illustrates

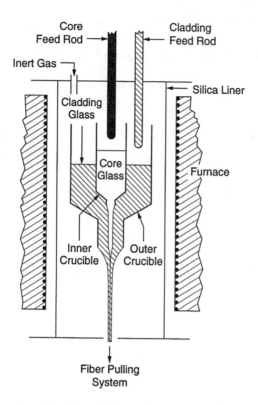

FIGURE 2.4 Examples of single-mode fiber refractive index profiles used to achieve minimum dispersion at specific wavelengths of operation.[7]

TABLE 2.1
ITU/TSS Recommendations for Single-Mode Fibers (1996)

Recommendation	Type of Fiber
G.652	Nondispersion-shifted fiber (1310 nm)
G.653	Dispersion-shifted fiber (1550 nm)
G.654	1550 nm loss-minimized fiber (pure silica core)
G.65x	Nonzero dispersion (new)

---

response to developing systems technology. When OFAs and dense wavelength division multiplexing (DWDM) is implemented over long spans, nonlinear effects including four wave mixing can limit performance. This is addressed by using fibers with a very small (nonzero) gradation of chromatic dispersion across the band of operation. In the case of Lucent TrueWave fiber, the zero dispersion wavelength is also tailored to be slightly below that of the operational band so that the dispersion in different channels is of the same sign. Another feature highlighted in Table 2.2 is that properties, including cutoff wavelength, of individual fibers can be modified

TABLE 2.2
ITU/TSS Recommendations for Single-Mode Fibers (1996)

Type of Fiber	G.652	G.653	G.654	G.65x
Outside diameter (μm)	125 ± 2	125 ± 2	125 ± 2	125 ± 2
Noncircularity of cladding (%)	<2	<2	<2	<2
Mode field diameter (μm) ±10%	9–10	7.0–8.3	10.5	8–11
Nonconcentricity of mode field (μm)	<1 (at 1310 nm)	<1 (at 1550 nm)	<1 (at 1550 nm)	<1 (at 1550 nm)
Attenuation (dB/km)				
1310 nm	<0.5	<0.55	—	—
1550 nm	<0.4	<0.35	<0.22	<0.35
1550 nm bend loss for 100 turns at 37.5 mm radius (dB)	<1	<0.5	—	<0.5
λ_c, fiber cutoff wavelength (nm)	<1280 (>1100)	—	1350–1600	—
λ_{cc}, cabled fiber cutoff wavelength (nm)	<1270	<1270	<1530	<1480
λ_o, zero dispersion wavelength (nm)	1300–1324	~1550	~1310	~1550
Chromatic dispersion [ps/(nm·km)]	<3.5 (1288–1339 nm); <5.3 (1271–1360 nm); <2.0 (1310 nm)	<3.5 (1525–1575 nm)	<20 (1550 nm) (1530–1560 nm)	>0.1, <6.0
Proof test level (G·Pa)	>0.35	>0.35	>0.35	>0.35
	(~0.69)	(~0.69)	(~0.69)	(~0.69)

when they are cabled (see Section 2.4.5). Attention must be paid to such effects when optimizing system, as distinct from fiber, performance.

While this chapter focuses on single-mode fibers for long-distance and high-capacity networks, multimode fibers, in which the core diameter is as much as 70% of the fiber diameter, continue to play a significant role in short-range communications systems. These include intra- and interoffice networks where, because of distance or application, dispersion is not a critical issue. Multimode fibers are more expensive to fabricate than single mode, because of higher materials and processing costs, but multimode systems typically cost less than single-mode systems, because fiber coupling and alignment are less critical and can be accomplished with lower-cost components and less-specialized training.

TABLE 2.3
Examples of Multicomponent Glass Fibers

Glass Composition (wt%)			NA	Loss (db/km)	λ (nm)	Ref.
I	CORE	SiO_2–B_2O_3–Na_2O	0.18	3.4	840	16
	CLAD	SiO_2–B_2O_3–Na_2O				
II	CORE	$55SiO_2$–18 GeO_2–$15Na_2O$–$6CaO$–$5Li_2O$–$1MgO$	0.23	4.2	850	11
	CLAD	$65SiO_2$–$7GeO_2$–$17Na_2O$–4 CaO–$5Li_2O$–$2MgO$				

2.3 MULTICOMPONENT GLASS FIBERS

Multicomponent glasses, specifically soda-lime silicate (Na_2O–CaO–SiO_2) and sodium borosilicate (Na_2O–B_2O_3–SiO_2) and related compositions in which silica comprises less than 75 mol% of the glass, were early candidate materials for optical communication fibers (Table 2.3). These were similar to commercial (nonfiber) glass compositions and amenable to conventional melting and processing. Drawing by the double crucible technique also offered a route for "continuous" fiber fabrication (Figure 2.6). Core-cladding index differences were typically achieved by varying the concentration or type of alkali in the respective glasses or by adding GeO_2 to the core glass. Graded-index profiles in the fiber could be tailored by using crucible designs which permitted more, or less, interfacial contact and interdiffusion between the core and cladding glasses during the fiber draw. The technology at the time focused on large-core, multimode fibers operating in the 840 nm optical window. Single-mode fibers could be fabricated with some difficulty, but operation at the 1310 and 1550 nm optical windows over significant lengths was precluded by high OH absorption in most fibers. Suitable lasers and detectors for these wavelengths were also not readily available.

Up to the mid 1970s significant efforts were made to fabricate low-loss multi-component telecommunication fibers in Europe,[8,9] the U.S.,[10] and Japan.[11,12] It was recognized that Rayleigh scattering in many multicomponent silicate glasses could be lower than in high-silica-content glasses[13-15] and that the achievable losses were largely determined by extrinsic impurities. Many innovative approaches were tried to minimize these impurities. The efforts yielded fibers with losses as low as 3.4 dB/km at 840 nm in 1977 (Figure 2.5).[16] Further loss reduction was dependent on reducing –OH contamination to the sub-parts-per-million (ppm) and transition metals to the low-ppb levels. It was clear that such progress would require major effort, investment, and further innovation, both in raw materials purification and glass processing.

At the same time, high-silica-content fibers made by vapor phase techniques were surpassing multicomponent glass fibers in terms of their minimum loss. In particular, low –OH levels opened up the telecommunication windows at 1310 and 1550 nm, where lower intrinsic loss was achievable. Tailoring of refractive index

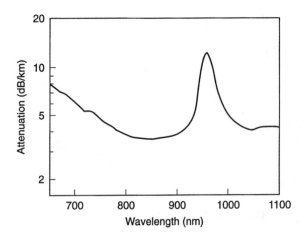

FIGURE 2.5 Lowest reported attenuation for a multimode, multicomponent (sodium boro-silicate) glass fiber.[16]

profiles for system applications was relatively easy, and their strength and durability was much higher than multicomponent fibers, primarily due to the higher degree of extended network formation in the high-silica glass structure. Difficulties in processing these high-temperature materials had also been satisfactorily addressed. In the circumstances, it is not surprising that all serious efforts on telecommunication fibers switched to high-silica compositions by the end of the 1970s. Since then, there has been occasional resurgence of interest in multicomponent glasses for fibers either as hosts for active ions or because they are predicted to have lower intinsic losses, at similar wavelengths, than high-silica fibers.[17,18] Any economic advantage of using ultra-low-loss multicomponent fibers for undersea or trunk routes has, however, been diminished by the development of OFAs. The intrinsically lower strength, reliability, and radiation hardness of the multicomponent fibers also present significant obstacles for their practical utilization. It is extremely unlikely that multicomponent fibers will ever replace high-silica-content fibers in telecommunication systems, except perhaps in relatively short lengths packaged in discrete devices for specialty applications.

An alternative fiberization route employing multicomponent glass which is worthy of mention is the Phasil process originally reported by Macedo and Litovitz.[19] In this technique an alkali borosilicate glass rod is phase separated by controlled heating into a silica-rich phase and an interconnected mixed-component phase. The mixed-component phase can be leached out in hot acid leaving a highly porous silica structure. The porous body is then completely doped with an index-increasing component (e.g., Cs^+) from solution. A core-cladding structure is then created by immersing the rod in deionized water to promote the out-diffusion of dopant from its surface. After careful drying, the resultant preform can be pulled directly into fiber. The fibers were typically multimode, graded index, in structure and something of a hybrid, having high-silica-content cladding and multicomponent core. An attempt was made to commercialize this process by Pilkington PLC in the U.K. in the late 1970s. The lowest reported loss for the fibers was 6.5 dB/km at 840 nm in 1979.[20]

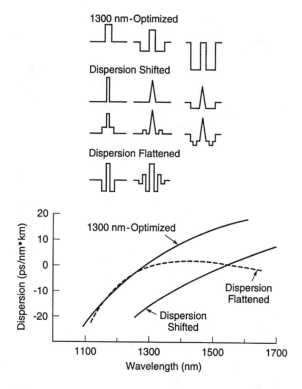

FIGURE 2.6 Schematic representation of a double crucible fiberization system with multicomponent glass feeds for core and cladding.

2.4 HIGH-SILICA-CONTENT FIBERS

The materials and fabrication procedures for high-silica fibers have been well documented in many publications (e.g., References 1 through 6). The following is intended as a brief summary of common practices.

2.4.1 MATERIALS AND PURITY

High-silica-content fibers are compositionally simple. In most instances, the cladding glass is 100% SiO_2 while the core glass is 90 to 95% SiO_2 with a few percent of dopants to increase the refractive index in order to achieve a guiding structure. The cladding glass in the vicinity of the core may also be doped to achieve specific refractive index profiles such as those shown in Figure 2.4. Fibers with high numerical apertures (NA) (e.g., for dispersion compensation) contain higher dopant levels (e.g., 20 to 30%) to achieve a desired refractive index profile for a targeted optical performance.

The range of dopants used in commercial fibers is relatively small. The concentration dependence of refractive index (n_D) and thermal expansion coefficient for some common dopants are shown in Figures 2.7 and 2.8.[21,22] GeO_2 and F are the

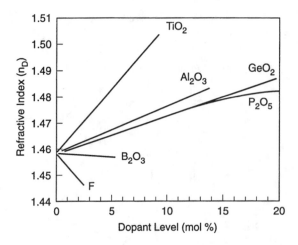

FIGURE 2.7   Refractive indexes (n_D) as a function of dopant level in bulk SiO_2. F levels are in atomic %.[21,22]

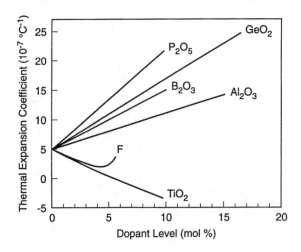

FIGURE 2.8   Linear thermal expansion coefficients as a function of dopant levels in bulk SiO_2. F levels are in atomic %.[21,22]

most widely used for elevating and depressing the index, respectively. In an extreme case (e.g., Sumitomo's Z-fiber) the core is undoped while the index of the cladding is substantially depressed by F doping. In some specialty fibers, nonstandard dopants such as Er (for OFAs) are also required for optical performance. These are typically introduced as codopants rather than as replacements for standard-index dopants.

It is advantageous to use core glasses with higher thermal expansion than the cladding. With such a combination, radial compressive stress is generated on cooling during fiberization which can increase the threshold for crack propagation (see Section 2.5.2).[1] The relative core/clad glass transition temperatures (T_g values) are also pertinent in this regard, since the linear thermal expansion of a glass typically increases fivefold above T_g. Fortuitously, most doped glasses (i.e., core) have lower T_g values than the parent composition. Consolidation during preform collapse (for tubular preforms) and fiber draw is also facilitated if the core glass has slightly lower viscosity than the cladding. If the viscosity difference is too great (e.g., by more than one or two orders of magnitude), however, surface tension effects make it difficult to achieve circular core geometry.

The ability to make extremely high purity glass preforms which can be drawn into low-loss fiber has been the key to the success of high-silica-content fibers for telecommunications. The route to high-purity glass is through vapor phase transport and reaction of high-purity metal halides. Typically, the desired metal halides are transported by bubbling carrier gases (e.g., O_2) through temperature-controlled reservoirs of liquid halides. The carrier gases are mixed and delivered with other reaction gases (e.g., O_2, H_2, etc.) to a controlled heat source where they are hydrolyzed (or oxidized) and deposited as glass. Vapor transport can impart significant purification as a consequence of large differences in vapor pressure between the glass-forming components and loss-increasing impurities (e.g., Fe, V, etc.). Vapor pressures as a function of temperature are shown in Figure 2.9 for a number of relevant metal chlorides.[23]

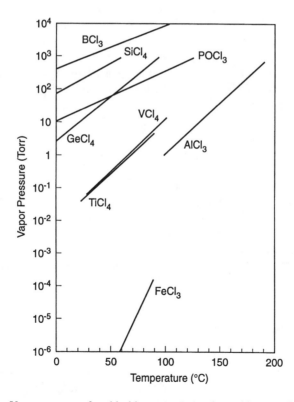

FIGURE 2.9 Vapor pressures for chloride precursors and transition metal impurities pertinent to preform fabrication.[23]

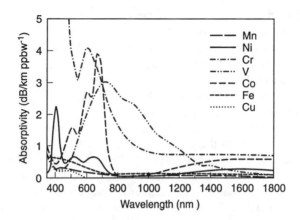

FIGURE 2.10 Absorbtivities of transition metal impurities in bulk SiO_2 made by a flame hydrolysis method.[25] Some elements are present in mixed oxidation states.

Other liquids and gases have been used as sources of both major constituents and dopants. Fluorine can be incorporated using SiF_4, BF_3, C_2F_6, or SF_6. Solid sources, such as $AlCl_3$, can be sublimed at high temperatures (e.g., 100 to 400°C) and delivered to the reaction zone in a heated carrier gas. $ErCl_3$ and other lanthanide halides can be delivered in a similar fashion.[24] The vapor pressure of such precursors may not be sufficiently different from that of absorbing impurity species to produce significant purification. Relatively high loss can be typical in some specialty fibers. Absorption loss contributions from transition metal impurities in bulk-fused silica are shown in Figure 2.10.[25] To achieve fiber loss approaching intrinsic levels at 1310 or 1550 nm, these and other metallic impurities must be reduced to sub-ppb levels in the core glass.

Losses due to hydroxyl (O–H) and hydrogen impurities are also of major concern. Overtone absorptions of hydroxyl, with peaks at 1240, 1380, and 1900 nm, can contribute of the order of 1 dB/km/ppm to loss at 1310 and 1550 nm in high-silica fibers.[26] Hydroxyl impurities are typically minimized by using pure reagents, dry gasses, and chlorine-based chemistry to dehydrate the deposited soot before densification. It is also beneficial to use the core dopants which are least susceptible to hydroxylation (e.g., GeO_2) and to minimize hydroxyl impurities in the cladding glass from where they may later diffuse into the core. Molecular hydrogen, which can readily diffuse into most glasses, can also contribute to loss, especially if it reacts with the glass to form hydroxyl species. More will be said on this topic in Section 2.5.1.

2.4.2 PREFORM FABRICATION

The preform route to fiberization, whereby a large glass body (typically 20 to 90 mm in diameter and 1 to 2 m in length) is fabricated with appropriate core/clad geometry and subsequently drawn to fiber dimensions, is the route of choice for high-silica-content fibers. Precise preform geometry is key to the production of high-quality fibers.

Three preform fabrication techniques are practiced commercially. These are outside vapor deposition (OVD), modified chemical vapor deposition (MCVD), and vapor axial deposition (VAD). All three share a common heritage in the work of Hyde[27] in the 1930s which established a flame process for delivering and decomposing vapors of "hydrolysable compounds of silicon" to deposit a fine silica soot and then sintering this to a dense glass. OVD and VAD are direct descendants of Hyde's technique, while MCVD is a close relative in which a flame is used indirectly to oxidize similar vapors which are flowing on the inside of a silica tube.

The three processes are illustrated schematically in Figure 2.11a, b, and c. In OVD, first core and then cladding soot is deposited by a traversing torch on a rotating silica mandrel, which is subsequently removed. The soot body is dehydrated in a chlorine-based atmosphere and sintered at high temperature to a dense glass, sometimes with hole closure. In VAD, core and cladding soot can be simultaneously deposited using multiple torches with different vertical orientations. The initial deposit is on a rotating seed rod and later on the soot boule itself as the process continues. There need be no hole in the soot body, and the dehydration and sintering steps are similar to OVD. In MCVD, the soot is produced by oxidation rather than hydrolysis. Minimal cladding and then core material is deposited, layer by layer, inside a rotating silica tube which ultimately becomes a major fraction of the cladding. Heating is provided by an external traversing torch. Soot deposition occurs near the leading edge of the flame, and consolidation occurs as the hotter section of the flame passes over the soot. There is no separate dehydration step, and the preform is eventually collapsed to a solid body by using a hotter and/or slower-traversing flame.

Many parameters must be established and controlled for uniform and reproducible preform manufacturing. Soot deposition is influenced by particle size and thermal gradients. The torch design plays an important role in establishing appropriate temperature profiles. As practiced, soot delivery rates are similar for OVD and VAD (10 to 25 g/min). Typically, only 50 to 60% of the soot produced ends up in the preform, while the remainder exits with the carrier and waste gases. MCVD deposition rates are much slower (2 to 3 g/min), but this limitation is offset by the smaller volume of deposit required since the substrate tube provides the majority of the cladding in MCVD.[28] The composition of the deposited glass depends on the rate of chemical delivery, relative concentrations, and reaction rates.[29] Fluorine doping can be accomplished in OVD and VAD by soaking the dehydrated, but unsintered, soot boule in a fluorine-containing atmosphere (e.g., SiF_4) at elevated temperatures.

High-quality, low-loss fibers are routinely drawn from preforms fabricated by each of the above techniques. All three processes can be commercially competitive, and currently there is an almost equal split in global manufactured volumes. Comparisons of throughput and cost efficiency depend on the system configuration, manufacturing volume, and fiber design. These should not be applied globally. Specific techniques may offer better routes for particular composition and/or index profiles. For example, the layer-by-layer deposition in MCVD can be advantageous for achieving trench and pedestal index profiles. The tubular MCVD configuration can also facilitate the selective codoping of the core (e.g., with Er^{3+}). Conversely,

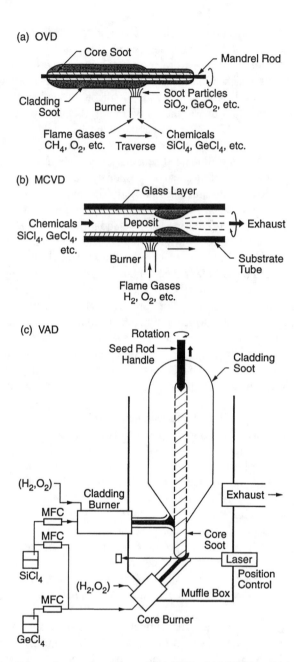

FIGURE 2.11 Schematic representation of (a) OVD, (b) MCVD, and (c) VAD preform fabrication routes.

the avoidance of a tubular stage in VAD prevents the occurrence of index dips at the center of a fiber core caused by the "burn off" of volatile dopants, such as GeO_2, during tube collapse. VAD is the route of choice for producing pure silica core fibers which are recognized to have the lowest transmission loss of any fibers currently manufactured. In this case, to facilitate dehydration and sintering, a silica core rod is first fabricated then overclad with fluorine-doped glass.

Typical preforms yield 50 km of fiber, or more. Larger preforms can give more fiber with potential cost reduction associated with the elimination of repetitive processing and handling. In OVD and VAD, larger preforms can be contrived by depositing more soot on the initial preform or by subsequent deposition. In MCVD, the initial preform size is limited by the availability of substrate tubes. However, these may also be overclad by a soot process or by using a "rod-in-tube" strategy in which the overcladding tubes are manufactured by a sol–gel or analogous route.[30]

The plasma chemical vapor deposition (PCVD) technique for preform manufacture should also be mentioned.[31] In configuration, PCVD resembles MCVD except that a low-pressure microwave plasma is generated inside the substrate tube and external heating can be provided by resistance heating. PCVD operates at lower temperature, and its main feature is the efficient deposition of thin vitreous layers without an intermediate soot stage. More layers are required to achieve core dimensions than in MCVD, but this can enable very precise tailoring of the refractive index profile. Also, the low-temperature processing can permit the incorporation of high fluorine levels in the deposited glass. These advantages have to be weighed against the operational complexity of a system that requires vacuum and microwave. PCVD is not practiced commercially at this time, primarily because of higher cost.

Prior to fiber draw, the surface of preforms is usually etched (e.g., in an HF solution), then flame polished to reduce the number of surface defects which can ultimately limit the fiber strength and reliability.

2.4.3 FIBER DRAWING

Draw towers for high-silica-content fibers typically range in height between 3 and 20 m and incorporate the functional components shown in Figure 2.12. Tall towers are used to access fast draw rates (>10 m/s) which can increase throughput and reduce manufacturing cost. The increased height is necessary to achieve increased spacing between the draw furnace and the coating system. This establishes the maximum time for fibers to cool to temperatures that will not compromise the thermal stability of coating materials. Forced air cooling devices may also be used to access high draw speeds.

In a typical draw, the preform, or fused extension handle, is clamped in a vertical feed mechanism above the draw furnace. It is lowered such that a small section of the preform protrudes below the hot zone of the draw furnace which is preheated to a temperature at which the cladding glass achieves a viscosity of 10^3 to 10^5 Pa·s (1950 to 2300°C for high-silica glasses). The preform necks-down in the hot zone and elongates under the weight of the protruding section. To start the draw, the elongated section is pulled to approximately fiber-dimensions then fed through the coating and curing systems to a drive capstan. The fiber diameter is controlled by the relative rate of preform feed and capstan rotation. Targeted dimensions are achieved by feedback control using in-line diameter monitoring. After the diameter control and coating processes are established, the fiber is fed to a take-up mechanism

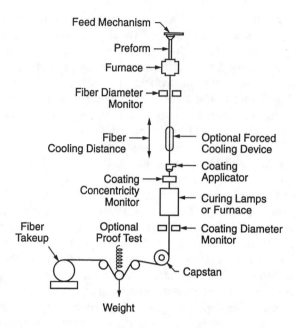

FIGURE 2.12   Schematic representation of a draw tower for high-silica preforms.

and the spooling rate is automatically adjusted to accept the fiber throughput under minimal tension. Precision alignment of the components and extremely tight feedback control have to be achieved to obtain fibers that meet the ITU/TSS recommendations. Cleanliness is also essential for maintaining high fiber strength and coating integrity. Many towers incorporate laminar flow capability for clean air.

Two types of draw furnaces are in common use: (1) graphite resistance furnaces, which are purged with inert gas, and (2) R.F. induction furnaces with zirconia susceptors. The latter are the more prevalent in manufacturing plants, although they must be carefully maintained and operated to limit the generation of zirconia particles, which can adhere to the softened glass surface. Such refractory defects promote crack propagation and can significantly reduce fiber strength.

On exiting the furnace, the optical fiber should have an almost pristine surface. To avoid mechanical surface damage and to preserve high strength, polymeric coatings are applied in-line immediately after drawing. A dual UV curable coating is often used, where the inner (primary) coating is soft and has a low elastic modulus, while the outer (secondary) coating is hard and has a high elastic modulus. This combination can alleviate stresses on the fiber during subsequent handling and makes it less susceptible to microbending loss. Coatings are applied by passing the fiber through specially designed applicators that promote good surface adhesion and centering of the fiber. The coatings are cured by passing through UV lamps before reaching the drive capstan. Coating thickness is typically in the range of 50 to 100 μm.

Alternative and/or additional coatings are sometimes applied to fibers for special applications. These include other UV or thermally cured polymeric coatings and

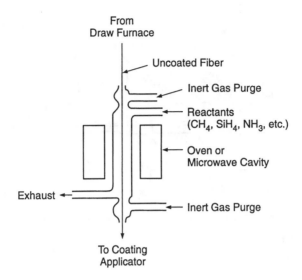

From
Draw Furnace

Uncoated Fiber

Inert Gas Purge

Reactants
(CH_4, SiH_4, NH_3, etc.)

Oven or
Microwave Cavity

Exhaust

Inert Gas Purge

To Coating
Applicator

FIGURE 2.13   Schematic representation of an in-line CVD reactor for producing hermetic coatings.

hermetic coatings. Hermetic coatings include low-melting metals (e.g., Sn), ceramics (e.g., silicon oxynitride), and amorphous carbon.[32-34] The hermetic coatings are typically applied in-line before the polymeric coatings to prevent environmental moisture or hydrogen penetration into the fiber. The metals can be applied by passing the fibers through molten metal in a heated coating applicator while ceramics and amorphous carbon can be applied by chemical vapor deposition (CVD). A schematic of an in-line CVD reactor is shown in Figure 2.13. An inert gas purge can be used to isolate the reaction zone where appropriate mixtures of vapors are reacted using resistance (or microwave) heating (200 to 500°C) to deposit coatings on the fiber. Amorphous carbon coatings 100 to 200 nm in thickness have been found to be effective in preventing the penetration of moisture and molecular hydrogen into the fiber. This will be discussed further in Section 2.5.1.

2.4.4   FIBER SPLICING AND INTERCONNECTION

The preform size limits the length of an individual fiber. This is seldom above 50 km. To span longer lengths, individual fibers have to be interconnected or joined. Fusion splicing is a robust technique for permanently joining fibers with minimal loss and strength penalties. It is used in the factory and also in the field to repair fiber breaks.

In a typical splicing procedure, the coating is first stripped from the tails of two fibers; the ends are cleaved to obtain fresh surfaces and positioned end to end into a splicing apparatus (e.g., Figure 2.14) with active or passive optical alignment and then fused with an electric arc or miniature torch. Current equipment and procedures routinely give losses <0.05 dB/splice (<1% loss) for standard single-mode fibers with no significant reduction in fiber strength. To achieve such losses the alignment of the cores has to be within 0.5 μm and their tilt angle less than 0.2°.[4,35] Procedures

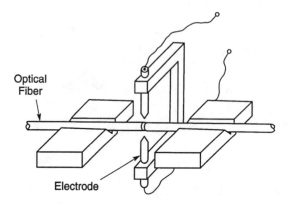

FIGURE 2.14   Schematic representation of an electric arc fusion splicer.[4]

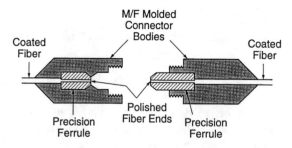

FIGURE 2.15   Schematic representation of a de-mountable fiber connector incorporating alignment ferrules for positioning the stripped fiber ends.

have also been developed for recoating polymeric and hermetic materials on the spliced regions.

Permanent mechanical splicing, in which fibers are butt-coupled within an alignment component (e.g., in a precision bore sleeve) then fixed in place (e.g., by epoxy resin), is conceptually simple but does not give the strength or low splice loss obtained by fusion. De-mountable mechanical connectors can be convenient for intraoffice applications, fiber termination points, and reconfigurable interconnection. A large variety of mechanical connectors for both individual and multiple fibers have been devised.[4,35] Most mechanical connectors (e.g., Figure 2.15) incorporate a lateral alignment component (e.g., a ceramic ferrule or a v-groove) on, or through, which the stripped end of a fiber is placed, or passed, then fixed in place (e.g., using an epoxy resin). Typically, the end of the fiber is ground or polished to attain a fixed position where it exits from the alignment component. Mating (e.g., with the aid of a biconic sleeve) of two matched connector bodies provides the fiber-to-fiber coupling. Coupling loss can be relatively high (0.1 to 1.0 dB) compared with a fusion splice but may be reduced by (1) optimizing the connector design, (2) using higher-precision components, (3) active alignment, (4) angle polishing of the fiber ends, or

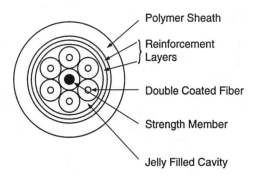

Polymer Sheath

Reinforcement Layers

Double Coated Fiber

Strength Member

Jelly Filled Cavity

FIGURE 2.16 Cross section of a cable design for a medium-long-haul terrestrial system.

(5) using index-matching elastomers to fill the air gap between the fiber faces. At some point the added cost associated with reducing coupling loss becomes prohibitive. The mechanical reliability is primarily determined by the connector design and should be adequate for the targeted applications and environments. Optical reliability associated with relative movement of the aligned fibers and the degradation of the optical interface is always a concern. The successful introduction of new connector designs depends on performance under rigorous environmental testing and meeting ITU/TSS and national (e.g., Bellcore) recommendations.

2.4.5 FIBER CABLES

Fibers must be adequately protected when deployed in real-world environments. A common practice is to incorporate the fibers into cables with multiple levels of protection. The technology of cable manufacturing is highly advanced, and various cable designs have been developed for both terrestrial and undersea environments. Important design considerations include mechanical strength, environmental stability, induced fiber loss, and hermeticity.[36] Fibers can be incorporated in cables as individual elements, bundles, or ribbons (e.g., for intraoffice or backplane interconnection) depending on the application.

An example of a long-haul terrestrial cable design is shown in Figure 2.16. It shows a number of fibers arranged around a central strength member (e.g., steel wire) inside a dense polymer sheath. Refinements that are frequently incorporated include (1) additional jacketing of the individual fibers, (2) filling of voids with "jelly" (a silicone or hydrocarbon grease loaded with fine glass particles) which can getter moisture and hydrogen, (3) positioning of fibers in plastic elements with axial grooves or conduits, and (4) additional buffer and/or armored layers.

As with any composite structure there is a potential for thermal expansion mismatch between the constituent elements. This is greatest between the fiber and the polymer components and can generate significant temperature-dependent stress on the fibers which may introduce microbending loss (see Section 2.5.1). The presence of a metallic strength member somewhat constrains the polymer expansion. Microbending loss is minimized through appropriate fiber, coating, and cable design.

2.5   PERFORMANCE AND RELIABILITY

Up until the early 1990s, issues of reproducible performance and long-term optical and mechanical reliability were prominent in the optical fiber industry. Many efforts have been directed toward understanding the mechanisms of optical loss and mechanical strength and their degradation with time. The success of these efforts and resulting innovations in fiber manufacturing and packaging has led to a diminution in the level of concern on reliability issues in recent years. This is especially true for mechanical strength and reliability, which depend on the initial quality of the fiber surface and subsequent stress and/or environmental exposure. Since the cladding glass is essentially identical in all fiber designs, common processing solutions, mechanical proof testing, and cabling practices can be implemented. Optical performance and reliability, on the other hand, depend on the composition and refractive index profile of the fiber core as well as environmental factors. Consequently, if the core structure and/or composition of specialty fibers is different from proven fiber designs, reliability qualification must still be performed.

To obtain estimates of long-term reliability commensurate with the projected system lifetime, accelerated testing is employed under conditions of elevated stress, temperature, humidity, pressure, and/or other pertinent environmental conditions. By assuming common mechanisms, activation energies and time constants can be obtained.

2.5.1   OPTICAL

Experimental single-mode fibers with losses close to the theoretical limits at 1310 and 1550 nm were first reported in the mid 1980s.[37,38] Similar losses are now routinely obtained in standard-production fibers (Figure 2.1).

The effects of different core dopants on the IR and UV absorption are illustrated in Figures 2.17 and 2.18.[39,40] For fibers operating near 1550 nm, GeO_2 doping provides an acceptable low-loss option. This is also the dopant of choice for 1310 nm operation although there is a measurable contribution to loss from the GeO defect edge in the UV. In specialty fibers with higher GeO_2 levels, substantial loss at 1550 nm has been observed (Figure 2.19).[41] This is attributed to the concentration dependence of the defect edge and enhanced scattering due to higher refractive index differences at the core/clad interface.

Macro- and microbending losses manifest themselves as rapid increases in attenuation beyond critical wavelengths (Figure 2.20).[42] Both are associated with changes in optical mode confinement due to stress-induced changes in refractive index profiles as a result of fiber bending. Dispersion-shifted, single-mode fiber designs are particularly sensitive to these effects. Macrobending is easily envisioned, while microbending is characterized by small amplitude variations of the order of a nanometer with a periodicity of the order of a millimeter. Microbending can result from the expansion mismatches between cable components. Bend loss sensitivity can be reduced by the prudent selection of fiber and cable design.

Several factors can produce increased fiber attenuation over a system lifetime including hydrogen and/or moisture penetration and optical defect creation due to

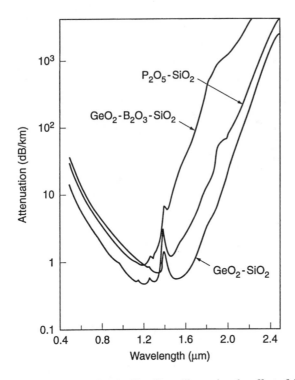

FIGURE 2.17 NIR attenuation of high-silica fibers illustrating the effect of different dopants on the intrinsic IR edge.[39]

natural background radiation. This increased attenuation can have a disproportionately high impact on system performance if it pushes the operation of such components as amplifiers and detectors out of their optimal range. Consequently, system specifications normally include maximum allowable increases in total attenuation (including all fiber and coupling losses) over the operational lifetime. Optical reliability depends on the prevention and/or management of all loss-inducing mechanisms.

Given the opportunity, moisture can diffuse from the surface to the core of a fiber with an attendant increase in attenuation at communication wavelengths due to overtone and combination absorptions of the –O–H vibration. This diffusion is relatively slow and can be minimized with appropriate selection of coatings and packaging. Dopant selection is also a factor, for example, GeO_2 is less susceptible to hydroxylation than P_2O_5. Hydrogen diffusion is more insidious because of small molecular size and slow reactivity under normal conditions. Diffusion occurs rapidly and is difficult to prevent. Figure 2.21 shows the result of an accelerated test for hydrogen-induced absorption in a standard SiO_2–GeO_2 fiber.[43] The effect is quite substantial and includes absorption due to hydrogen and hydroxyl reaction products. Some hermetic coatings, including amorphous carbon, can limit hydrogen diffusion (Figure 2.22).[44] These are recommended for use in environments conducive to hydrogen generation. Sources of hydrogen can include polymeric cabling materials and electrolytic reactions involving power supplies and seawater.

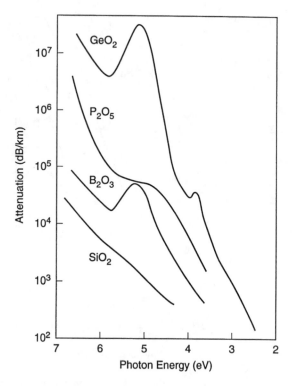

FIGURE 2.18   UV attenuation of doped SiO_2 glass, including intrinsic and defect absorptions.[40]

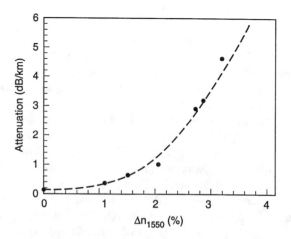

FIGURE 2.19   Fiber attenuation at 1550 nm for a series of SiO_2–GeO_2 core fibers with increasing dopant level, given as percent change in core vs. clad index (see Figure 2.7), incorporating data from Reference 41.

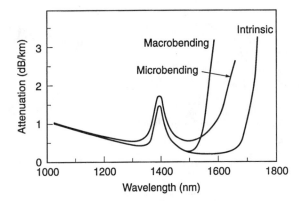

FIGURE 2.20 Phenomenological illustration of the effects of macro- and microbending on attenuation of dispersion-shifted single-mode fibers.[42]

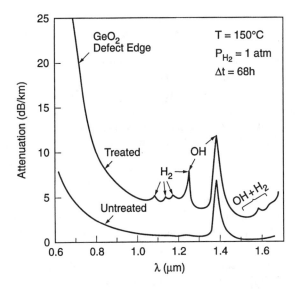

FIGURE 2.21 Increased fiber attenuation resulting from the exposure of an SiO_2–GeO_2 core fiber to hydrogen at 150°C for 68 h.[43]

Exposure to ionizing radiation can produce defect centers in fibers which also contribute to optical loss. Natural radiation, which is approximately 0.1 to 1 rad/year, can be sufficient to produce significant degradation over system lifetimes. Radiation "hot spots" on the ocean floor can also be an additional problem for submarine cables. Standard dopants including P_2O_5 and GeO_2 have been identified as progenitors of such defects, and even undoped silica is not immune. Spectral losses for GeO_2–P_2O_5–SiO_2 and GeO_2–SiO_2 fibers before and after exposure to a ^{60}Co gamma ray source are shown in Figure 2.23.[45] The fiber without P_2O_5 is seen to be somewhat

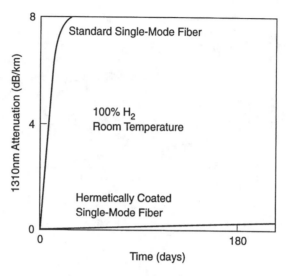

FIGURE 2.22 Effect of an amorphous carbon hermetic coating on the 1310 nm attenuation of a SiO_2–GeO_2 core single-mode fiber after exposure to hydrogen at room temperature.[44]

more resistant, which is another reason for selecting GeO_2 as the dopant of choice. The induced loss in an SiO_2–GeO_2 fiber at 1310 nm as a function of cumulative dose is shown in Figure 2.24.[46] The effect is seen to be linear up to extremely high doses. At present, there is no complete solution to radiation damage. Radiation hardness is enhanced by avoiding the more-sensitive dopants and shielding from some types of radiation by the cable structure.

2.5.2 MECHANICAL

For practical purposes the strength of a fiber should be sufficient to withstand initial handling stresses including those generated in cabling and deployment. Furthermore, this strength should not degrade, during the system lifetime, below the level at which fracture can occur either under as-deployed stress or anticipated handling. Both aspects of strength have been investigated as a function of fiber length and environment. In accord with the statistical probability of encountering a flaw, longer fiber lengths exhibit a higher failure probability at lower stress.

The intrinsic strength of a flawless silica fiber is estimated to be between 7 and 14 GPa. Such strengths have only been approached for short lengths of fiber at very low temperatures (e.g., 77 K) where degradation mechanisms are kinetically inhibited. In practice, fiber strength is compromised by the presence of flaws which serve to concentrate the applied stress and promote fracture. These flaws include abrasions, microcracks, inclusions, diameter fluctuations, and/or particles (particularly those on the surface of the fiber). As an example, surface flaws of only 1 μm in depth are sufficient to reduce the strength of a standard fiber by a factor of 100 or more.[4]

The initial fiber strength depends on the probability of encountering a flaw of sufficient size to concentrate the applied stress above the critical fracture level. The long-term reliability is determined by the existence of subcritical flaws and their

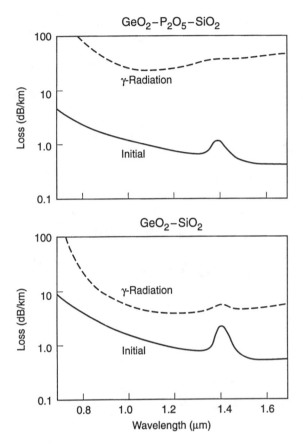

FIGURE 2.23 Increased fiber attenuation resulting from ^{60}Co gamma ray irradiation (10^4 rad) of GeO$_2$–SiO$_2$ and GeO$_2$–P$_2$O$_5$–SiO$_2$ core fibers.[45]

potential to grow to a critical size for the stress levels subsequently encountered. As previously noted, the initial fiber strength can be considerably enhanced by (1) etching and fire polishing preforms to heal microcracks, (2) maintaining a low particle count and low humidity in the draw furnace and precoating environments, and (3) applying of protective coatings in-line. The latter limits subsequent fiber abrasion and particle contamination.

As a quality control, in-line proof testing is commonly employed during fiber draw to screen out gross fiber flaws. As shown in Figure 2.12, a load is applied to the fiber to produce a transient stress immediately before the fiber is spooled. According to ITU/TSS recommendations, standard fibers should pass proof testing of at least 0.35 GPa (typically twice this level is adopted). This level is selected to establish a practical initial strength while minimizing the possibility of introducing or growing subcritical flaws which may later accelerate long-term fatigue.

Environmental moisture is recognized as a major promoter of flaw growth through corrosion. This is consistent with the results presented in Figure 2.25, where fibers with a hermetic SiON coating exhibit slower fatigue in normal atmosphere than those

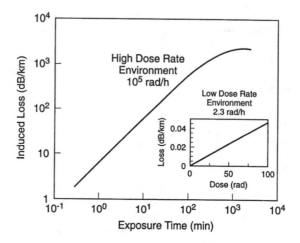

FIGURE 2.24 Induced 1310 nm loss of a single-mode GeO_2–SiO_2 core fiber as a result of cumulative ^{60}Co gamma ray irradiation.[46]

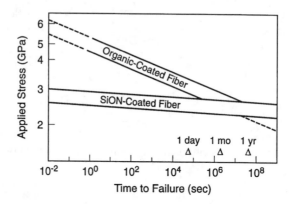

FIGURE 2.25 Time to failure of comparable single-mode fibers with standard polymeric and SiON coatings.

with polymeric coatings which are permeable to moisture. The lower initial strength of the fibers with an SiON (and also for amorphous carbon) coating is illustrative of the influence of surface chemistry and reactivity on strength and flaw generation.

Enhanced fatigue resistance has also been achieved by employing compositional modification of the outer cladding glass to induce surface compression. The resulting compression inhibits subcritical flaw growth at lower-magnitude stress.[47] One example is Corning's Titan fiber, in which titanium doping is used to produce the compressive cladding (by virtue of the high glass transition temperature and low thermal expansion coefficient relative to the core and standard cladding). To date, the strength and fatigue resistance of standard fibers have proved to be adequate for most applications, and the market demand for compressive cladding fibers has been limited.

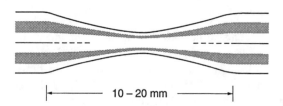

FIGURE 2.26   Schematic representation of a fused fiber coupler in which two fibers are fused and stretched to promote evanescent coupling between the fiber cores.[48]

2.6   SPECIALTY FIBERS

Specialty is a generic descriptor frequently used for fibers that differ from *standard* components in design, composition, coatings, or functionality. In the present context, it is used to describe fibers for advanced optical components that have enabled implementation of the revolutionary changes in design and operation of telecommunication systems in the 1990s. The most recognized are Er-doped fibers for OFAs. These and several other specialty fibers are described in this section.

Serendipitously, germanosilicate core fibers are appropriate, and in some cases enabling, for many specialty applications, including the examples given below. This has accelerated implementation, since the chemistry, fabrication, and reliability are essentially identical to that of standard fibers. Propagation loss can be relatively high (at most a few decibels per kilometer) in specialty fibers because their designs are optimized for other functionalities. In most instances, however, relatively short lengths of specialty fiber are required and the impact on the total loss budget of most systems is minimal.

2.6.1   FUSED FIBER COUPLERS

These are basic, but key, components in many systems which enable the low-loss interconnection of multiple fibers. They are a preferred route for launching pump light into fiber lasers and amplifiers. Their operation depends on bringing fiber cores into close proximity such that all or part of the light can be transferred between the fibers by evanescent coupling. The efficiency of this transfer depends on several factors, including the wavelength of operation, the NA of the fibers, the proximity of their cores, and the distance, or coupling length, over which they are in proximity.

One example of a fused fiber coupler is shown schematically in Figure 2.26.[48] Fabrication in this case involves stripping the coating from sections of two fibers, twisting them together to ensure good contact, heating and stretching the twisted section, then recoating. The stretched region is typically 10 to 20 mm long, and the degree of coupling may be tailored by actively monitoring the output from the two fibers during fabrication. Standard splicing procedures can be used to incorporate the fused fiber couplers into a fiber span, and individual couplers can be concatenating for larger-scale interconnection.

2.6.2 TAPERED CORE FIBERS

These are useful elements in devices that require coupling between standard and small-core, high-NA fibers. The mode field mismatch between such fibers can introduce significant loss. This mismatch may be reduced by downtapering or uptapering the standard or high-NA fiber cores, respectively. Downtapering can be accomplished by stretching. This also reduces the fiber OD which can introduce other problems. Uptapering may be achieved during fusion splicing, without affecting the OD by prolonged heating of the high-NA fiber end to promote thermal outdiffusion of index-increasing dopants from the core.[49]

2.6.3 POLARIZATION-MAINTAINING FIBERS

As noted in Section 2.2, the propagation constants of normally degenerate polarized modes in a single-mode fiber can be separated by introducing birefringence in the radial refractive index. If the birefringence is sufficiently large, the fibers may be used to transmit polarized light without significant cross talk. Such fibers are particularly useful for interferometric sensors (e.g., fiber gyroscopes) and other polarization-sensitive devices. The birefringence is typically engineered at the preform stage by mechanical deformation (flattening, grinding, and/or core drilling) of solid preforms or by selective etching and reflow of the inner cladding in tubular preforms (or overcladding tubes) at an appropriate stage in their processing. As illustrated in Figure 2.27 for several polarization-maintaining fiber (PMF) designs, the origin of birefringence is asymmetry in either the core or cladding. Core ellipticity can produce birefringence directly by altering the effective refractive index on different axes. This may be considered as an extreme example of PMD. Cladding asymmetry primarily influences the effective refractive index through differential stress effects on the core.[50-52] Boron codoping is frequently used as a means of producing the stress-inducing cladding.

2.6.4 DISPERSION-COMPENSATING FIBERS

A significant fraction of established telecommunications systems still have embedded optical fibers optimized for 1310 nm operation and use electronic repeaters. These fibers exhibit substantial positive dispersion in the 1550 nm range which precludes the direct replacement of repeaters with Er–doped OFAs and system upgrades to all-optical operation. One strategy that can facilitate such upgrades is to incorporate additional sections of dispersion-compensating fibers (DCF), which have negative dispersion that balances the dispersion of the embedded fiber. In practice, DCFs are designed with large negative dispersion. Fiber lengths are customized for the application and are much shorter (e.g., <10%) than the embedded span. DCF design for single-wavelength operation is relatively straightforward, while design for broadband operation is more involved.

High-NA, small-core (e.g., <5 µm diameter), germanosilicate fibers exhibit large negative dispersion near 1550 nm, primarily because a larger fraction of the mode field is in the cladding (slower propagation). This can be accompanied by higher attenuation and greater sensitivity to PMD. The magnitude of the dispersion is

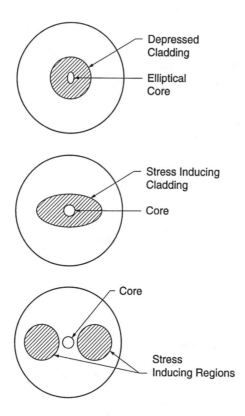

Depressed
Cladding

Elliptical
Core

Stress Inducing
Cladding

Core

Core

Stress
Inducing Regions

FIGURE 2.27 Core/cladding cross sections for several PMFs.[50-52]

extremely sensitive to core diameter. One figure of merit for benchmarking DCF designs is obtained by dividing the magnitude of the dispersion by the attenuation. This is plotted against the percent of elevation of core index for a series of germano-silicate core fibers in Figure 2.28.[53]

To illustrate DCF performance on the coherence of 1550 nm transmission over 1310 nm fibers, Figure 2.29 shows two time-based "eye-diagrams" taken on each side of a section of DCF (note that an additional OFA is included to compensate for the attenuation of the DCF). Before the DCF, the effect of dispersion on trans-mitted signals is evident. After the DCF, the signals overlie one another as in the original transmission.

2.6.5 ERBIUM-DOPED FIBER FOR OPTICAL AMPLIFIERS

More than any other specialty fiber these components have spurred the development of advanced optical telecommunication systems in recent years. Prior to their imple-mentation, electronic repeaters were used to boost signal intensity every 20 to 100 km along a fiber span. These were appropriate for single-wavelength, fixed bit-rate operation. With the advent of OFAs, all-optical transmission could be realized which enabled multiwavelength, bit-rate-transparent operation. The available capacity in

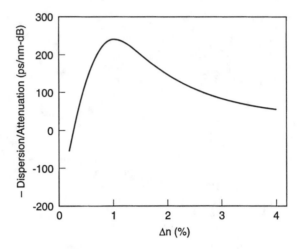

FIGURE 2.28 Figure of merit for DCFs plotted against core/clad index difference for a series of germanosilicate fibers.[53]

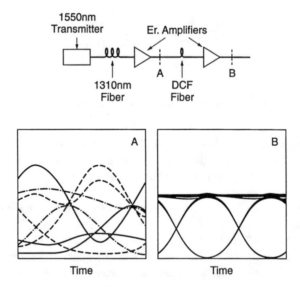

FIGURE 2.29 Effect of incorporating DCF on the coherence of transmitted 1550 nm signals over 1310 nm fiber. The eye diagrams represent data at points A and B on the span.[53]

new and existing systems operating in the 1550 nm window was thus multiplied manyfold.

Much has already been written on Er-doped fibers for OFAs and rare earth–doped fibers in general, including some excellent texts.[54,55] Only key features will be outlined in this section. The basis of operation is stimulated emission in the 1550 nm optical window between the $^4I_{13/2}$ and $^4I_{15/2}$ electronic energy levels of Er^{3+} ions in a

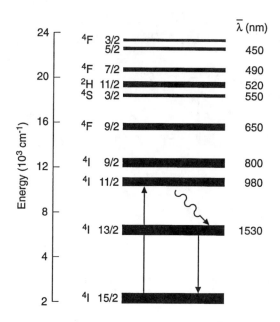

FIGURE 2.30 Electronic energy level diagram for Er^{3+} ions in a silica host.

glass host (see Figure 2.30). The $^4I_{13/2}$ level can be populated directly by exciting the Er^{3+} ions with light of appropriate energy or indirectly by pumping to a higher level (e.g., $^4I_{11/2}$) followed by nonradiative decay to the $^4I_{13/2}$ level. Subsequent relaxation to the $^4I_{15/2}$ level occurs radiatively, with a 1/e lifetime of approximately 10 ms in high-silica glasses. To exploit these transitions in an OFA, the fiber core is codoped with Er up to a few thousand ppm. Pump light, typically 980 or 1480 nm, is coupled into the fiber, and emission between the $^4I_{13/2}$ and $^4I_{15/2}$ levels is stimulated by, and transfers power to, transmitted signals in the 1550 nm range.

Rare earth–doped fibers have been fabricated by liquid and vapor impregnation of soot bodies which are subsequently sintered to provide core rods or by using vapor sources in an MCVD configuration, as described earlier. Additional codopants such as Al are frequently included to increase the solubility of the rare earth ions in the core glass and/or to modify the profile of the Er emission spectra for better WDM performance. To utilize pump power efficiently it is advantageous to use high-NA, small-core fibers as shown in Figure 2.31. Typically, 10 to 30 m of Er-doped fiber are used in an OFA.

Many factors must be considered when optimizing an OFA for a given application. These include the performance of sources and detectors, operational bit rate, span length, amplifier location, pump wavelength, pump efficiency, noise figure, gain saturation, amplified spontaneous emission (ASE), excited state absorption (ESA), nonlinear effects, broadband requirements, and system reliability.[55] Amplifiers may be located near the source (power amplifier), near the detector (preamplifier), or remote from both (in-line amplifier). Pumping can also be implemented near to, or remote from, the doped fiber in the same (copropagation) or opposite (contrapropagation) direction to the signal or in both (codirectional). The optimal

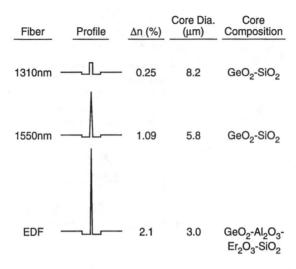

Fiber	Profile	Δn (%)	Core Dia. (μm)	Core Composition
1310nm		0.25	8.2	GeO_2-SiO_2
1550nm		1.09	5.8	GeO_2-SiO_2
EDF		2.1	3.0	GeO_2-Al_2O_3-Er_2O_3-SiO_2

FIGURE 2.31 Comparison of fiber refractive index profiles and properties of erbium-doped (EDF) and standard single-mode fibers for 1310 and 1550 nm operation.

configuration is system dependent, although copropagation usually gives lower noise while codirectional pumping improves reliability. It is notable that failure of an OFA leaves an absorbing component in-line. In codirectional pumped systems both pumps would have to fail before this occurs. Long-haul systems may employ a series of similar in-line OFAs. In other situations, enhanced performance can be achieved by concatenating amplifiers with different performance characteristics.[56] Such configurations may be used to incorporate gain-equalizing components for WDM applications without adding excessive noise.

Any of several pump wavelengths corresponding to electronic absorptions of the Er^{3+} ions can be used, although 980 and 1480 nm are usually preferred. Among other features these are least susceptible to ESA. Yb^{3+} codoping also facilitates pumping in the 800 to 1000 nm range through absorption to the $^2F_{5/2}$ level of Yb^{3+} followed by radiative transfer to the $^4I_{11/2}$ level of Er^{3+}.[57] Most amplifiers for undersea links currently employ 1480 nm pumping, because of established source reliability and negligible ESA.

The noise figure is the ratio of system noise after and before an amplifier. It is always greater than 1. ASE is a major contributor to noise, generating a broad rather than selective emission at signal frequencies. The noise figure is minimized by prudent selection of the operating parameters for the amplifier and system configuration.[55,56]

Gain as a function of pump power is plotted in Figure 2.32 for an amplifier pumped at 1480 nm.[58] Gain coefficients >10 and >6 dB/mW have been demonstrated for 980 and 1480 nm pumping, respectively.[59] The gain at a given pump power eventually diminishes with increasing signal power as a result of the competing dynamics of population inversion and stimulated emission (see Figure 2.33). Preamplifiers are usually operated below this threshold to minimize noise, while power amplifiers can be operated at higher levels of gain saturation.[55]

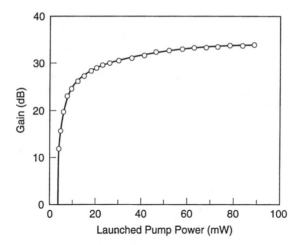

FIGURE 2.32 Example of signal gain obtained at 1544 nm, in a 40 m length of Er-doped fiber, as a function of 1490 nm pump power.[57]

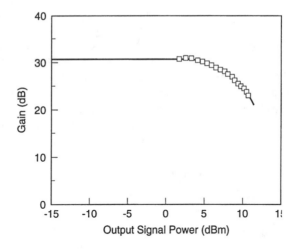

FIGURE 2.33 Gain saturation for the same fiber represented in Figure 2.32, at a pump power of 24.5 mW. The saturation output power is defined as the point at which the gain is 3 dB below its unsaturated value.[58]

Broadband operation requires amplifiers that provide almost constant gain over a defined wavelength range. Without employing external gain flattening, this depends on the emission (or, more specifically, the gain) spectra of the active dopants in a given host. The emission spectrum is influenced both by homogeneous broadening, due to phonon-induced transitions within the Stark levels of the dopant ions, and inhomogeneous broadening, due to variations in the nature and symmetry of different sites occupied by these ions. For Er^{3+} in silica and germanosilicate glasses, codoping with Al and/or P significantly enhances the inhomogeneous broadening (see

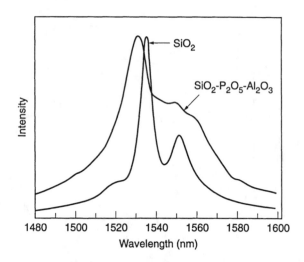

FIGURE 2.34 Normalized emission profiles for the $^4I_{13/2}$ to $^4I_{15/2}$ transition of Er^{3+} in fused silica and Al–P codoped silica.[54]

Figure 2.34) and can provide relatively flat gain between 1535 and 1555 nm. Wider bandwidths (>35 nm) have also been demonstrated in the same fibers by using long-period photodefined gratings to achieve gain equalization by wavelength-selective attenuation. This provides a practical alternative to ZBLAN and other nonsilicate glass fibers for broadband operation.[60] Span length is also a factor since it introduces wavelength-dependent attenuation, dispersion, and increasing nonlinear effects. In the real world, span lengths are not constant and amplifier performance has to be tuned for each application. This can be achieved by selecting the operating parameters of the amplifier or the tuning performance of ancillary components.

2.6.6 FIBERS WITH PHOTOINDUCED GRATINGS

The development of techniques whereby reflective gratings can be directly photodefined in the core of germanosilicate fibers has also been one of the most-enabling factors for developing advanced optical telecommunication systems in recent years. With this functional integration, many optical components and devices became practical, including short-cavity and single-frequency fiber lasers, Raman lasers and amplifiers, dispersion compensators, ultrasharp filters, multiplexers and demultiplexers, gain equalizers, and optical taps. Most of these functions can only be implemented in all-optical transmission systems, and the advent of OFAs was a chronological prerequisite.

The basis of the technology is the phenomenon whereby the refractive index of a germanosilicate glass can be increased by exposure to UV light.[61] This is attributed to the creation of electronic defects at germania sites by photoexcitation. The defects exhibit absorption in the far UV, including a peak at 195 nm, which generate an increase in index throughout the optical window of the material. Two subsequent innovations made this effect of practical importance for telecommunications. The

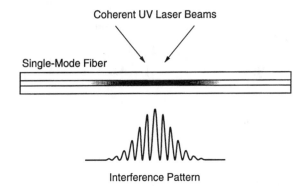

Coherent UV Laser Beams

Single-Mode Fiber

Interference Pattern

FIGURE 2.35 Schematic representation of photoinduced grating writing using UV side illumination. A phase mask is used to generate the interference between the coherent beams which defines the grating intensity and periodicity.

first was the writing of gratings in germanosilicate core fiber using side illumination with coherent ArF or KrF laser sources and phase masks as illustrated in Figure 2.35,[62] and the second was the implementation of "hydrogen loading" to increase the photosensitivity and thereby the magnitude of the refractive index change.[63] KrF (248 nm) illumination is preferred because it matches an absorption peak for oxygen-deficient germanium sites present in most germanosilicate glasses. This permits efficient absorption of the radiation and facilitates the creation of other defects through multiphoton absorption.[64] Hydrogen loading entails the impregnation of fiber with molecular hydrogen (or deuterium) under high pressure (e.g., 20 to 750 atm) for as long as several days. After impregnation, the fibers exhibit enhanced photosensitivity which diminishes after several hours when the hydrogen diffuses out of the fiber core. The mechanism of the enhancement is not fully understood, but it is postulated that hydrogen stabilizes intermediates in the photosensitive reaction. Achieved index changes can be of the order of 0.02 (>1%) as shown in Figure 2.36 for a commercial dispersion-shifted fiber.[63] Most of the index change is permanent under ambient conditions, but is annealed out at temperatures above a few hundred degrees Celsius. The photosensitivity generally increases with germania content, but useful gratings can be written in low- through high-NA germanosilicate core fibers.

Gratings such as those shown in Figure 2.35 can be defined with precise periodicity and can be used to achieve full or partial reflection of signals with subnanometer wavelength resolution. Such filtering can be used for demultiplexing signals of different wavelength. Also, by selecting appropriate fiber designs, the reflected light may be partially transferred to cladding modes, where it is attenuated. This can be exploited to flatten gain profiles in OFAs for enhanced broadband operation. Gratings can be angled for off-axis monitoring or chirped (tailored gradations of periodicity) for tuning device performance. Tuning can be accomplished through constructive and destructive interference and reflection.

The fabrication and implementation of photodefined components continues to be an area of considerable activity. As performance demands increase, concerns with

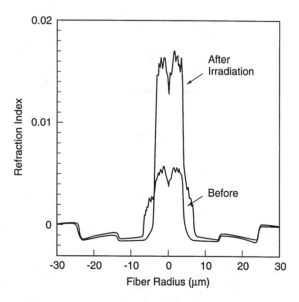

FIGURE 2.36 Effect of UV irradiation (248 nm KrF, 600 mJ/cm^2 for 1 h) on the refractive index profile of a standard single-mode germanosilicate fiber after hydrogen loading.[63]

manufacturability and reliability naturally emerge. There is scope for materials optimization of the photosensitive fibers to reduce the reliance on hydrogen loading and also to minimize index degradation with time. Concern over the latter is addressed at present by partially annealing the gratings (e.g., at 100°C); this reduces subsequent decay in index at ambient conditions to a few percent over the projected system lifetime.[65] Also, because the periodicity of the gratings can change with temperature, many devices are operated isothermally. Temperature control provides an additional mechanism for tuning performance. Applications incorporating gratings and photodefined components are far from exhausted.

2.6.7 FIBER LASERS

Narrow-band, single-mode fiber lasers have been demonstrated using rare earth–codoped fibers with cavities defined by Bragg gratings. Device lengths of only a few centimeters are practical as shown in Figure 2.37.[66] The operational wavelength can also be tuned by using temperature to adjust the periodicity of the gratings. Such devices are robust and have potential use as sources and pumps both in telecommunications and other areas. Pump light can be launched external to the cavity using fused fiber couplers.

Superfluorescent fiber sources including cladding pumped fiber lasers have also been demonstrated.[67] These offer high output power with the possibility of inexpensive diode array pumping. Applications include pumps for OFAs and Raman amplifiers. One fiber structure that has been employed is shown in Figure 2.38. It comprises a single-mode core codoped with rare earths (e.g., Nd or Yb) and a flattened silica "pump" cladding which supports multimode propagation when surrounded by

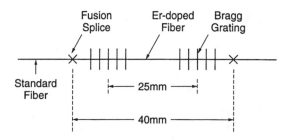

FIGURE 2.37 Schematic of a short-length fiber laser incorporating Er-doped fibers and a laser cavity defined by Bragg gratings.[66]

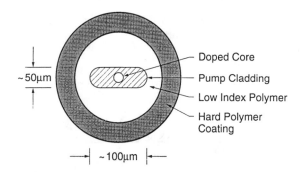

FIGURE 2.38 Schematic of a cladding pumped fiber laser.[67]

a lower-index polymer cladding (e.g., $n_D = 1.39$). Pump light, launched at an angle or from the side, becomes trapped in the pump cladding and can be efficiently absorbed by the core as it propagates along the fiber. Output powers >5 W have been demonstrated using externally defined cavities. The flattened geometry of the pump cladding facilitates better coupling to output of the diode array.

2.6.8 FIBERS FOR RAMAN LASERS AND AMPLIFIERS

The concept of using stimulated Raman scattering to fabricate lasers in germanosilicate fibers has been of interest since the 1980s.[68] The development of photodefined gratings and high-power fiber lasers has made this a practical option in the 1990s. The basis of operation is phonon (or multiphonon) scattering, which is offset (Stokes and anti-Stokes) from the pump frequency. This scattered light can be optically confined and may propagate in a fiber structure. The Stokes spectrum of a germanosilicate fiber, pumped at 1064 nm, is shown in Figure 2.39.

The effect can be used to fabricate lasers for different wavelengths by employing configurations such as that depicted in Figure 2.40. This is composed of high-NA germanosilicate fiber, typically 0.1 to 1.0 km in length, nested within cascaded photodefined Bragg gratings and employs a high-power pump source (e.g., a cladding pumped fiber laser). The gratings selectively confine and recirculate the pump and Raman frequencies up to the $(n - 1)$th order, while permitting lasing from the nth

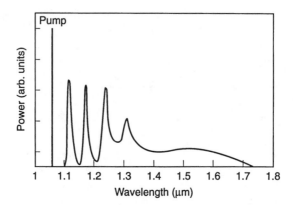

FIGURE 2.39   Raman (Stokes) spectrum of a high-NA germanosilicate core fiber. The pump in this case was a Nd:YaG laser at 1064 nm.

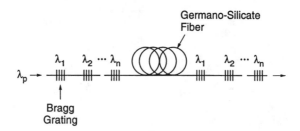

FIGURE 2.40   Schematic of a Raman fiber laser, incorporating high-NA germanosilicate fiber and a series of nested Bragg gratings to produce progressive laser cavities.

order. Such lasing can be contrived at almost any wavelength within the optical window of silica-based fibers by using different pumps and/or accessing different-order Raman peaks. Power conversion can be quite efficient at high pump power as shown in Figure 2.41 for 1240 nm (third order) laser emission from a fiber pumped at 1064 nm.[69]

Amplification of signals in the frequency range of nth-order Raman peaks can also be achieved by using gratings that confine Raman frequencies up to the $(n-1)$th order. Gain is obtained through nth-order stimulated Raman emission as the signal propagates through the same fiber. Using this method, amplification of 1310 nm signals was demonstrated for the first time in germanosilicate fibers using a 1064 nm (Nd^{3+}) cladding pumped laser and cascaded gratings for 1117, 1175, and 1240 nm.[69] Gains up to 25 dB have subsequently been achieved with only 350 mW of pump power in high-NA fibers.[70] An alternative strategy, which has also been demonstrated for 1310 nm amplification, is to establish a recirculating germanosilicate fiber ring in which power can be directly transferred from higher-order Raman emissions without the use of cascaded gratings.[71]

The 1310 nm Raman amplifiers offer a conventional fiber alternative to rare earth–doped fluoride, and other nonsilicate fiber amplifiers. In analogy to Er-doped OFAs this can enable the conversion of embedded systems to all-optical and broadband

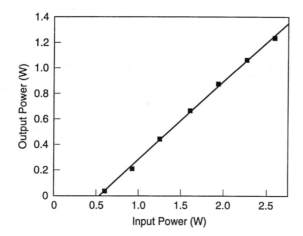

FIGURE 2.41 Example of the performance of a Raman fiber laser.[69]

operation. Broadband operation is feasible since the Raman peaks are relatively broad (see Figure 2.39) and the wavelength of cladding pumped fiber lasers can be varied by as much as 100 nm. Gain equalization can also be implemented as in Er-doped OFAs.

Lasers and amplifiers for the 1550 nm optical window are also accessible via cascaded Raman configurations. High-power 1480 nm lasers are considered practical for pumping of some Er-doped OFAs.[72] Raman amplifiers have intrinsically higher noise figures than Er-doped OFAs; however, experiments have shown that this may be adequately addressed through prudent configurational design.

2.6.9 TRUEWAVE FIBERS

TrueWave and analogous fibers are fast becoming an industry standard, replacing existing dispersion-shifted fiber designs for 1550 nm broadband operation especially in long-haul systems. The index profile of TrueWave fiber is triangular on a raised pedestal (inner cladding) and is similar to some of the designs shown in Figure 2.4. The purpose of the new design is to introduce a small gradation, typically <0.1 ps/nm^2·km, in dispersion across the 1550 nm band. This minimizes interference between discrete wavelength channels due to nonlinear effects including four-wave mixing. These effects can be limiting in long-haul systems, where power densities can be high and the interaction lengths are long, particularly if the channel spacing is small and there is no time delay (dispersion) between them. TrueWave fiber makes DWDM a practical option in long-haul systems, although the fiber attenuation is marginally increased (typically <0.3 dB/km).

2.7 FUTURE DIRECTIONS

High-silica-content fibers are firmly established as the interconnection media of choice for optical telecommunication networks. They have met, and continue to

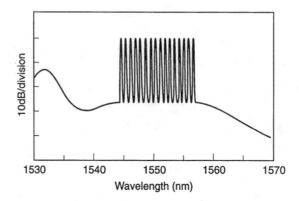

FIGURE 2.42 Amplification in a WDM system with 16 channels using high-silica-based Er/Al codoped fiber. In this case, there is excellent gain equalization over the wavelength band of the channels.[60]

meet, system needs, and their potential is far from exhausted. The manufacturing of such fibers is a mature technology. For more than a decade, refinements in standard fiber design, composition, and fabrication have been incremental and geared to performance optimization or cost reduction. Existing processes and compositions have also proved appropriate, with some modification, for manufacturing specialty telecommunication fibers. This is an area of continuing activity, which is developing synergistically with the demand for increased network functionality. Alternative glass fibers that offer some aspect of enhanced performance may be accepted for use in discrete components if they are able to satisfy both real and perceived reliability issues.

In contrast to the fibers, advances in system design, capacity, and operation in the 1990s have been radical and largely attendant on the implementation of all-optical networks. Areas of continuing interest include (1) higher transmission capacity, (2) longer-range transmission, (3) network distribution, (4) multimedia transmission, and (5) hybrid and intelligent networks.

Higher transmission capacity can be accessed through higher bit-rate operation or by increasing the number of transmission channels (i.e., DWDM). Both routes are being aggressively pursued. Some systems now operate at 10 Gb/s/channel, and this could increase to 40 Gb/s in the near future based on demonstrated advances in sources and detectors. The maximum number of channels that can be employed depends on the usable bandwidth, channel spacing, and gain spectra of the OFAs. Systems with 16 channels have been demonstrated using conventional high-silica OFAs with a 15 nm bandwidth as shown in Figure 2.42.[73] By implementing gain equalization, 32- and 64-channel operation should be achievable with similar components. For Raman gain amplifiers, the bandwidth could span the optical window of the fiber. In this extreme case, the wavelength-dependent attenuation would be substantial and the gain profile would have to be tailored to equalize signal intensity. System transmission capacities (bit rate × number of channels), exceeding 1 Tb/s have already been demonstrated experimentally.[74]

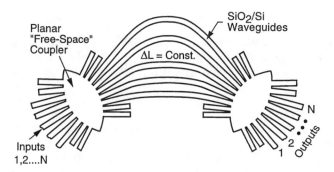

FIGURE 2.43 Schematic of a high-silica planar light guide circuit for routing optical signals. These passive components are typically a few centimeters in width and less than 10 cm in length.[75]

The practical range for coherent all-optical transmission with in-line OFAs can be limited by dispersion, nonlinear phenomena, cross talk between channels, and/or noise. These factors are enhanced in systems operating at high bit rate and/or narrow channel spacing. New fiber designs, including nonzero dispersion-shifted fiber, can offer some relief while advances in data handling, and system operation (both largely software driven) and noise reduction should also extend the practical transmission range.

Networks are now being designed and constructed that require large-scale routing and distribution of data. As the level of interconnection required increases (e.g., in metropolitan networks), it becomes practical to incorporate planar waveguide components in the networks. This will reduce the number of individual fiber components that has to be concatenated to achieve large-scale optical interconnection. High functionality, including multiplexing (see Figure 2.43), can be achieved using passive components only a few centimeter in length. At present, planar waveguide technology is analogous to optical fiber technology at an earlier stage of its development.[75] There is considerable scope for advances in waveguide design, materials, processing, packaging, and interconnection.

In terms of network management, it may be advantageous, from the point of view of cost, reliability, ease of reconfiguration and/or multimedia transmission, to construct "passive optical networks" in which the active network components, including switches, are located only at termination points. Performance and reliability may also be improved by implementing "intelligent" networks in which transmission channels are dedicated for active monitoring and control.

New network architectures have been proposed to address specific applications. One example is the "ring around Africa," comprising of an offshore fiber-optic ring with strategic onshore drops for intracontinental interconnection. Also under development are hybrid networks where fiber and wireless systems will be integrated to provide efficient solutions for both local and large-area interconnection.

The vision of a high-speed global information network, substantially composed of optical fiber telecommunication links, that is interactive and supports multimedia data transmission is well on the way to becoming a reality.

REFERENCES

1. S. E. Miller and A. G. Chynowyth, Eds., *Optical Fiber Telecommunications*, Academic Press, New York (1979).
2. S. E. Miller and I. P. Kaminow, Eds., *Optical Fiber Telecommunications II*, Academic Press, New York (1988).
3. T. Li, Ed., *Optical Fiber Telecommunications: Vol 1. Fiber Fabrication*, Academic Press, New York (1985).
4. H. Murata, *Handbook of Optical Fibers and Cables*, Marcel Dekker, New York (1988).
5. J. C. Daly, Ed., *Fiber Optics*, CRC Press, Boca Raton, FL (1984).
6. T. Izawa and S. Sudo, *Optical Fibers: Materials and Fabrication*, D. Reidel, Boston (1987).
7. L. G. Cohen, *J. Lightwave Technol.*, LT3, 958 (1985).
8. K. J. Beales and C. R. Day, *Phys. Chem. Glasses*, 21, 521 (1980).
9. H. M. J. M. Van Ass, R. G. Gossink, and P. J. W. Severin, *Electron. Lett.*, 12, 369 (1976).
10. A. D. Pearson and W. R. Northover, *Am. Ceram. Soc. Bull.*, 57, 1032–1039 (1978).
11. S. Takahashi, S. Shibata, and M. Yasu, *Rev. NTT Electr. Commun. Lab.*, 27, 123 (1979).
12. T. Yamazaki and M. Yoshiyagawa, *Proc. IOOC'77*, 617–620, Tokyo (1977).
13. H. N. Daglish, *Glass Technol.*, 11, 30 (1970).
14. J. Schroeder, R. Mohr, P. B. Macedo and C. J. Montrose, *J. Am. Ceram. Soc.*, 56, 510 (1973).
15. D. A. Pinnow, L. G. Van Uitert, T. C. Rich, F. W. Ostermeyer, and W. H. Grodkiewicz, *Mater. Res. Bull.*, 10, 133 (1975).
16. K. J. Beales, C. R. Day, W. J. Duncan, and G. R. Newns, *Electron. Lett.*, 13, 755 (1977).
17. M. E. Lines, *J. Non Cryst. Solids*, 103, 279 (1988).
18. A. J. Bruce, W. A. Reed, A. E. Neeves, M. Chui-Sabourin, L. G. Van Uitert, W. H. Grodkiewicz, and M. E. Lines, *Am. Ceram. Soc. Bull.*, 68, 1297, Paper 18GP-89F (1989).
19. P. B. Macedo and T. A. Litovitz, *Method of Producing Optical Waveguide Fibers*, U.S. Patent #3938974 (1976).
20. C. R. Bamford and D. G. Loukes, *Glass Technol.*, 20, 166 (1979).
21. P. C. Schultz, *Wiss. Z. Friedrich-Schiller-Univ. Jena, Math. Naturwiss. Reihe*, 32, 215 (1983).
22. H. Takahashi, A. Oyobe, M. Kosuge, and R. Setaka, *Tech. Dig., ECOC'86*, 3 (1986).
23. R. C. Weast, Ed., *Handbook of Chemistry and Physics*, 70th ed., CRC Press, Boca Raton, FL (1990).
24. S. B. Poole, D. N. Payne, and M. E. Ferman, *Electron. Lett.*, 21, 737 (1985).
25. P. C. Schultz, *J. Am. Ceram. Soc.*, 57, 309 (1974).
26. D. B. Keck, R. D. Maurer, and P. C. Schultz, *Appl. Phys. Lett*, 22, 307 (1973).
27. J. F. Hyde, *Making a Transparent Article of Silica*, U.S. Patent #2272342 (1942).
28. J. R. Simpson, J. B. MacChesney, K. L. Walker, and D. L. Wood, *Adv. Ceram.*, 2, 8 (1981).
29. J. B. MacChesney and D. J. DiGiovanni, *Materials Science and Technology*, Vol 9: *Glasses and Amorphous Materials*, J. Zarzycki, Ed., 753–780, VCH, New York (1991).
30. J. B. MacChesney, D. W. Johnson, D. A. Fleming, and F. W. Walz, *Electron. Lett.*, 23, 1005 (1987).

31. P. K. Bachmann, P. Geittner, and H. Lydtin, *Tech. Dig., OFC'86,* 76 (1986).
32. J. Wysocki, *Appl. Phys. Lett.,* 34, 17 (1979).
33. K. J. Beales, D. M. Cooper, W. J. Duncan, and J. D. Rush, *Tech. Dig., OFC'84,* paper W15 (1984).
34. P. J. Lemaire, K. S. Kranz, K. L. Walker, R. G. Huff, and F. V. DiMarcello, *Electron. Lett.,* 24, 1323 (1988).
35. C. M. Miller, S. C. Mettler, and I. A. White, *Optical Fiber Splices and Connectors,* Marcel Dekker, New York (1986).
36. P. S. Oh, *Ceram. Trans.,* 49, 125 (1995).
37. R. Csencsits, P. J. Lemaire, W. A. Reed, D. S. Shenk, and K. L. Walker, *Tech. Dig. OFC'84,* 54 (1984).
38. H. Yokato, H. Kanamori, Y. Ishiguro, G. Tanaka, S. Tanaka, H. Takada, M. Wantanabe, S. Suzuki, K. Yano, M. Hoshikawa, and H. Shimba, *Tech. Dig. OFC'86,* 11 (1986).
39. H. Osanai, T. Shioda, T. Moriyama, A. Arki, M. Horiguchi, T. Izawa, and H. Takata, *Electron. Lett.,* 12, 549 (1976).
40. N. Shibata, *Optical Characteristics of High Silica Glass Fibers for Communication,* Ph.D. Thesis (Nagoya University, Japan, 1983).
41. S. T. Davey, D. L. Williams, D. M. Spirit, and B. J. Ainslie, *Proc. SPIE,* 1171, 181 (1989).
42. D. Marcuse, *Light Transmission Optics,* Van Nostrand Reinhold, New York (1982).
43. A. Tomita and P. J. Lemaire, *Electron. Lett.,* 21, 71 (1985).
44. K. E. Lu, G. S. Glaesmann, M. T. Lee, D. R. Powers, and J. S. Abbott, *J. Opt. Quantum Electron.,* 22, 227 (1990).
45. H. Itoh, Y. Ohmori, and M. Nakahara, *J. Lightwave Technol.,* LT4, 473 (1986).
46. S. R. Nagel, *Proc. SPIE,* 717, 8 (1986).
47. P. S. Oh., P. H. Prideaux, and X. G. Glavas, *Opt. Lett.,* 3, 241 (1982).
48. V. J. Tekippe, *Proc. SPIE,* 1085, 72 (1989).
49. T. Haibara, T. Nakashima, M. Matsumoto, and H.Hanafusa, *IEEE Photonics Technol. Lett.,* 3, 348 (1991).
50. R. B. Dyott, J. R. Cozens, and D. G. Morris, *Electron. Lett.,* 15, 380 (1979).
51. R. H. Stolen, W. Pleibel, and J. R. Simpson, *IEEE J. Lightwave Technol.,* LT2, 639 (1984).
52. M. P. Varnham, D. N. Payne, R. D. Birch, and E. J. Tarbox, *Electron. Lett.,* 19, 146 (1983).
53. A. M. Vengsarkar and A. E. Miller, *Tech. Dig.: Symp. Optical Fiber Measurements,* NIST Special Publication, 864, 175–180 (1994).
54. M. J. F. Digonet, Ed., *Rare Earth Doped Fiber Lasers and Amplifiers,* Marcel Dekker Inc, New York (1993).
55. E. Desurvire, Ed., *Erbium Doped Fiber Amplifiers: Principles and Applications,* Wiley Interscience, New York (1994).
56. J.-M. P. Delavaux and J. A. Nagel, *IEEE J. Lightwave Technol.,* 13, 703 (1995).
57. J. E. Townsend, W. L. Barnes, K. P. Jedrzejewski, and S. G. Grubb, *Electron. Lett.,* 27 (1991).
58. E. Desurvire, C. R. Giles, J. R. Simpson, and J. Zyskind, *Opt. Lett.,* 14, 1266 (1989).
59. M. Nakazawa, Y. Kimura, and K. Suzuki, *Proc. IEE/LEOS/OSA Topical Meeting on Optical Amplifiers and Applications,* Paper PDP1 (1990).
60. B. Clesca, D. Ronach, D. Bayart, Y. Sorel, L. Hamon, M. Guibert, J. L. Beylat, J. F. Kerdiles, and M. Semenkoff, *IEEE Photonics Technol. Lett.,* 6, 509 (1994).
61. K. O. Hill, Y. Fujii, D. C. Johnson, and B. S. Kawasaki, *Appl. Phys. Lett.,* 32, 647 (1978).

62. G. Meltz, W. W. Morey, and W. H. Glenn, *Opt. Lett.*, 14, 823 (1989).
63. P. J. Lemaire, R. M. Atkins, V. Mizrahi, and W. A. Reed, *Electron. Lett.*, 29, 1191 (1993).
64. R. M. Atkins, V. Mizrahi, and T. Erdogan, *Electron. Lett.*, 29, 385 (1993).
65. P. J. Lemaire, A. M. Vengsarkar, W. A. Reed, and D. J. Digiovanni, *Appl. Phys. Lett.*, 66, 2034 (1995).
66. G. A. Ball, W. W. Morey, and W. H. Glenn, *IEEE Photonics Technol. Lett.*, 3, 613 (1991).
67. H. Po, E. Snitzer, R. Tumminelli, L. Zenteno, F. Hakima, N. M. Cho, and T. Haw, *Tech. Dig. OFC'89*, paper PD7 (1989).
68. A. Hasegawa, *Appl. Opt.*, 23, 3302 (1984).
69. S. G. Grubb, T. Erdogan, V. Mizrahi, T. Strasser, W. Y. Cheung, W. A. Reed, P. J. Lemaire, A. E. Miller, S. G. Kosinski, G. Nykolak, and P. C. Becker, *Tech. Dig. OSA: Optical Amplifiers and Their Applications*, Washington, D.C., paper PD3 (1993).
70. E. M. Dianov, A. A. Abramov, M. M. Bubnov, A. V. Shipulin, A. M. Prokhorov, S. L. Semjonov, and A. G. Schebunjaev, *Opt. Fiber Technol.*, 1, 236 (1995).
71. S. V. Chernikov, Y. Zhu, R. Kashap, and J. R. Taylor, *Electron. Lett.*, 31, 472 (1995).
72. P. B. Hansen, L. Eskilden, S. G. Grubb, A. M. Vengsarkar, S. K. Korotky, T. A. Strasser, J. E. J. Alphonsus, J. J. Veselka, D. G. Digiovanni, D. W. Pecham, E. C. Beck, D. Truxal, W. Y. Cheung, S. G. Kosinski, D. Gasper, P. F. Wysocki, V. L. Da Silva, and J. R. Simpson, *Electron. Lett.*, 31, 1460 (1995).
73. S. Yoshida, S. Kuwano, and K. Iwashita, *Electron. Lett.*, 31, 1765 (1995).
74. A. H. Gnauck, A. A Chraplyvy, R. W. Tkach, J. L. Zyskind, J. W. Sulhoff, A. J. Lucero, Y. Sun, R. M. Jopson, F. Forghieri, R. M. Derosier, C. Wolf, and A. R. McCormick, *Tech Dig. OFC'96*, paper PD20-2 (1996).
75. H. M. Presby, Ed., *Selected Papers on Silica Integrated Optical Circuits*, SPIE Optical Engineering Press, Bellingham, WA (1996).

3 Fluoride Glass–Based Fibers

Jasbinder S. Sanghera, Lynda E. Busse,
Ishwar D. Aggarwal, and Charles Rapp

CONTENTS

0-8493-2489-0/98/$0.00+$.50
© 1998 by CRC Press LLC

3.1 INTRODUCTION

Fluoride glasses based on ZrF_4 are predicted to have a minimum optical loss of less than 0.01 dB/km, which is more than an order of magnitude lower than the 0.12 dB/km predicted and practically realized for silica fibers. This phenomenon is related to the fact that these are low-phonon-frequency glasses and, hence, the multiphonon energy is shifted to longer wavelengths. In addition, fluoride glasses possess low nonlinear refractive indices and, in some cases, a negative thermo-optic coefficient (dn/dT). Furthermore, these glasses are excellent hosts for rare earth elements. As a result, there are many applications for optical fibers, such as low-loss repeaterless links for long-distance telecommunications, fiber lasers and amplifiers, as well as infrared (IR) laser power delivery. More recently, there has been interest in using fluoride fibers in remote fiber-optic chemical sensor systems for environmental monitoring using diffuse reflectance and absorption spectroscopy.

It is quite apparent that the fluoride glasses are an important class of materials since there are numerous practical applications for both the glasses and fibers. Therefore, this chapter describes the glass compositions and properties, glass and fiber fabrication processes, and optical and mechanical properties of fluoride glass fibers. Some applications using fluoride fibers are also described.

3.2 FLUORIDE GLASS AND FIBER COMPOSITIONS

3.2.1 FLUOROBERYLLATE GLASSES

Historically, the first fluoride glasses were based on beryllium fluoride and have been known since at least 1926.[1] These glasses are very stable and have many unique properties including good ultraviolet (UV) transparency,[2] low index of refraction, low optical dispersion, and a low nonlinear refractive index. In fact, BeF_2 has the lowest refractive index ($n_D = 1.2747$) and the highest Abbe number ($v = 106.8$) of any inorganic material.[3] However, the combination of the high toxicity, volatility, and hygroscopic nature of BeF_2 poses serious problems in melting, forming, handling, and disposal of these glasses. Despite this, these glasses were investigated extensively in the 1970s for use as laser materials.[4] The primary interest was in the development of materials with a high resistance to damage for use in high-power laser systems. Since it is the nonlinear component of the refractive index that results in "self-focusing" and damage to the optical components in a high-power laser system, beryllium fluoride glasses should have an exceptional resistance to damage.[5-7]

Since BeF_2-based systems have not been fabricated into optical fibers, we shall not discuss them further. However, for those interested, the compositions and properties of the fluoroberyllates are discussed in more detail by Baldwin et al.[8] The rest of this chapter deals with heavy metal fluoride (HMF) glasses, so named since they contain significant quantities of heavier cations compared with Be.

3.2.2 FLUOROZIRCONATE GLASSES

Fluorozirconate glasses have been the most extensively studied HMF glasses and are so named since they contain ZrF_4 as the major component. The first ZrF_4-based glass was reported in 1975 and was obtained quite by chance while Poulain et al.[9] were trying to fabricate crystals from the ZrF_4–BaF_2–NaF system doped with NdF_3. However, X-ray examination revealed that the product was mostly amorphous. The glass-forming region for this system is shown in Figure 3.1 and encompasses the binary ZrF_4–BaF_2 (ZB) system when rapid quenching is employed. Even though these glasses were not very stable, this prompted a great deal of interest and it was not long before other metal fluorides were added to facilitate stable glass formation. Since then, many other glass-forming systems have been reported.[10] For example, the addition of ThF_4[11] and LaF_3[12] to the binary ZB system to form ZBT and ZBL glasses increases the glass-forming region and leads to more-stable glasses. Although these glasses are easily formed, they are not sufficiently stable for practical applications. In addition, thorium is radioactive (weak α emitter), and therefore its use is discouraged in many laboratories and institutions.

It has been demonstrated that the addition of only a few percent AlF_3 to the fluorozirconate glasses greatly enhances the glass stability.[13] In fact, the ZrF_4–BaF_2–LaF_3–AlF_3 (ZBLA) glasses can be formed in pieces weighing at least several hundred grams and can be used for practical applications such as small bulk optics and fibers.[14] Also, substitution of La with Gd gives rise to the ZBGA glasses

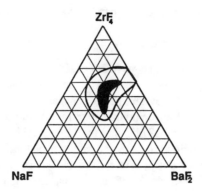

FIGURE 3.1   Glass-forming region in the ZrF$_4$–BaF$_2$–NaF system. Solid region corresponds to glasses obtained by slower cooling. (From Poulain, M., in *Fluoride Glass Fiber Optics,* I. D. Aggarwal and G. Lu, Eds., Academic Press, New York, 1991, 15. With permission.)

which have reasonable stability.[15] Aside from AlF$_3$, it was shown that the addition of NaF to the ZBL system also resulted in significant improvement in stability.[16] Although these glasses were stable, their crystallization tendency was still high enough to cause problems during fabrication of low-loss fibers. This prompted Ohsawa and Shibata[17] to add both AlF$_3$ and NaF to the ZBL system to fabricate ZBLAN glasses. Generally speaking, these glasses are very stable, can be formed into pieces weighing in excess of 40 kg,[18] and are the compositions most frequently used in bulk optics and IR fibers.

The increase in the glass stability with the addition of various fluorides can be seen in the series of glasses listed in Table 3.1. Critical cooling rates without crystallization, as determined by differential scanning calorimetry (DSC), are given for each glass. The glasses with the slowest critical cooling rates are those with the greatest stability with respect to crystallization. The stabilizing effect of AlF$_3$ and

TABLE 3.1
Critical Cooling Rates, R_c (°C/min), for Several ZrF$_4$–BaF$_2$–Based Glasses

Glass ID	ZrF$_4$	BaF$_2$	GdF$_3$	LaF$_3$	YF$_3$	AlF$_3$	LiF	NaF	R_c
1	63	33	4	—	—	—	—	—	370
2	60	32	4	—	—	4	—	—	70
3	52	20	—	5	—	3	20	—	26
4	49	22	—	3	3	3	20	—	26
5	48	23.5	—	2.5	2.5	2.5	—	20	4
6	47.5	23.5	—	2.5	2	4.5	—	20	1.1
7	53	20	—	4	—	3	—	20	0.7

After Kanamori, T. and Takahashi, S., *Jpn. J. Appl. Phys.,* 24, L758, 1985 and Iwasaki, H., *SPIE,* 618, 2, 1986.

TABLE 3.2
Typical HMF Glasses Used in the Preparation of Optical Fibers

	ZrF_4	BaF_2	LaF_3	GdF_3	AlF_3	LiF	NaF	PbF_2	HfF_4	n_D	Ref.
Clad	59.6	31.2	—	3.8	5.4	—	—	—	—	1.5132	21
Core	61	32	—	4	3	—	—	—	—	1.5162	
Clad	53	19	5	—	3	20	—	—	—	1.512	21
Core	51	16	5	—	3	20	—	5	—	1.525	
Clad	53	20	4	—	3	—	20	—	—	1.4991	21–24
Core	53	15	4	—	3	—	20	5	—	1.5106	
Clad	39.7	18	4	—	3	—	22	—	13.3	1.4925	21
Core	53	20	4	—	3	—	20	—	—	1.4991	

the alkali fluorides can be seen. One particular composition that has shown unusual stability consists of 53 mol% ZrF_4, 20% BaF_2, 4% LaF_3, 3% AlF_3, and 20% NaF and is commonly called ZBLAN.[21-24] Few compositions are known which show an equal or lesser tendency to crystallize, and, therefore, this or similar compositions have been commonly used in the fabrication of optical fibers. However, for an optical fiber it is necessary to have a second glass of lower or higher refractive index to act as the cladding or core glass with the ZBLAN, respectively. It is also desirable that the two glasses have similar softening or glass transition temperatures and similar viscosity profiles vs. temperature. This minimizes both the crystal growth rate in both glasses and the induced mechanical stress due to the fiber-drawing process. Formulation of the second glass is usually accomplished by the partial substitution of other fluorides into the base glass. Examples of core and cladding glass pairs are shown in Table 3.2. The numerical apertures for the third and fourth pairs of glasses are 0.18 and 0.13, respectively. In addition to the substitutions of HfF_4 for ZrF_4 and PbF_2 for BaF_2, many other partial or full substitutions can be made in the ZBLAN composition to increase or decrease the refractive index. Some of these include other rare earth fluorides for LaF_3; GaF_3 or InF_3 for AlF_3; and PbF_2, LiF, or CsF for NaF. Some of these partial substitutions have little effect on the glass stability, while others can significantly decrease the glass stability. Unfortunately, both the HfF_4 and PbF_2 substitutions described above result in decreases in the stability of the glass with respect to crystallization.

One measure of the glass stability is the difference between the glass transition temperature (T_g) and the first indication of crystallization (T_x) of the glass on heating using differential thermal analysis (DTA) or DSC. The decrease in glass stability ($T_x - T_g$) for the PbF_2- and HfF_4-substituted glasses is shown in Table 3.3. A decrease in $T_x - T_g$ by 20°C results in a large melt, such as 500 to 1000 g, to exhibit crystallization.[25] Also shown is a new glass composition based on a Li–Ga substitution which exhibits good stability ($T_x - T_g = 102$°C) and may be a candidate for fiber drawing.

TABLE 3.3

Thermal and Optical Properties of ZBLAN and Modified ZBLAN Glasses for Core/Clad Optical Fibers

	ZBLAN	Hf	2Pb	4Pb	Li & Ga
ZrF_4	53	39.7	53	53	53
BaF_2	20	18	20	20	25
LaF_3	4	4	4	4	4
AlF_3	3	3	3	3	1
NaF	20	22	18	16	8
HfF_4	—	13.3	—	—	—
PbF_2	—	—	2	4	—
LiF	—	—	—	—	7
GaF_3	—	—	—	—	2
n_D	1.4991	1.4934	1.5074	1.5148	1.5104
T_g (C)	259	259	260	262	256
T_x (C)	359	340	348	339	358
$T_x - T_g$ (C)	100	81	88	77	102
NA^a	—	0.13	0.16	0.22	0.18

[a] Numerical aperture with ZBLAN.

After Nice, M. L. et al., Final Report, Contract number N00014-91-C-2361, 1993.

It is often found in glass composition research that the properties of a glass are linearly dependent on the amount of each constituent in the glass. This is true for many physical, thermal, and optical properties. Therefore, linear models or equations can be derived which relate the properties of a glass to the composition. These often take the form of

$$B = a_1 p_1 + a_2 p_2 + a_3 p_3 + \cdots + a_n p_n \qquad (3.1)$$

where B is the value of the property, p_1, p_2, p_3, ..., p_n are the percentages of the various glass constituents, and a_1, a_2, a_3, ..., a_n are empirically derived constants. Properties such as the density, refractive index, and coefficient of thermal expansion show a linear dependence on the composition over a very large compositional range. Other properties such as the liquidus temperature and glass stability may show a linear dependance on the composition over a very limited range. Nice et al.[25] have used this approach to derive coefficients which can be used to calculate n_D, T_g, and $T_x - T_g$ for ZBLAN types of glasses (Table 3.4). These coefficients can be very useful in formulating a glass composition with the desired combination of n_D and T_g. The positive effects of LaF_3, AlF_3, and GaF_3 and the negative effect of PbF_2 on the glass stability can be seen in the $T_x - T_g$ coefficients.

TABLE 3.4

Coefficients for Calculating Changes in n_D, T_g, and $T_x - T_g$ from the Composition of Modified ZBLAN Glasses

	$\Delta n_D \times 10^{-2}$	ΔT_g	$\Delta(T_x - T_g)$
ZrF_4	1.5237	2.8079	1.028
BaF_2	1.5314	2.9220	−0.120
NaF	1.4132	1.0936	0.222
LaF_3	1.5398	4.3731	6.563
AlF_3	1.3758	4.0545	3.963
LiF	1.4682	−0.4496	1.530
GaF_3	1.4379	3.5455	3.882
PbF_2	1.7387	1.8486	−3.433
HfF_4	1.4989	3.0221	0.603
CsF	1.3987	1.1250	0.685
$NaCl$	1.5875	−0.8398	—
YF_3	1.4486	4.1101	—
InF_3	1.5107	3.2486	—

After Nice, M. L. et al., Final Report, Contract number N00014-91-C-2361, 1993.

3.2.3 FLUOROALUMINATE GLASSES

Aluminum fluoride glasses were first reported in 1949 by Sun[26] in the AlF_3–PbF_2–SrF_2–MgF_2 system. Since these glasses were not very stable, there was not much interest in them. However, after discovery of the fluorozirconate glasses, attention also turned into making more-stable fluoroaluminate glasses. Initial glass compositions based on AlF_3–BaF_2–CaF_2 (ABC)[27] were soon followed by glasses in the YF_3–AlF_3–BaF_2–CaF_2 (YABC)[28] and BaF_2–AlF_3–ThF_4–YF_3 (BATY)[29] systems. Although the latter compositions have been fabricated into rods and fibers, they are still prone to crystallization, and are therefore inappropriate for low-loss fiber fabrication. Glasses have been fabricated in the CdF_2–LiF–AlF_3–PbF_2 (CLAP) system,[30] but it appears that these glasses exhibit phase separation.[31]

More recently, it has been demonstrated that small quantities of ZrF_4 (<15 mol%) can stabilize the fluoroaluminate glass system.[31] These glasses are more chemically durable against attack by water and possess higher glass transition temperatures (typically around 400°C) than ZBLAN. Despite this, the glasses are not as stable as ZBLAN since their critical cooling rates are higher and also because AlF_3 shifts the multiphonon edge to shorter wavelengths. Nevertheless, appropriate core and cladding glass compositions have been developed[32] and are shown in Table 3.5. The numerical aperture of this pair is equal to 0.3.

TABLE 3.5
Fluorozircoaluminate Core and Cladding Glass Compositions (mol%)

	AlF$_3$	ZrF$_4$	YF$_3$	MgF$_2$	CaF$_2$	SrF$_2$	BaF$_2$	NaF	Cl	Index	T_g (C)
Clad	30.2	10.1	6.3	3.5	20.3	13.2	7.6	8.8	—	1.4313	369
Core	25.1	12.8	11.1	3.7	15.4	13.6	12.6	5.7	1.2	1.4629	362

After Miura, K., *Mater. Sci. Forum,* 67–68, 335, 1991.

3.2.4 OTHER FLUORIDE GLASSES

InF$_3$-based glasses have received attention since they transmit to slightly longer wavelengths than ZBLAN. Reasonably stable glasses have been fabricated from the BaF$_2$–InF$_3$–ZnF$_2$–YbF$_3$–ThF$_4$ (BIZYT) system.[33] Although numerous other glass-forming systems have been reported in the literature, e.g., glasses based on divalent cations,[34] ThF$_4$,[35] UF$_4$,[36] and oxyfluorides,[37] these are predominantly unstable and unsuitable for fabrication into optical fibers.

3.3 FABRICATION OF FLUORIDE GLASSES

High-purity chemicals are required for the preparation of ultralow-loss optical fibers. The two main chemical purification techniques employed have been based on dry and wet processing. Dry processing techniques include examples such as sublimation[38] and chemical vapor processing.[39] Examples of wet processing techniques include reprecipitation[40] and solvent extraction.[41] While the purification techniques are outside the scope of this chapter, a detailed review of purification procedures and analytical techniques is given by Ewing and Sommers.[42]

Certain precautions have to be taken when melting HMF glasses since both the fluoride precursors and the melt are readily attacked by moisture. Maze et al.[43] have defined three reaction mechanisms between water and fluorides which are dependent upon the temperature regimes shown below:

$$MF_n + xH_2O \leftrightarrow MF_n \cdot xH_2O \qquad T < 150°C \qquad (3.2)$$

$$MF_n + xH_2O \leftrightarrow MF_{n-x}(OH)_x + xHF \qquad 150°C < T < 250°C \qquad (3.3)$$

$$MF_n + xH_2O \leftrightarrow MF_{n-2x}O_x + 2xHF \qquad T > 600°C \qquad (3.4)$$

Furthermore, any hydroxyl groups present in the melt will react with the fluoride ions according to

$$OH^- + F^- \leftrightarrow O^{2-} + HF \qquad (3.5)$$

Overall, hydroxyl groups are liberated from the melt via formation of HF and formation of oxide. Since the hydroxyl levels are expected to be in the sub-parts-per-million (ppm) range, then only small levels of oxide will be formed. As will be shown in a following section, the presence of oxide species can have a detrimental effect on the glass stability and optical quality. Therefore, it is critical to use anhydrous chemicals and melt under dry and inert conditions. Unfortunately, this can lead to the formation of reduced Zr (3+ and 2+), which has a detrimental effect on the glasses. A partially reduced glass may show increased light scatter, a dark color, and/or an uncontrolled nucleation and crystallization of the glass. Therefore, a variety of melting processes have been developed that minimize these types of problems. One of the earliest, and probably still the most commonly used, is the ammonium bifluoride (ABF) process.[11]

3.3.1 AMMONIUM BIFLUORIDE PROCESSING

The primary advantage of ABF, NH_4HF_2 processing is that the glass can be formulated from most available fluorides, oxides, or hydroxides. During the infancy years of fluoride glass fiber-optic development, high-purity fluoride precursors were either not commercially available or difficult to obtain. Therefore, fluorozirconate glasses were fabricated using available fluorides and ZrO_2 as a precursor but with the addition of ABF.[11] Typically, the batches were mixed with ABF and heated slowly (and/or stepwise) to the glass melting temperature (about 800 to 850°C). This was usually done in a Pt or Au crucible in an inert atmosphere such as Ar. During the heating cycle, the ABF decomposes to produce NH_3 and HF gases. It is the HF gas that fluorinates the oxide precursors and converts them to their respective fluorides. The reaction between oxides and ABF can be represented as follows:

$$MO_n + (n - x/2)NH_4HF_2 \rightarrow MF_{2n} \cdot xNH_4F + nH_2O + (n - x/2)NH_3 \quad (3.6)$$

Water formed from the reaction with any oxides, along with any excess HF and NH_4F, is volatilized during the heating cycle. Ar and NH_3 atmospheres are reducing and tend to leave the glass in a chemically reduced state which leads to increased light scatter.[44] Therefore, later stages of melting usually include the use of an O_2-containing atmosphere in order to completely oxidize the glass.[45]

The dissociation rate of ABF is extremely high above 150°C and, therefore, its supply may be exhausted before complete fluorination has occurred. To overcome this problem, several modifications of the ABF process have been developed to produce low-oxide-containing glasses. For example, Takahashi et al.[46] pretreated high-purity precursors at 400°C with a mixture of ABF and HF gas followed by Ar/O_2 at 900°C. For comparison, a glass was made without using HF. The oxide contents were 84 and 290 ppm, respectively.

More recently, it was demonstrated that ABF undergoes a solid-state room-temperature chemical reaction with specific hydrated and anhydrous precursors used in ZBLAN types of glasses.[47] The products of this reaction were defined as ammonium fluoride–metal fluoride complexes. The kinetics of their formation and subsequent dissociation were followed by thermal gravimetric analysis (TGA) and the products identified using X-ray diffraction analysis. It was observed that these complexes decomposed with the evolution of HF above temperatures at which ABF decomposed. An example of one of the reaction schemes is shown in Equation 3.7. These materials were used to melt fluorozirconate glasses. In this case, the batches were mixed with ABF for about 18 h for complete prereaction. Then the batches were heated up to 500°C for several hours to dissociate completely and remove any residual ammonium fluoride complexes. In this manner, the oxygen contents were reduced from the range of 100 to 400 ppm for unprocessed chemicals down to 10 to 50 ppm for the new process.

$$
\begin{aligned}
ZrF_4 \cdot H_2O + 3NH_4FHF &\rightarrow (NH_4)_3 ZrF_7 + 3HF + H_2O \\
&\downarrow \\
&(NH_4)_2 ZrF_6 + NH_4F \\
&\downarrow \\
&NH_4 ZrF_5 \quad + NH_4F \\
&\downarrow \\
&ZrF_4 \qquad + NH_4F
\end{aligned}
\tag{3.7}
$$

Advantages of ABF processing include the wide selection of batch materials. Disadvantages include the large reduction in volume between the batch (including excess ABF) and the final glass, the necessity for an all-precious-metal (e.g., Pt) melting chamber, problems with unreacted oxides, control of the oxidation state, traces of NH_3, and possible contamination of the glass with precious metal particles. Furthermore, ABF is a source of contamination since it is typically wet and contains many impurities that have a detrimental effect on the glass quality.

3.3.2 DRY PROCESSING

Another common melting process used for the preparation of zirconium fluoride glasses is to directly melt the fluoride batch materials which are relatively free of oxide or hydroxide contamination. These high-purity chemicals are now commercially available from several vendors, such as BDH Corp. and Air Products, Inc., although distillation of some of the chemicals may still be required to lower further the oxide contents. Hence, it is possible to melt these precursors without ABF, but an oxidizing atmosphere is necessary to avoid possible reduction of the glass. The most common oxidizing atmospheres used are O_2 and mixtures of O_2 with N_2 or Ar.[48] These atmospheres must be very dry to avoid reaction with the water and thus increase the oxide content of the glass. Reaction of the fluoride batch materials and/or glass melt with the O_2 in the atmosphere is not thermodynamically favorable and so the increase in the oxide content of the glass by this reaction is negligible. However, if the glass contains chloride ions, reaction with O_2 in the melting atmosphere is favorable and the oxide content of the glass can increase significantly by

FIGURE 3.2   A large fluorozirconate glass window with a diameter of 21 in. and a thickness of 1.5 in.

this reaction.[18] For example, Rapp[18] has shown that remelting a fluorozirconate glass doped with 0.5 wt% Cl in a 100% O_2 atmosphere for 1 h at 800°C will increase the oxide content by about 300 ppm. No measurable increase in oxide was apparent after 3 h for the pure fluoride glass.

Direct melting or remelting of fluoride glasses in an O_2-containing atmosphere is usually done in a precious metal crucible (Pt or Au). A resistance-heated furnace with a platinum or fused silica liner to control the atmosphere is suitable for small melts. Typical melting temperatures and times for small melts (several hundred grams) can be 800°C for several hours. Large melts of many kilograms may take 24 h or more for complete melting and homogenization.[18] Figure 3.2 shows a large glass window fabricated in this manner. Small or medium-sized melts can be cooled and annealed in Pt crucibles since the glass separates from the crucible on cooling. This is not true for gold crucibles since the fluoride glass adheres to the gold crucible on cooling, much like oxide glasses do to platinum.

Advantages of the direct melt process using an O_2-containing atmosphere include the simplicity of the melting process, a relatively wide choice of batch materials, a highly oxidized glass with low concentration of scattering centers, and no carbon contamination. The melting process is quite simple compared with the ABF process since a fused silica tube can be used to control the atmosphere and because the processing times are shorter. Also the amount of toxic or corrosive gases produced during melting are very small compared with the ABF process. The primary disadvantages of this process are the necessity for low-oxide-content chemicals and the possible contamination of the glass with precious metal particles.[49] However, precious metal contamination should be no more of a problem in this process than for glasses prepared by the ABF process.

3.3.3 REACTIVE ATMOSPHERE PROCESSING

Reactive atmosphere processing (RAP) of fluoride glasses is based on the method used for halide crystal growth. In this case, the fluoride glass batches are melted in the presence of a reactive gas which removes hydroxyl and oxide impurities from the melt. Usually an inert gas (such as Ar or nitrogen) is mixed with a reactive gas such as CCl_4,[50] chlorofluorocarbons (CFCs),[51] SF_6,[52] and NF_3.[53] These halogen-containing compounds decompose at the melting temperatures of the glass to give free halogen. This free halogen provides the oxidizing conditions necessary for removing hydroxyl and oxide impurities and, therefore, for producing a good fluoride glass. For example, the overall effect of CCl_4 can be represented as follows:

$$O^{2-} + CCl_4 \rightarrow 2Cl^- + COCl_2 \qquad (3.8)$$

$$OH^- + CCl_4 \rightarrow Cl^- + COCl_2 + HCl \qquad (3.9)$$

While the by-products are gaseous and readily removed from the melt, it is evident that Cl^- ions are introduced into the fluoride melt. As pointed out in Section 3.3.2, this can lead to increased oxide levels in the glass during remelting in an oxygen environment. This has an adverse effect on the glass stability, as will be shown later. Therefore, it is advisable to use non-chlorine-containing gases such as SF_6 and NF_3, and, in this case, the reactions can be represented as follows:

$$O^{2-} + F_2 \rightarrow 2F^- + \tfrac{1}{2}O_2 \qquad (3.10)$$

$$2OH^- + F_2 \rightarrow 2F^- + H_2O + \tfrac{1}{2}O_2 \qquad (3.11)$$

These gases are strong oxidizing agents, and so they prevent the formation of reduced Zr species.

Melting is usually done in vitreous carbon crucibles to avoid precious metal contamination of the melt. The RAP gases react with Pt or Au crucibles to produce high levels of dissolved metal. Moderate levels of metal produce a coloration of the glass, while high levels can precipitate and give rise to significant light scattering in the glass. Therefore, a suitable melting process is to use RF induction heating of the vitreous carbon crucibles containing the batch materials inside a fused silica tube. A flowing atmosphere containing several percent SF_6 is sufficient to maintain oxidizing conditions for the glass. Melting times and temperatures are similar to the other processes (800°C for several hours). Since the silica tube is cool, little reaction occurs between the tube and the reactive atmospheres. The melting chamber can be as simple as a tube with end caps, or as complex as some systems developed where the fused silica tube is attached to the bottom of a glove box. One such system consisting of modified glove boxes[54] is shown in Figures 3.3 and 3.4. Systems such as this allow for the complete handling, weighing, and mixing of the batch materials under a controlled dry atmosphere. After melting is complete, the crucible can be

FIGURE 3.3   A dry box system used for handling fluoride glasses and precursors located inside a class 100 clean room. The RAP furnaces are attached to the underside of the dry box.

FIGURE 3.4   A RAP furnace showing the RF coils used for heating. The white discoloration represents the buildup of sublimation product after numerous melting cycles.

raised into the glove box while the glass is still molten for casting or annealing in the controlled atmosphere. Annealing can be done in the carbon crucible since the glass separates from the crucible on cooling. However, a thin layer of carbon or fluorocarbon can be present at the glass surface which had been in contact with the carbon crucible. This layer can be removed by grinding, aqueous etching, or reactive plasma etching.[55]

Glasses prepared in vitreous carbon crucibles with a reactive halide atmosphere have exhibited the lowest scattering losses of any melting process (see section on glass properties). Disadvantages can include a high chloride level in the glass if CCl_4 or other CFC atmospheres are used. Other problems can include toxic waste gases and the deposition of decomposition products of the reactive gases on the crucible walls during melting. These decomposition products make it difficult to directly cast preforms from glass melted in vitreous carbon crucibles. However, if the glass is remelted in Pt or Au under an O_2 atmosphere for a short time, the contamination is oxidized and excellent quality glass is obtained which can be easily processed into fibers. This two-step melting process can be used to produce high-quality optical fibers.

3.3.4 OTHER PROCESSING TECHNIQUES

3.3.4.1 Chemical Vapor Deposition

The chemical vapor deposition (CVD) technique is an alternative process to melting and casting of fluoride glasses into preforms since this process is routinely used to fabricate low-loss silica preforms and subsequently long lengths of low-loss fibers. Therefore, it is not unexpected that this route has been explored for fluoride glass fabrication. For example, Sarhangi and Thompson[56] have used organometallic precursors to deposit soot which has been consolidated into BeF_2 glass as well as BeF_2 glass, doped with AlF_3 (to increase the refractive index of BeF_2). The measured Rayleigh scattering loss of 2.4 dB/km at 488 nm was lower than the value of 11.6 dB/km for silica. Although BeF_2 is a good glass former, numerous problems were encountered during the fabrication process, such as boiling of BeF_2 layers during the preform collapse process.

The deposition of a soot or glassy powder, as is done in the silica CVD preform fabrication process, results from the precursors decomposing and reacting in the gas phase and condensing on a substrate as a particulate or powder, often called soot. This process is referred to as a homogeneous gas phase reaction. The soot is subsequently sintered into a dense glassy layer. In the case of HMF, the sintering process is not favorable since the surface tension, which is the driving force for the densification, is four times lower than for silica. Hence, sintering must be performed at elevated temperatures which subsequently leads to crystallization. However, the microwave plasma-assisted CVD process is a relatively low temperature deposition process suitable for forming glassy materials in a dense morphology for preform fabrication.[57] In this process, volatile fluorinated metal β-diketonate organometallic precursors were chosen as source materials based upon their thermal stability and sublimation

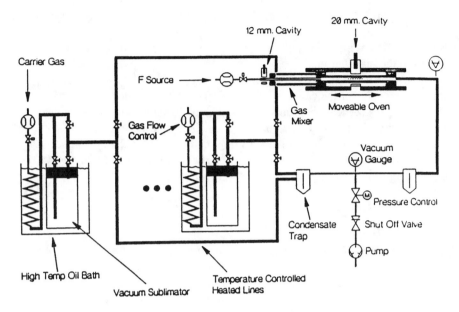

FIGURE 3.5   Schematic of microwave CVD system.

temperatures. The solid source precursors included $Zr(HFA)_4$, $Ba(FOD)_2$, $La(FOD)_3$, $Al(HFA)_3$, and $Na(HFA)$. These precursors were volatilized by vacuum sublimation and transported to the plasma reactor in an argon carrier gas. Figure 3.5 shows a schematic diagram of the microwave plasma-enhanced MOCVD (metal organic CVD) system. The microwave operated at a frequency of 2.45 GHz and at a power level of 75 W. In the presence of SF_6 gas, the films deposited on ZBLAN substrates were dense, glassy, transparent, and colorless with good adhesion. However, optical transmission data indicated slightly reduced transmission in the UV compared with the ZBLAN glass substrate. Furthermore, secondary ion mass spectroscopy indicated that the carbon content in the films was two to three orders of magnitude higher than the substrate. This value was reduced by the addition of 1% O_2 gas to the deposition chamber. Table 3.6 shows that the composition of the films was close to the target values. However, spatial uniformity was not investigated.

TABLE 3.6
Plasma-Enhanced Microwave CVD Deposition of ZBLAN Glass

Composition	mol%				
	ZrF_4	BaF_2	LaF_3	AlF_3	NaF
Target	53.0	20.0	4.0	3.0	20.0
Experimental	50.0	20.8	4.0	5.7	19.5

Fujiura et al.[58] have obtained similar results for 30-μm-thick ZBLAN films which were deposited inside ZBLYAN-cladding glass tubes at a deposition rate of about 60 μm/h. However, the films did not exhibit uniform composition along the entire length of the cladding tube.

While CVD processing is a mature technology for fabricating silica fiber optics, this is obviously not the case for HMF glasses. Certain questions still remain, such as whether or not the carbon impurities can be removed without compromising the HMF glass properties and whether or not coatings with uniform composition can be deposited. Also, can thicker glass coatings be fabricated with high optical quality and without crystallization? If these issues can be resolved, then low-loss HMF fiber optics via plasma-enhanced CVD processing will become a reality.

3.3.4.2   Sol–Gel Process

Only a few reports have been published regarding sol–gel processing of HMF glasses.[59,60] Typically, the metal alkoxides are used to form an oxide gel which is then partly dried and fluorinated with HF to produce the fluoride glass powder. This powder is subsequently remelted to form a monolithic fluoride glass sample. Such glasses have not been fabricated into optical fibers, presumably because of the high levels of unwanted oxide, hydroxyl, and carbon impurities.

3.4   FLUORIDE GLASS PROPERTIES

3.4.1   Intrinsic Optical Properties

The main reason for developing fluoride glasses and fibers is their optical properties. For example, greater than 50% transmission can be achieved between 0.25 and 7 μm (Figure 3.6).[61] While fluoroaluminate glasses do not transmit as far as fluorozirconate glasses, non-ZrF_4-based glasses based on heavy metal cations can transmit to even longer wavelengths. As a result, fluoride glasses are candidate materials for both bulk window and fiber IR optical applications.

The total intrinsic optical absorption (α_t) as a function of wavelength in an ideal vitreous solid can be expressed by[62]

$$\alpha_t = \left(A/\lambda^4\right) + B_1 \exp\left(B_2/\lambda\right) + C_1 \exp\left(-C_2/\lambda\right) \qquad (3.12)$$

where A, B_1, B_2, C_1, and C_2 are constants. The first term represents the Rayleigh scattering loss due to microscopic density and composition fluctuations in the material. The second and third terms describe the losses due to UV absorption from the electronic band edge (Urbach tail) and the IR losses arising from multiphonon absorption, respectively. The combination of these three terms gives rise to the V-shaped transparency curve[63] shown in Figure 3.7. In fluoride glasses, the contribution from the electronic edge is negligible in the IR, and therefore intersection of the Rayleigh and multiphonon curves can be used to predict the minimum loss and wavelength at which this occurs. Theoretical estimates for fluorozirconate glasses

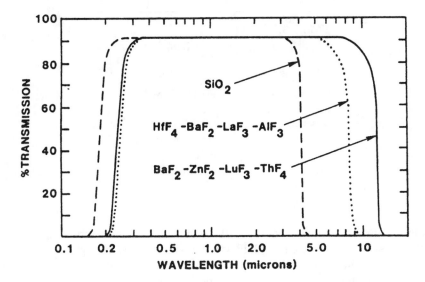

FIGURE 3.6 Transmission curves for 5-mm-thick samples of fused silica and two heavy metal fluoride glasses. (From Drexhage, M. G., in *Treatise on Materials Science & Technology,* Vol. 26, *Glass IV,* M. Tomozawa and R. Doremus, Eds., Academic Press, New York, 1985, 208. With permission.)

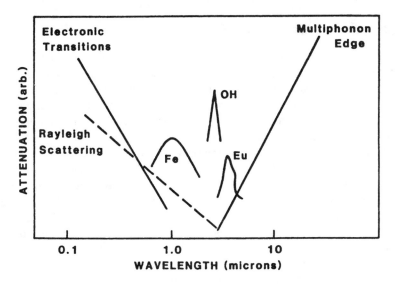

FIGURE 3.7 Schematic representation of the electronic absorption, Rayleigh scattering, and multiphonon absorption in a fluoride glass. Some extrinsic absorption bands are also shown. (From Miyashita, T. and Minabe, T., *IEEE J. Quantum Electron.,* QE-18, 1432, 1982. ©1982 IEEE. With permission.)

TABLE 3.7
Absorbing Impurities in Fluoride Glasses

Type of Species	Examples of Species	Ref.
Transition metal ions	Fe^{2+}, CO^{2+}, Ni^{2+}	64, 65
Rare earth ions	Nd^{3+}, Pr^{3+}	64, 65
Dissolved gases	CO, CO_2	66
Molecular ions	OH^-	67
	NH_4^+	68
	SO_4^{2-}, PO_3^-	69
Reduced Species	Zr^{3+}, Zr^0	70
Oxides	Zr–O	71

lead to values of approximately 0.01 dB/km at 2.55 μm. For comparison, the minimum loss of silica is about 0.12 dB/km at 1.55 μm. In fluoride glasses the actual values for minimum loss and the corresponding wavelength are dependent upon the glass composition since this dictates the position of the multiphonon edge and the contribution from Rayleigh scattering. Therefore, shifting the edge to longer wavelengths can lead to lower theoretical losses at slightly longer wavelengths. A more-detailed analysis of the parameters in Equation 3.12 for fluorozirconate fibers is given later in this chapter. Unfortunately, the intrinsic loss limit has not been attained in fluoride glasses as a result of several factors, namely, extrinsic absorption and scattering, and these are discussed in the following sections.

3.4.2 EXTRINSIC ABSORPTION

While the transmission window of fluoride glasses extends from about 0.2 μm to beyond 8 μm depending upon composition, there are numerous impurities which possess absorption bands within this transmission window and these are listed in Table 3.7. Some of the major contributors are transition metal and rare earth ions.[64,65] A complete review of the position of the absorption bands and the absorption coefficients of these species in fluoride glasses can be found in Reference 45. The contributions from the rare earth and transition metal ions to the loss at 2.55 μm for some fluorozirconate glasses are listed in Table 3.8. The oxidation states of the transition metal and rare earth ions are +2 and +3, respectively. The oxidation state of Fe can be +2, +3, or a mixture of +2 and +3 depending upon whether an oxidizing or reducing atmosphere was used during melting. The detrimental effect of Fe^{2+} at 2.55 μm is highlighted by its high absorption coefficient of 15 dB/km/ppm compared with <0.1 dB/km/ppm for Fe^{3+}. Practically, it is envisioned that all the Fe is in the +3 state due to melting under oxidizing conditions and, therefore, has little effect on the loss at 2.55 μm. On the other hand, Co^{2+} and Nd^{3+} have a significant contribution to the loss at 2.55 μm. Table 3.9 shows the measured concentration of several ions in a fluoride glass along with the contribution to the total loss at 2.55 μm. The transition metal concentrations were measured using graphite furnace atomic absorption (GFAA) spectroscopy while the Nd concentration was determined using

TABLE 3.8
Absorption Coefficients (dB/km-ppm) for
Cationic Impurities in Fluorozirconate Glasses

Ion	Ref. 64[a]	Ref. 65[b]
Fe	28.0[c]	15.0[d]
Co	31.0	17.0
Ni	6.0	2.4
Cu	0.14	3.0
Ce	—	—
Pr	1.8	0.01
Nd	20.0	22.0
Sm	2.6	3.3
Eu	1.4	2.5
Tb	—	0.2
Dy	0.7	1.6

[a] At 2.5 µm.
[b] At 2.55 µm.
[c] Mixtures of +2 and +3 states.
[d] All in +2 state.

TABLE 3.9
Transition Metal and Rare Earth Ion Content and Loss
Contribution at 2.55 µm in a Fluoride Glass

	Fe	Co	Ni	Cu	Nd
Ion content (ppb-wt)	108	<7	<7	<6	5
Loss contribution (dB/km)	0.016	<0.119	<0.017	<0.018	0.110

photoluminescence spectroscopy.[72] The values of the absorption coefficients quoted by France et al.[65] were used to calculate the contribution to the loss at 2.55 µm. The concentration of Fe^{2+} was assumed to be approximately 1/100th of the total Fe content since an oxidizing atmosphere was used during melting. In this case, the extrinsic absorption loss is equivalent to <0.280 dB/km at 2.55 µm.

The presence of OH^- ions has been observed in both bulk glass and fiberglass.[67] The fundamental OH band is centered at about 2.87 µm and has an absorption coefficient of approximately 5000 dB/km/ppm.[73] In addition, there are two bands observed at 2.24 and 2.42 µm which are attributed to combination bands with Zr–F and Ba–F, respectively.[74] The ratio of these bands to the fundamental OH band are 1:1.06:100, respectively. The OH spectrum in a fluoride glass is shown in Figure 3.8. The OH band is broad and asymmetric with high absorption on the long wavelength side. This is attributed to hydrogen bonding to fluorine ions. The high electronegativity

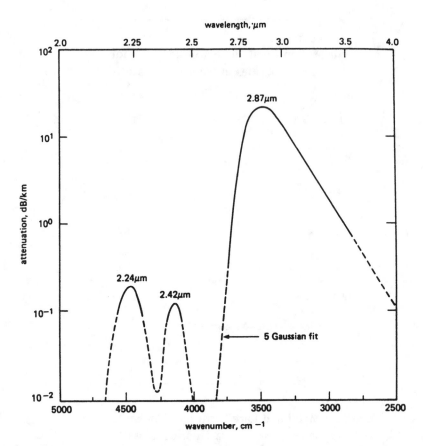

FIGURE 3.8 The OH spectrum in a ZBLALiPb glass rescaled to 20 dB/km peak height. Broken curve is an empirical multiple Gaussian fit. (From France, P. W. et al., *Electron. Lett.,* 20, 608, 1984. With permission.)

of the fluorine increases the strength of the hydrogen bond and therefore shifts the OH absorption to longer wavelengths. A low-loss window exists around 2.55 μm which coincides with the predicted wavelength of minimum loss for a fluorozirconate glass. In addition, the OH overtone occurs at 1.44 μm. The intensity of the overtone band is a factor of 10^{-3} lower than the fundamental band.

It has already been shown that HF is liberated by the reaction of water with a fluorozirconate melt. Unassociated and associated HF have fundamental absorption bands at 2.61 and 2.9 μm, respectively.[75] Therefore, any dissolved HF is likely to be hydrogen bonded to the fluorine in the glass and subsequently contribute to the fundamental OH band at 2.87 μm.

Such dissolved gases as CO and CO_2 have sharp fundamental absorption bands at 4.762 and 4.255 μm with absorption coefficients of 2.7×10^4 and 9×10^3 dB/km/ppm, respectively.[66] There are other absorption bands associated with these gases in the 2.6 to 4.7 μm region. For example, CO_2 has an absorption band at 2.675 μm, but the absorption coefficient is a factor of 100 times lower than the fundamental band. The formation of these gases is attributed to the presence of carbon and oxide impurities. However, selection of better-quality chemicals results in a significant reduction in their presence. While the molecular ion NH_4^+ possesses an absorption band near 3 µm, SO_4^{2-} and PO_3^- possess bands beyond 4.5 µm and therefore do not contribute to the loss at 2.55 µm. The NH_4^+ ion can arise from ABF processing of the chemicals. Its concentration is typically low and can be further reduced by prolonging the melting schedule. The reduced species such as Zr^{3+} and Zr^0 are somewhat different since they both absorb and scatter light. Since their origin is attributed to melting under reducing conditions, they can be eliminated by melting under an oxidizing atmosphere. The oxide bands attributed to Zr–O are centered at 7.4 µm and therefore have negligible effect at 2.55 µm. However, high oxide contents can cause crystallization and lead to extrinsic scattering losses. This will be discussed in the next section. Nevertheless, higher-quality chemicals and RAP melting can be used to form glasses with lower oxide contents.

3.4.3 EXTRINSIC SCATTERING

3.4.3.1 Glass Stability

Glass stability is an extremely important parameter for successful fabrication of crystal-free glasses, preforms, and fibers. It is well known that crystals can grow homogeneously (intrinsic) or heterogeneously (impurity related). If the predominant mechanism of crystal growth in fluoride glasses is homogeneous nucleation, then glass stability is considered to be poor, and this places a severe restriction on the attainment of ultralow-loss fiber. However, if the predominant mechanism is impurity related, namely, heterogeneous nucleation, then with appropriate control of both chemicals and processing one should be able to eliminate the impurities on which the crystals grow and therefore successfully produce ultralow-loss glasses and fibers.

The question arises of how to define and measure glass stability. One parameter, namely, the critical cooling rate to avoid crystallization (R_c), is routinely used to define glass-forming ability but this does not give information on what might happen to the glass during reheating for preform fabrication or fiber drawing. Another parameter, namely, ($T_x - T_g$), which is defined as the temperature difference between the glass transition (T_g) and crystallization temperature (T_x), is commonly used to describe the thermal stability of a glass during reheating. However, both techniques are insensitive to small levels (<1%) of crystallinity. Furthermore, even though the value of ($T_x - T_g$) is large, the fiber-drawing temperature may be relatively close to the crystallization temperature and therefore lead to crystallization during fiber drawing. Practically, it is desirable to draw fibers in the viscosity range of 10^3 to 10^6 P and preferably above the liquidus temperature to minimize crystallization. Table 3.10 lists the viscosities of some glass-forming systems at the liquidus. While typical oxide glasses can be drawn into fibers from above the liquidus temperature due to their high melt viscosities, fluoride glasses possess low melt viscosities almost like water. While low melt viscosities are good for fining and homogenization of the melts, it is not possible to draw fluoride fibers from above the liquidus. A low viscosity at the liquidus means that the fluoride glasses will possess relatively low

TABLE 3.10
Viscosities of Oxides and Fluorides at Their Liquidus Temperatures

	Melting Point or Liquidus Temperature (°C)	Viscosity (P)
SiO_2	1734	1×10^7
Na_2O-SiO_2	878	6×10^3
$51ZrF_4-20BaF_2-3LaF_3-2AlF_3-20NaF-4GaF_3$	450	2.7×10^0
$53ZrF_4-20BaF_2-4LaF_3-3AlF_3-20NaF$	600	3×10^{-1}
H_2O	0	2×10^{-2}

viscosities at temperatures slightly below the liquidus. This is disadvantageous since a low viscosity goes hand-in-hand with an increased likelihood of crystallization due to ionic diffusion. Figure 3.9 shows the viscosity–temperature profile for a fluorozirconate glass.[76] Work has been performed to lower the liquidus temperature and thereby increase the viscosity at the liquidus temperature.[48] Since the primary and secondary phases which crystallize on cooling are AlF_3 and LaF_3, small modifications to the ZBLAN composition, namely, substitution of GaF_3 for AlF_3 and LaF_3, have led to a reduction in the liquidus temperature from 600°C to about 450°C.

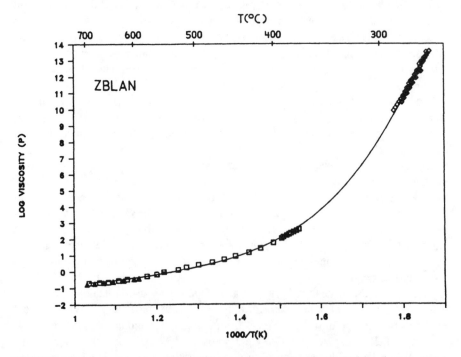

FIGURE 3.9 Arrhenius plot of ZBLAN viscosity from the glass transition temperature to above the liquidus temperature. (From Hasz, W. C. et al., *Mater. Sci. Forum,* 32–33, 592, 1988. With permission.)

TABLE 3.11
Nucleation Rates and Crystal Densities
of Fluorozirconate Glasses

Glass Sample	Nucleation Rate (nuclei/cm^3/s)	Crystal Density (crystals/cm^3)	Ref.
ZBLAN	10^4–10^3	10^9–10^8	77
ZBLAN	10^1–10^3	10^6–10^2	78
ZBLAN	$<10^{-4}$	10–0	79

While this is a significant change, the viscosity still remains too low for fiber drawing from above the liquidus.

Other approaches, such as the nucleation rate, have been used to define glass stability. However, studies have shown that the number of crystals formed on reheating the glass is independent of the nucleation rate. This indicates that homogeneous nucleation is not the predominant mechanism causing crystallization, but rather it is heterogeneous nucleation. Therefore, it would appear more appropriate to define glass stability in terms of crystal density as determined by the number of crystals formed after reheating the bulk glass. The rate of crystal growth has been measured at various temperatures. For instance, by heating the glass at 340°C for 10 min, the ZB crystals grow to approximately 100 μm and are therefore easily detected by optical microscopy. The crystal density is subsequently defined as the number of crystals detected per cubic centimeter. Table 3.11 shows the crystal density data of some glasses[77-79] along with the estimated nucleation rates. Glasses have been prepared that do not exhibit crystallization on rigorous heating.[79] Hence, bulk crystallization is not a problem limiting ultralow loss. However, knowledge of the nature and origin of the heterogeneities responsible for crystallization is very important so that these scattering centers can be eliminated in order to obtain ultralow-loss glasses and fibers.

3.4.3.2 Origin and Elimination of Crystals and Defects

Table 3.12 summarizes the source of different types of crystals and defects found in fluorozirconate glasses, preforms, and fibers. Typically, the fluoride crystal phases grow predominantly on submicron platinum, zirconium oxide, and zirconium oxyfluoride particles. ZrO_2 has the lowest free energy of formation of any oxide in a fluorozirconate glass, and so ZrO_2 crystals are preferentially formed after glass melting.[84] However, zirconium oxyfluoride microcrystals have also been identified in fluorozirconate glasses after the addition of Al_2O_3 or La_2O_3.[85] Figure 3.10 shows the adverse effect of oxide content on the stability of fluorozirconate glasses, as noted by the increase in the crystal density on reheating.[79] This figure also demonstrates that the nucleation step is not important in determining the overall number of crystals on reheating. Figure 3.11 shows the presence of an oxyfluoride crystal nucleating another fluoride crystal. Mitachi et al.[89] have suggested that about 100 ppm oxide is soluble in a fluorozirconate glass. Thereafter, any excess oxide

TABLE 3.12

Scattering Centers Found in Fluorozirconate Glasses, Preforms, and Fibers

Crystals/Defects	Source	Ref.
ZrF_4	Contamination from sublimed particles	80
$NaF-ZrF_4$	Surface growth (bubbles)	
ZrF_4-BaF_2 and $2ZrF_4-BaF_2$	Oxyfluorides	80
	Pt nuclei	81
LaF_3	Incomplete dissolution and/or surface growth (container walls)	80
$LaF_3 \cdot 2ZrF_4$		82
AlF_3, Al-rich phase	Solubility limit exceeded and/or incomplete dissolution	83
ZrO_2	Moisture, hydroxides, and oxide impurities	84
ZrO_xF_y		80, 85
Fe, Ni phosphides	Phosphate impurities in precursors	86
Pt (Au, Ir, Rh)	Crucible reactions	87
Carbon	Organic impurities/crucible reactions	80, 88
Bubbles	Contraction, cavitation, gas precipitation	80

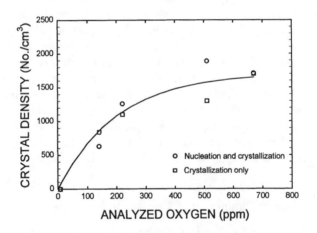

FIGURE 3.10 Crystal density vs. oxygen content in fluorozirconate glasses. One set of glasses was given an initial nucleation treatment at 290°C for 24 h prior to the crystal growth treatment at 340°C for 10 min. For comparison, another set of glasses was only heated at the crystal growth temperature.

precipitates out and leads to increased scattering. This is highlighted in Figure 3.12 which shows the increase in the measured scattering loss with oxide content in a ZBGA glass. Mitachi and Tick[90] observed an increase in the critical cooling rate of fluorozirconate glasses with increasing oxide content indicating more rapid cooling rates were required to inhibit crystallization. They also observed anomalous minima in the critical cooling rates for non-ZrF_4-based glasses, such as fluorohafnate, CLAP, and BCYA glasses.

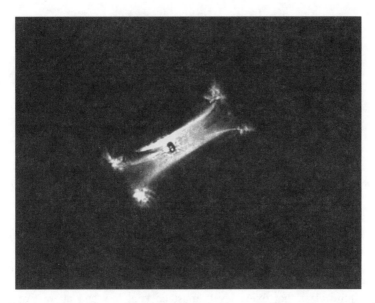

FIGURE 3.11 A zirconium oxyfluoride crystal at the center of a ZB fluoride crystal (~100 μm). Dendritic growth of 2ZB fluoride crystals is evident at the corners of the ZB fluoride phase.

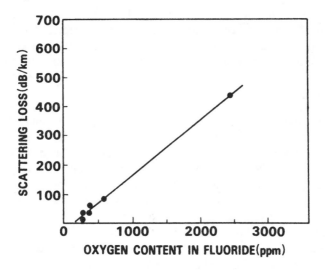

FIGURE 3.12 Relationship between scattering loss and oxygen content in a ZBGA glass. (From Mitachi, S. et al., *Jpn. J. Appl. Phys.,* 24, L827, 1985. With permission.)

Since oxides are impurities in fluorozirconate glasses, their concentration may not be uniform in the glass and regions can be enriched with respect to oxide causing localized crystal precipitation. Practically, it may be advisable to fabricate glasses with oxide contents well below 100 ppm.

TABLE 3.13
Typical Properties of Fluorozirconate Glasses

Property	Value	Ref.
Glass transition temperature (°C)	240–455	92
Expansion coefficient (°C $\times 10^{-7}$)	100–187	92
Density (g/cm^3)	4.5–5.3	92
Leach rate in water (g/cm^2/day)	10^{-2}–10^{-3}	93
Young's modulus (GPa)	50–60	94
Vickers hardness (kg/mm^2)	220–270	94
Fracture toughness, K_{Ic} (MPa·m$^{1/2}$)	0.25–0.3	94
Poisson's ratio	0.25–0.30	94
Refractive index, n_D	1.48–1.54	92
Nonlinear index, n_2 (10^{-13} esu)	0.9	92
Abbe number	68–80	92
dn/dT (K^{-1})	-1×10^{-5}	96

It is quite clear that oxide levels must be kept low and alternative crucible materials such as vitreous carbon must be used in conjunction with an oxidizing gas, such as SF$_6$ gas. This glass processing leads to the fabrication of stable and high-optical-quality glasses.[91] For example, the oxide content of these glasses was below 10 ppm and the glasses did not exhibit bulk crystallization on reheating at 340°C from 10 min up to 24 h. Furthermore, the measured scattering loss of the glasses at 0.63 μm was 4.5 dB/km which extrapolated to 0.02 dB/km at 2.55 μm. This is very close to the theoretically predicted value of about 0.01 dB/km.

3.4.4 GENERAL PROPERTIES

Some typical properties of fluorozirconate glasses are listed in Table 3.13. The glass transition temperatures of fluorozirconate glasses are lower than for typical silicate glasses.[92] The expansion coefficients and densities are quite large and are attributed to the ionic bonding character and presence of heavy cations.[92] These properties are dependent upon the components in a fluorozirconate glass as well as the type of glass. For example, fluoroaluminate glasses possess higher glass transition temperatures and lower densities and expansion coefficients.

The chemical durability of fluorozirconate glasses is quite low.[93] For instance, the leach rates of fluorozirconate glasses in water are about six orders of magnitude higher than Pyrex glass. Fluoroaluminate glasses possess slightly lower leach rates, but they are still more than three orders of magnitude higher than Pyrex.

The values for Young's modulus of fluorozirconate glasses are ³/₄ that of silica and comparable with that of Pyrex glass. The hardness and fracture toughness are lower than silica, but higher than chalcogenide glasses. The fracture toughness, which is an approximate measure of intrinsic strength, is about one third that of silica, implying that the ultimate strength may be approximately one third that of silica.[61,94]

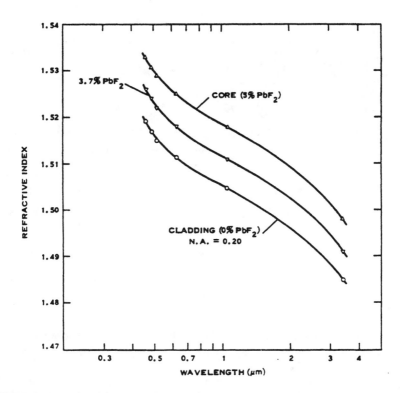

FIGURE 3.13 Refractive index specta of ZBLALi glass, a potential cladding composition, and two Pb-doped glasses which are potential core compositions. (From Levin, K. H. et al., *Glass Technol.*, 23, 143, 1983. With permission.)

The fluorozirconate glasses possess a relatively low refractive index (n_D), a low nonlinear index (n_2), and a high Abbe number.[92] The values of n_D and n_2 for fluoroaluminate glasses tend to be slightly lower than for the fluorozirconate glasses. Figure 3.13 shows the change in refractive index vs. wavelength for three fluorozirconate glasses.[95]

The information-carrying capacity or the bandwidth of a fiber is related to the material dispersion parameter $M(\lambda)$ of the core glass given by[63]

$$M(\lambda) = (\lambda/c)(d^2n/d\lambda^2) \tag{3.13}$$

where c is the velocity of light, λ is the wavelength, and n is the refractive index. The maximum bandwidth in an optical waveguide is achieved at a wavelength λ_0, where $M(\lambda)$ goes to zero. For optimum properties, it is desirable for λ_0 to coincide with the wavelength of minimum absorption for a given fiber material. This is practically the case for silica. Figure 3.14 shows a plot of material dispersion vs. wavelength for two fluoride glasses and silica.[63] In the case of the fluoride glasses, the value of λ_0 is slightly lower than the predicted wavelength for minimum loss.

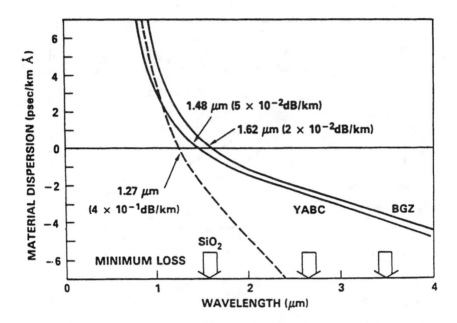

FIGURE 3.14 Material dispersion for silica and two heavy metal fluoride glasses. (From Miyashita, T. and Minabe, T., *IEEE J. Quantum Electron.,* QE-18, 1432, 1982. ©1982 IEEE. With permission.)

Nevertheless, high bandwidths, in excess of 1 GHz/Å·km should be attainable in fluoride fibers since the dispersion curves are relatively flat above λ_0. By contrast, silica exhibits a much steeper change in dispersion with wavelength. Hence, significant dispersion occurs if the operating wavelength is shifted from 1.3 μm in silica fiber.

Laser windows and other optical components that are subjected to high-energy laser beams will fail optically at power levels far below those sufficient to cause melting or fracture. The failure is a combined result of spatially inhomogeneous laser heating and electronic photorefraction that causes the component to act as a lens. A consequence of the lenslike behavior is a significant distortion of the laser intensity in the far field. The optical path distortion (OPD) caused by laser heating in an isotropic window is given by

$$\text{OPD} = \Delta Tl\big[(n-1)(1+v)\alpha + dn/dT + 0.25n^3Y\alpha(q_{11} + q_{12})\big] \qquad (3.14)$$

where l is the laser path length through the window, n is the refractive index at the laser wavelength, v is Poisson's ratio, α is the linear coefficient of thermal expansion, dn/dT is the thermo-optic coefficient, Y is Young's modulus, and q_{11} and q_{12} are stress-optic coefficients. The temperature rise ΔT is given by

$$\Delta T = P\beta/\rho C_p \qquad (3.15)$$

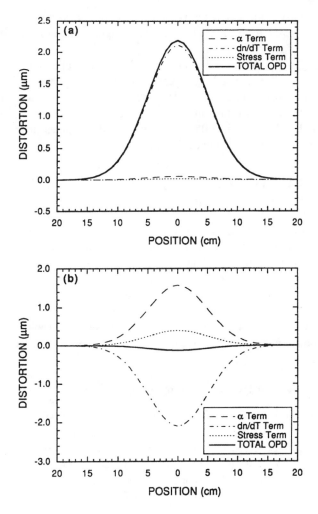

FIGURE 3.15 Contribution to optical path distortion for 5-cm-thick (a) silica and (b) ZBLAN windows exposed to a 1 MW laser for 10 s and assuming the absorption coefficients for both are 10^{-4} cm^{-1}.

where P is the laser intensity, β is the absorption coefficient, and ρ and C_p are the density and heat capacity of the window material, respectively. The OPD of a high-energy laser window can be calculated for any material using these equations. Figures 3.15a and b show the contribution from the terms represented by the linear coefficient of thermal expansion, the thermo-optic coefficient, and the stress-optic coefficients and the total theoretical OPD for silica glass and ZBLAN calculated by using property data from the literature.[91,96] These curves were calculated assuming a 10-cm-diameter beam from a 1-MW laser and 5-cm-thick windows with absorption coefficients of 1×10^{-4} cm^{-1} for both glasses. In Figure 3.15a, the terms for the coefficient of expansion and dn/dT for silica have the same sign and result in a large positive OPD. In Figure 3.15b, the advantage of using ZBLAN for high-energy laser windows is clearly evident. The terms representing the linear coefficient of thermal expansion and *dn/dT* for ZBLAN have opposite signs and nearly cancel, resulting in a near-zero OPD.

Fluoride glasses appear well suited for UV laser applications. For instance, fluoride glasses exhibit reversible changes in absorption with power which are similar to silica when exposed to an excimer laser operating at 248 nm (KrF laser).[97] This has been attributed to a two-photon absorption mechanism. The two-photon absorption coefficient of ZBLAN is 3.05×10^{-9} cm/W and compares favorably with the value of 1.82×10^{-9} cm/W obtained for silica.

3.5 FABRICATION OF FLUORIDE GLASS OPTICAL FIBERS

There are two basic techniques that have been used to produce core/clad optical fibers from silicate glasses, namely, the redraw of preforms into fibers and the direct draw of fibers from a crucible. Many variations of these two basic techniques have been developed, including preparation of preforms using CVD, "rod-in-tube," ion exchange, leaching, sol–gel processing, and updraw from a melt. The same two basic techniques of preform redraw and direct draw from a crucible have been used to prepare optical fibers from fluoride glasses. However, because of the unique properties and problems of the fluoride glasses, fiber preparation techniques have been modified accordingly.

3.5.1 PREFORM FABRICATION TECHNIQUES

The various techniques that have been used to fabricate fluoride glass preforms can be divided into two main classes, namely, the bulk cast preforms and the tubular preforms. The former represent preforms which possess the desired core/clad geometry and are drawn directly into optical fibers. The second class represents preforms which are formed into tube shapes and subsequently need to be collapsed prior to or during fiber drawing. The following is a brief description of these techniques.

3.5.1.1 Clad-Over-Core Process

Mitachi et al.[98] described the fabrication of the first core/clad fluoride glass preform using the clad-over-core process. In this case, the clad melt was poured around a core rod placed inside a metal mold. While the resulting preform possessed a uniform clad-to-core-diameter ratio, there were numerous bubbles trapped at the core/clad interface. Reheating at the core/clad interface led to crystallization. Fibers fabricated from these preforms had high transmission losses.

3.5.1.2 Built-In-Casting Process

Another variation of this method, known as the built-in-casting[99] process is shown in Figure 3.16. The cladding melt is poured into a metal mold held at the glass transition temperature. After waiting a few seconds, the mold is inverted and the melt from the central region of the mold flows out, leaving behind a tubular-shaped

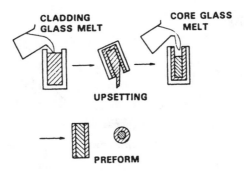

FIGURE 3.16 Built-in-casting process. (From Myashita, T. and Minabe, T., *IEEE J. Quantum Electron.*, QE-18, 1432, 1982. ©1982 IEEE. With permission.)

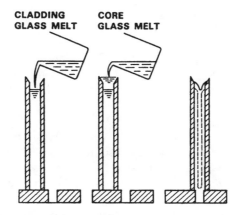

FIGURE 3.17 Modified-built-in-casting process.

void in the center. Next, the core melt is poured in to fill up the void and the preform subsequently annealed. While this is a simple technique, interfacial and core bubbles still remain a problem. Furthermore, this process is not appropriate for long fiber lengths since the preforms are tapered, i.e., clad-to-core diameter ratio varies along the length of the mold assembly. Nevertheless, a loss of about 1 dB/km has been quoted on short lengths using this method.[100]

3.5.1.3 Modified-Built-In-Casting Process

The modified-built-in-casting process, developed by Sakaguchi and Takahashi,[101] reduced the taper associated with the built-in-casting method. This is shown in Figure 3.17. In this process, a precise amount of the cladding melt was poured into a specially designed mold. Prior to the central part of the cladding melt solidifying, the core melt was cast onto the cladding melt. The mold assembly was then shifted over to a section where the bottom of the plate was missing. Next, the cladding melt from the central section is drained from the mold, and, as it does so, the core melt

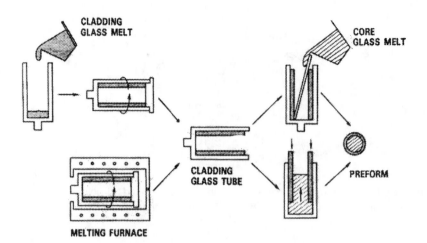

FIGURE 3.18   Rotational casting process. (From Tran, D. C. et al., *Electron. Lett.,* 18, 657, 1982. With permission.)

is drawn into the cladding glass along the mold axis. While a slight taper still exists, typical fiber losses of about 10 dB/km can be obtained for 200 m lengths of fiber.

3.5.1.4   Suction Casting Process

The suction-casting process[102] is a somewhat more refined version of the modified-built-in-casting technique which has been used to fabricate single-mode fiber. In this process, the cladding melt is poured into a specially designed cylindrical mold having a reservoir at the bottom and is preheated close to the glass transition temperature. Next, the core melt is cast onto the cladding melt before it solidifies. On cooling, volume constriction of the melt occurs, which results in the formation of a tubular cladding tube. This results in a suction effect on the core glass melt down to the bottom, thereby filling up the cladding tube. One can control the preform length and core/clad diameter ratio by appropriate choice of reservoir volume and mold diameter. The single-mode geometry was obtained by inserting the preform inside an outer cladding tube of similar composition to the inner cladding tube and subsequently drawing this configuration into a single-mode fiber. The lowest loss obtained was 8.5 dB/km at 2.2 μm. Despite the clean interface and the possibility of single-mode fibers, the preforms still possess some taper, which limits the fabrication of long lengths of low-loss fiber.

3.5.1.5   Rotational Casting Process

The problem of tapered preforms was solved by Tran et al.[103] using the rotational casting process shown in Figure 3.18. In this technique, the cladding glass is first melted in a platinum crucible and then poured into a gold-plated mold preheated at about the glass transition temperature. This is subsequently spun at several thousand rpm to form a highly concentric cladding tube whose diameter is precisely controlled by the amount of cladding glass. Next, the tube is set vertically and the core glass

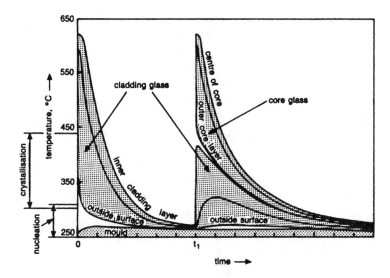

FIGURE 3.19 Predicted temperatures of core and cladding glasses and mold during the rotational casting process. (After Williams, J. R. et al., *Mater. Sci. Forum,* 67–68, 318, 1991. With permission.)

melt is either cast into the tube or the tube is dipped into the core melt. The cast preform is then annealed to prevent cracking. Fibers produced by this technique have been made with a total optical loss of less than 1 dB/km.[104] It is critical that the entire fabrication process be done in a very clean and dry environment, such as a dry box with laminar flowing inert gas. Otherwise, contamination at the interface between the core and cladding will produce crystals and cause scattering of light. Williams et al.[105] have calculated the temperature–time profile for the rotational casting process from which it becomes quite clear that crystallization can occur at the core/clad interface (Figure 3.19). At time zero the cladding glass is cast into the preheated mold at the glass transition temperature (T_g), is spun at high speed, and subsequently begins to cool to near T_g to form a solid tube. At time t_1 the core melt is cast into the cladding tube which leads to reheating of the inner surface of the clad (adjacent to the core). Then, both the core and cladding glasses cool to T_g and are annealed. It is apparent that since the inner surface of the cladding is reheated into the critical crystallization temperature region, then crystals will grow in the presence of nuclei. Hence, it is critical to control the glass pouring and mold temperatures to minimize crystal size and therefore reduce scattering losses.

Another problem associated with the rotational casting process is the presence of bubbles. The bubbles can be either trapped at or near the core/clad interface during the pouring process or be present down the centerline of the core and are attributed to contraction of the glass during cooling. The latter tend to be low-pressure bubbles and often collapse during the fiber-drawing process. Recently, Harbison and Aggarwal[106] showed that hot isostatic pressing can be used to remove small bubbles from preforms without inducing crystallization or altering the geometry of the fluoride preforms. However, the removal of large bubbles, for example, a few

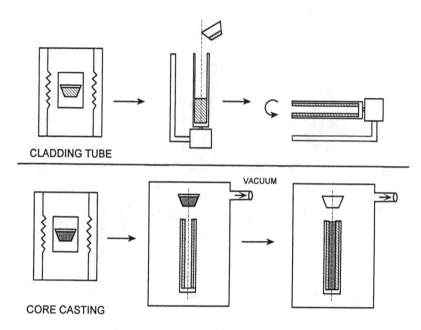

FIGURE 3.20 Casting core under vacuum.

millimeters in diameter, is accompanied by a noticeable geometric change in the glass dimensions. A typical treatment consisted of 1 h at 270°C under 20 kpsi, although less pressure can also be used.

3.5.1.6 Vacuum-Casting Process

In an attempt to eliminate the bubbles encountered with rotational casting, the process was modified to first pull a vacuum on the core melt and the cladding glass tube before casting the core glass under reduced pressure.[91] This process is shown in Figure 3.20, where the cladding glass tube is formed by rotational casting and placed into a vacuum chamber with the molten core glass prior to casting the core glass. The vacuum leads to outgassing from the core melt and removes physically adsorbed impurities on the cladding tube before the core glass is cast. The most important advantage of this technique over conventional rotational casting is the reduction of bubbles formed at the core/cladding interface. However, mixing of the core and cladding glasses while the core is cast is evident with this technique and the standard rotational casting technique, but can be minimized by casting the core glass at the lowest temperatures possible. Contraction bubbles in the center of the core are also reduced by casting at the lowest possible temperature; however, the tendency for crystallization of the core glass is more common at lower temperatures. Measured losses on fiber drawn from preforms made by this technique were <1.0 dB/km at 2.5 μm on short lengths of fiber (30 m).

The total-vacuum-casting technique represents a modified and slightly improved variation of the above process.[91,107] In this process, both the cladding and core glasses

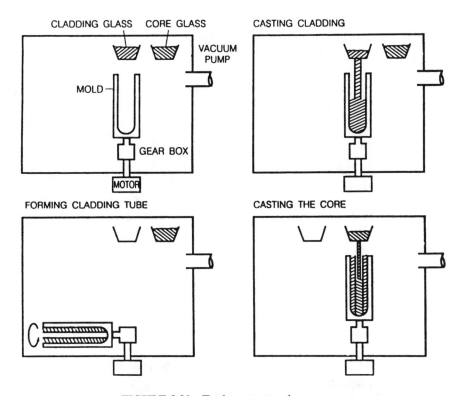

FIGURE 3.21 Total-vacuum-casting process.

are cast under vacuum. The main advantage of this technique over the previously described technique is that the core/cladding interface is not exposed to atmosphere and the core glass can be cast much sooner after the cladding tube is formed. A schematic drawing of this process is shown in Figure 3.21. Both core and cladding glass crucibles are simultaneously placed into a vacuum chamber containing the preform mold that is equipped with a direct drive motor to spin the mold. Results using this casting technique show that the absorption peak at 2.9 μm due to hydroxyl impurities is greatly reduced to <1 dB/km corresponding to an OH concentration of <10 ppb.[91] The advantage of using this casting technique is that it allows the cladding glass to remain above T_x, the crystallization temperature. This avoids reheating the cladding glass to T_x after pouring the core melt which usually results in crystallization near the core/cladding interface. Losses of <1 dB/km have been achieved by this technique on fiber lengths longer than 100 m.[107]

3.5.1.7 Rod-in-Tube Process

The rod-in-tube process requires a high-quality fluoride core glass rod which is usually fabricated by melting, casting, and subsequently polishing. This is inserted into a cladding tube which was obtained by the rotational casting method. During fiber drawing, the interface is evacuated so that the cladding tube collapses around the core rod. While bubbles in the core can be avoided with this method, the fibers

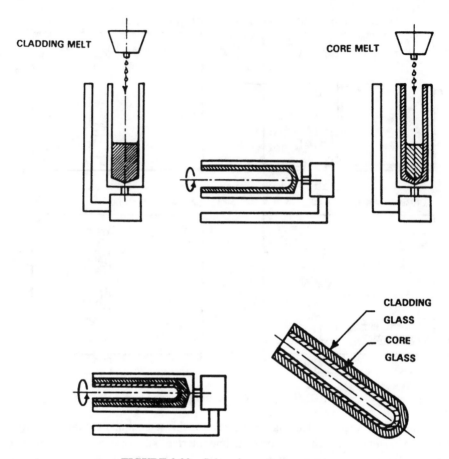

FIGURE 3.22   Spin-spin-casting process.

typically exhibit high scattering due to interfacial defects, particularly trapped gas, airborne particulate, and embedded particles of the polishing media.

3.5.1.8   Spin-Spin-Casting Process

Another method of casting fluoride glass preforms, termed spin-spin casting,[91] involves making a cladding tube by rotational casting and then spin-casting a thin layer of core glass inside the cladding tube as shown in Figure 3.22. The resulting tube-in-tube preform can be either directly drawn into fiber or collapsed and then drawn into fiber.

Several elaborate models have been developed for the viscous collapse of silica preforms prepared by the CVD process which can be used for collapse of fluoride tubes.[107-109] The driving force for the collapse process can be viewed in a simplistic way by the following relationship:

$$dr/dt \propto \frac{P_i - P_o - \sigma f(\text{ID, OD})}{\eta}$$

(3.16)

where dr/dt is the collapse rate, P_i is the internal tube pressure, P_o is the external tube pressure, σ is the surface tension, η is the glass viscosity, ID is the inside diameter of the tube, and OD is the outside diameter of the tube. This relationship demonstrates the competing forces of differential pressure $(P_i - P_o)$ and surface tension on the rate of collapse at a given temperature or viscosity. Experimental results indicate that high temperatures or large differential pressures lead to core eccentricities as a result of the rapid collapse rate. These should be avoided by optimization of pressure and temperature control. The collapse rate is expressed by an Arrhenius relationship:

$$\text{Rate} = Ae^{-Q/RT} \tag{3.17}$$

where Q is the activation energy for the collapse process. A differential pressure of 25 torr leads to an activation energy of 427 kJ/mol for the temperature range of 310 to 325°C. This value is similar to the values obtained for viscous flow of fluoride glasses. This is not unexpected from Equation 3.16, since the viscosity is a stronger function of temperature than surface tension. Fiber scattering losses measured at 0.63 μm were observed to be related to the core eccentricity. Measured losses on short sections (12 cm) of fibers were 0.07 ± 0.04 dB/km at 2.55 μm. This technique is capable of being scaled up to produce large preforms, since it is a rapid quench process capable of producing both multimode and single-mode fiber. Consequently, long lengths of low-loss fiber may be achieved from large spin-spin casting preforms. However, the process is currently limited by the poor concentricity control during the collapse of the tubular preform into a solid.

3.5.2   Crucible Techniques

Even though the preform method has yielded the lowest fiber losses, it has limitations on length and core size. It is difficult to produce a preform which can be drawn into a fiber much longer than a few hundred meters. Also, it is difficult to produce single-mode fibers. The core/clad ratio in a single-mode fiber is too small to produce in a single casting. Therefore, it is necessary to stretch and overclad the preform several times to produce a single-mode fiber by this technique. These limitations can be overcome by the double-crucible process, which was initially developed for low-melting-temperature oxide glasses. The first attempts to draw fluoride fiber using the single-crucible technique were not completely successful, probably due to the difficulty of controlling the viscosity of the glass in the crucible nozzle.[111,112]

3.5.2.1   Double-Crucible Process

Preliminary efforts were made to produce optical fibers by the double crucible technique, but from above the liquidus.[113] These attempts were largely unsuccessful since the melts were so fluid that only poor-quality unclad fibers could be made, and those with significant difficulty.

Tokiwa et al.[114] used a double-crucible technique to draw fibers from below the liquidus. The core and clad glasses were first melted in gold crucibles at the normal

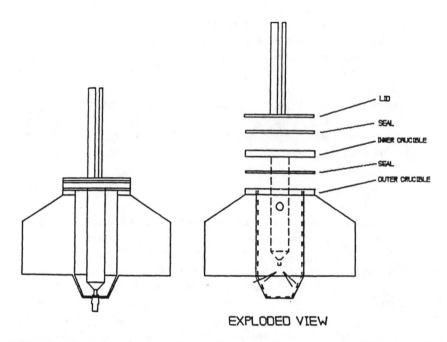

EXPLODED VIEW

FIGURE 3.23   Schematic of double crucible used to melt, quench, and draw core-clad
fluoride glass fibers.

glass-melting temperatures (about 800°C). The glasses were then poured into the
double crucible. The glass and double-crucible assembly was equilibrated at about
320°C which is at a viscosity of about $10^{5.2}$ P. Fibers can still be drawn at this
viscosity, but the crystal growth rate is very low so that quite long fibers can be
drawn before crystallization becomes a problem. Single-mode fibers were drawn
with core and cladding diameters of 13 and 180 μm, respectively, and cutoff wavelength of 2.36 μm.

In this technique it is critical that the core and cladding glasses have the same
softening temperatures. This will minimize the crystal growth rate in both glasses.
Therefore, an appropriate choice for the core glass is the ZBLAN composition
consisting of 53 mol% ZrF_4, 20% BaF_2, 4% LaF_3, 3% AlF_3, and 20% NaF. The
cladding glass can be a similar composition with HfF_4 partially substituted for the
ZrF_4 (see Table 3.3).

Remelting cullet can give reasonably good results. For example, multimode
fibers over 700 m long and single-mode fibers over 500 m long were produced.[25] A
modified version of this technique was used to produce fibers which had a loss of
only a few decibels per kilometer.[25] In the modified technique, the two glasses were
melted in the double crucible, and then the entire crucible assembly was quenched
rapidly to the fiberizing temperature. This preserved the quality of the glass without
introducing bubbles or crystal nuclei during the casting process. A sketch of this
modified double-crucible system is shown in Figure 3.23. A photograph of the
components is shown in Figure 3.24. The system consists of an inner crucible that
is concentrically located within an outer crucible. Core and cladding glasses are

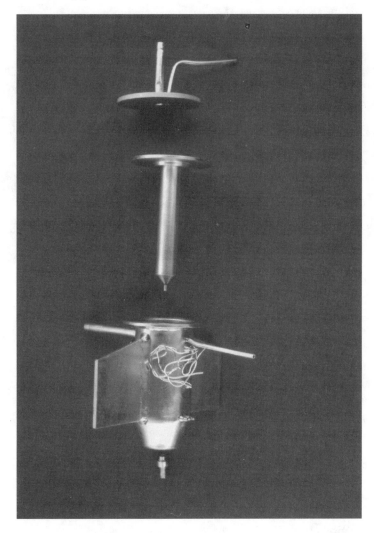

FIGURE 3.24   Photograph of double crucible used to draw core-clad fluoride fibers.

placed in the two crucibles and the tips of the crucibles are plugged during melting. The crucibles were heated electrically by passing a current through the outer crucible. The absence of insulation allowed for rapid cooling through the most critical temperatures associated with high crystal growth rates.

The best results to date have been achieved using a platinum–gold alloy (such as 95% Pt–5% Au) as the crucible material. This appears less efficient in nucleating crystals at the crucible/glass interface. The minimum total loss achieved by this process was 5 dB/km measured on a length of 125 m of multimode fiber.[25] In addition, single-mode fibers were also produced by this technique. The cross section of a single-mode fiber produced by the double-crucible technique is shown in Figure 3.25. No losses were reported for the single-mode fibers.

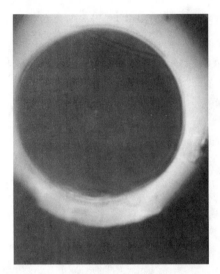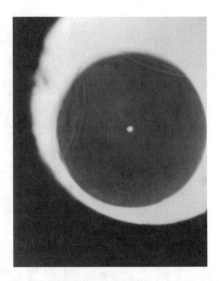

FIGURE 3.25   Cross section of a single-mode fluoride glass optical fiber produced by the double-crucible process.

3.5.2.2   Triple-Crucible Process

The triple-crucible system has also been explored.[25] In this case, a higher-viscosity glass was placed in the outer crucible and the core and cladding glasses were placed in the inner two crucibles. The purpose of this system was to provide a means for fiberizing fluoride glasses above their liquidus. Subliquidus drawing can only be performed for a finite time before crystallization becomes an issue. If the glasses can be held above the liquidus, there is no constraint on the time of fiberization, since there is no crystallization to worry about, and very long lengths of fiber can be drawn. Although preliminary three-layer fibers have been produced, they were not of sufficient quality to measure loss spectra.

3.5.3   OTHER TECHNIQUES

Some alternative techniques have been used to fabricate preforms, especially for single-mode fibers. This is because casting techniques are unable to give core diameters small enough for single-mode dimensions. The picture is further complicated by the necessity for a clad-to-core diameter ratio greater than 5 to eliminate extrinsic losses at the polymer jacket interface. It is difficult to fabricate uniform and small-diameter core structures. A modified version of the rod-in-tube process can be used to make single-mode fibers as shown in Figure 3.26. The first step consists of forming a preform with a core/clad structure. This can be done using any of the techniques previously described. In the second step, the preform is stretched so that the diameter is reduced uniformly to some predetermined size. In the third step, this stretched preform is placed inside a cladding glass tube which can be fabricated by the rotational casting technique. The inner diameter of the tube

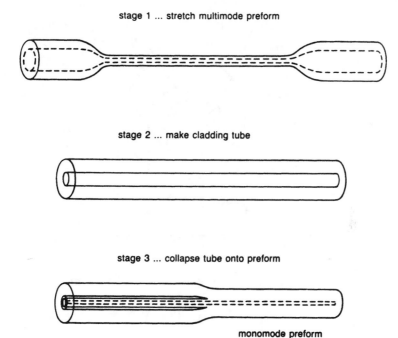

stage 1 ... stretch multimode preform

stage 2 ... make cladding tube

stage 3 ... collapse tube onto preform

monomode preform

FIGURE 3.26 Fabrication of a single-mode fluoride fiber by the modified rod-in-tube method.

is slightly larger than the stretched preform, and the composition of the preform cladding and the cladding tube should be identical. The cladding tube is then collapsed onto the preform to remove the gap between the two structures by applying a vacuum at the interface. This new preform, which possesses a larger clad-to-core ratio, can be subsequently stretched and reinserted into another tube and the whole process repeated until the correct cladding-to-core diameter ratio is achieved for a single-mode fiber.

The reactive vapor transport (RVT) process developed by Tran et al.[115] is suitable for making single-mode fluoride glass preforms. In this process, a reactive vapor originating from metal halides or halogenated gases is carried inside a fluoride glass cladding tube. Depending on the processing parameters, chlorine, bromine, or iodine ions can be exchanged and incorporated in the glass, thus increasing the refractive index of the layer of the internal cladding tube. Figure 3.27 shows a schematic of the cross-sectional view and the corresponding index profile. The index profile is not a step function since this is a diffusion process. Nevertheless, ion exchange was seen to a depth of over 0.5 mm. The advantage of this process is that it avoids casting of the core glass which is responsible for defects at the core/cladding interface. But a disadvantage is that this tubular configuration has to be collapsed prior to or during fiber drawing. As pointed out previously, this is not trivial and may account for why there is no optical loss data reported for fibers.

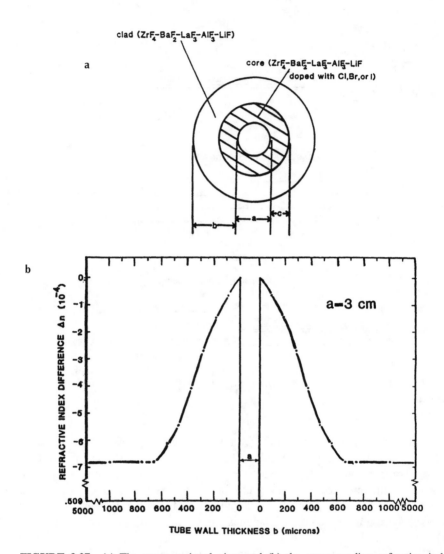

FIGURE 3.27 (a) The cross-sectional view and (b) the corresponding refractive index profile using the RVT process. (From Tran, D. et al., in *Tech. Dig. Conf. Optical Fiber Communications,* New Orleans, 1984. With permission.)

A modified version of this process was developed[116] since displacement of fluoride ions by chloride, bromide, or iodide ions is less favorable than the reverse displacement and since collapse of the tubular preform can present problems. In this case, a core rod made with the nominal ZBLAN composition but containing several percent chloride ions was placed in a chamber containing HF. The F ion readily displaced the Cl ion such that penetration to 2 mm was evident. Although no optical data were reported for fibers, this process should be capable of producing good-quality preforms since there is no casting or collapse required prior to fiber drawing.

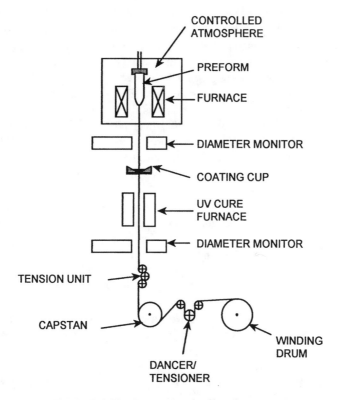

FIGURE 3.28 Schematic of a fiber draw tower.

The CVD process for fluoride glass fabrication has already been described in Section 3.3.4.1. While preliminary coatings on the inside of cladding tubes have been applied, it is apparent that thicker, more-uniform, and higher-optical-quality coatings are needed. Despite this, the CVD process, as is the case for silica, still remains a strong candidate for fabrication of long lengths of low-loss multimode and single-mode fluoride glass fibers. However, more work is required to accomplish these objectives.

3.6 FIBER DRAWING

As pointed out in the previous sections, the fluoride fibers are obtained from either the crucible or the preform routes. Fluoride glass preforms are drawn into fibers using a resistively heated furnace or an RF induction furnace. Since the fluoride glass viscosity shows a strong temperature dependence, precise control of temperature is critical to minimize dimensional fluctuations and crystallization. In addition, it is important to have a dry inert or reactive atmosphere during fiber drawing to prevent crystallization at the preform neck-down region as a result of the action of moisture. A schematic of a draw tower is shown in Figure 3.28. The preform is usually lowered into the draw furnace at a few millimeters per minute at the draw temperature, which depends upon composition and which in turn defines the viscosity–temperature profile. For a ZBLAN preform, the fibers are drawn at 312°C, which corresponds to a viscosity of approximately $10^{5.5}$ P. The fiber travels down through a diameter monitor and into a coating cup which contains a UV curable acrylate. The polymer is cured as the fiber passes through the UV furnace. The thickness of the coating is monitored with another diameter monitor. The fiber then passes through a tension unit, a capstan unit, a dancer/tensioner, and is finally attached to a winding drum and drawn at about 10 m/min under a 20 g load. The overall diameter of the fiber can be controlled to within ±2 μm. While the clad-to-core diameter ratio of the fiber is fixed and defined by the preform dimensions, the overall diameter can be changed in a controlled manner by appropriate changes in the furnace temperature, the feed rate, and the fiber draw rate.

The polymer coatings are applied to prevent mechanical abrasion as the fiber travels through the various sections during fiber drawing. Aside from UV curable acrylate, heat-shrinkable Teflon tubing can also be applied to the preform. During fiber drawing the Teflon softens and flows with the fiber. The only limitation with using Teflon is the upper operating temperature of around 320°C. Therefore, it may not be appropriate for drawing some of the glasses with higher drawing temperatures, such as the fluoroaluminates. Teflon coatings are not as popular as acrylates since they have to be removed by scraping with a razor blade and, therefore, require a sensitive touch. On the other hand, the acrylates soften and swell in the presence of methylene chloride and are simply pulled off.

The crucible fiber-drawing techniques have already been described in the previous section. The same draw tower can be used as for the preforms, except that Teflon cannot be used as a coating material.

It is critical that the preforms and double-crucible glasses be handled in a clean-room environment to minimize contamination, which not only affects the optical properties but also the fiber strength.

3.7   MULTIMODE AND SINGLE-MODE FIBER DESIGN

The majority of fluoride fibers fabricated were of multimode design because of ease of fabrication. Initially, it was believed that a multimode design (i.e., choice of fiber core/clad diameters and numerical aperture) similar to that of silica fiber would be appropriate. Experimentally, however, it was found that due to the longer propagation wavelength, different parameters were necessary to reduce the loss caused by the polymer coatings used to protect the fiber surface. It was determined[117] that a thicker cladding and a higher numerical aperture were needed to reduce such losses. Table 3.14 shows results for calculated values of the ratio of cladding radius (b) to core radius (a) needed such that 90% of the modes propagating in the fiber have loss less than 0.01 dB/km. The ratio is strongly dependent on the numerical aperture (NA) and the V parameter, defined as

$$NA = \left(n_{co}^2 - n_{cl}^2\right)^{1/2} \tag{3.18}$$

TABLE 3.14

Parameters for Fluoride and Silica Multimode Fibers

2a (μm)	NA[a]	V	b/a[b]	2b (μm)	Minimum Bend Radius R (cm)
Parameters for Fluoride Multimode Fiber: $\lambda = 2.5$ μm					
80	0.14	14	3.25	259	13
70	0.16	14	3.25	226	9
56	0.20	14	3.25	181	5
114	0.14	20	2.5	284	17
99	0.16	20	2.5	249	12
80	0.20	20	2.5	199	6
136	0.14	24	2.0	273	20
119	0.16	24	2.0	239	13
95	0.20	24	2.0	191	7
Typical Silica Multimode Parameters: $\lambda = 1.3$ μm					
50	0.20	23.6	2.5	125	4

---

[a] Numerical aperture.

[b] Minimum values such that 90% of modes have loss <0.01 dB/km.

After Busse, L. E. and Aggarwal, I. D., *J. Lightwave Technol.*, 9, 828, 1991.

$$V = (2\pi a)\,\mathrm{NA}/\lambda \qquad (3.19)$$

where n_{co} and n_{cl} are the refractive indices of the core and clad glasses, respectively, and λ is the propagation wavelength.

As the results in Table 3.14 suggest, the parameters for low-loss fluoride and silica multimode fibers are quite different. The calculations showed that use of fiber with numerical aperture of at least 0.20 allows for reduction in the cladding thickness, which is favorable because of the unstable nature of the glass. The final column gives the calculated minimum bend diameter that can be tolerated such that less than 10% of the modes propagating in the fiber will be unguided as a result of radiative losses. Increasing the numerical aperture allows for smaller radius bends, which in addition results in lower microbending loss.

For practical use of fluoride optical fibers in a communications link, single-mode fiber is needed. Thorough analysis of single-mode fluoride fiber design with propagation wavelength equal to 2.55 μm was carried out by Tokiwa and Mimura.[118] They used the theory established for silica fibers with parameters adjusted for fibers of ZBLAN core and HZBLAN clad composition. Although the criterion of $V < 2.405$ is necessary for single-mode propagation, there are other important factors that the authors considered. These included the minimization of excess losses (due to microbending, splicing, coating material, and macrobending), as well as the zero dispersion limit. Their analysis showed that the excess losses caused by microbending

TABLE 3.15
**Recommended Parameters for Ultralow-Loss Fluoride
Single-Mode Fibers**

Fiber Diameter (μm)	Operating Wavelength λ (μm)	Core Radius a (μm)	Relative Refractive Index Difference	Dispersion (ps/km/Å)
150	2.5	6.9	0.43	1.784
150	3.5	8.1	0.60	3.288

After Tokiwa, H. and Mimuran, Y., *J. Lightwave Technol.*, LT4, 1260, 1986.

and coating materials are effectively reduced as the fiber outer diameter is increased. The authors also predicted that, although fusion splicing can be applied to fluoride fibers, it is expected to give the most excess loss. In addition, the analysis showed that the zero-dispersion limit (material plus waveguide dispersion) can be achieved in the wavelength region less than 3.25 μm. For their optimal choice of parameters at operating wavelengths of 2.5 and 3.5 μm, although the zero-dispersion limit is not attainable, the dispersion changes little with λ. Table 3.15 lists the parameters and the calculated dispersion for fluoride single-mode fiber at the two operating wavelengths, where the relative refractive index difference Δ is defined as

$$\Delta = \left(n_{co}^2 - n_{cl}^2\right)/\left(2n_{co}^2\right) \tag{3.20}$$

Thus, a typical single-mode fiber design for propagation at a wavelength of 2.5 μm would have a numerical aperture of 0.14, core diameter of 13.8 μm, and clad diameter of 150 μm.

3.8 FIBER ATTENUATION

3.8.1 PREDICTED INTRINSIC LOSS

Predictions of the intrinsic optical attenuation in fluoride fiber were made by France et al.[45] using the empirical equation:

$$\alpha_{Intrinsic} = A \exp\left(-a/\lambda\right) + B/\lambda^4 \tag{3.21}$$

where A and a are coefficients of the multiphonon absorption and B is the Rayleigh scattering coefficient. (The UV absorption edge is not included since the loss is negligible at wavelengths above 1 μm.) The IR edge was measured over five orders of magnitude for ZBLAN-based glass and fibers of varying path length, and the results of a fit to the IR edge yielded the parameters:

$$\ln A = 23.2 \quad a = 71.6 \tag{3.22}$$

such that the resultant IR edge absorption at 2.55 μm was predicted to be 0.004 dB/km.

The Rayleigh scattering coefficient B was determined from fiber-scattering measurements to yield a minimum value of $B = 0.72$ μm^4 dB/km. From the combined multiphonon and Rayleigh scattering contributions the predicted minimum loss at 2.55 μm was calculated to be 0.024 dB/km. This loss is slightly higher than original predictions (0.01 dB/km at 2.55 μm), but was believed to be representative of the practical minimum attenuation for a low-scattering fiber. This minimum attenuation in fluoride fiber was still eight times lower than the minimum loss of SiO_2 fiber and would result in significant increase in repeater spacings for a telecommunications system. There are problems, however, associated with extrinsic losses in the fibers, such as scattering due to defects and absorption due to OH, transition element, and rare earth impurities; these limitations will be described below.

3.8.2 EXTRINSIC SCATTERING LOSS PREDICTIONS

Extrinsic scattering centers in fluoride fibers result from defects introduced during the glass, preform, or fiber fabrication processes or may be present in the residual chemicals. Many studies of these defects and their origin were made for fluoride glass, as described earlier in Section 3.4.3.2. In addition, France et al. performed an analysis of what defects were most likely to cause extrinsic scattering in fluoride fiber, as well as how much scattering they would contribute. A complete analysis can be found in Reference 45 and will only be summarized here. The authors found that the scattering in a uniform dielectric medium of refractive index n_0 containing spherical defect particles of radius a and refractive index n_1 (where n_0 is real and n_1 may be complex) is dependent on m, the relative refractive index given by

$$m = n_1/n_0 \qquad (3.23)$$

on x, a measure of the relative particle size:

$$x = 2\pi a n_0/\lambda \qquad (3.24)$$

and on Q, the scattering efficiency, or ratio of the true scatter cross section to the geometric cross section. The scattering coefficient ε is determined to be

$$\varepsilon = Q\pi a^2 N/L \qquad (3.25)$$

where N = number of scattering centers per unit volume and L is the length of fiber. The loss α for the fiber of core radius r containing N defects is given by

$$\alpha = (N/L)10(\log_{10}\varepsilon)(a/r)^2 Q \quad (dB/km) \qquad (3.26)$$

or

$$\alpha = 10(\log_{10}\varepsilon)(a/r)^2 Q \quad (dB/defect) \qquad (3.27)$$

TABLE 3.16
**Extrinsic Scattering Defects and the Defect Radius vs. Wavelength
Dependence of Their Scattering**

Particle	n_1	m	Radius (μm) for λ^{-4} Scattering	Radius (μm) for λ^{-2} Scattering	Radius (μm) for λ^0 Scattering
Platinum	$n - ik$	$A - iB$	N/A	N/A	>0.1
ZrO_2	2.18	1.453	<0.3	N/A	>0.3
Gas bubbles	1.00	0.667	<0.3	N/A	>0.3
LaF_3 crystals	1.60	1.067	<0.3	$0.3 < a < 4$	>4.0
$\beta BaZrF_6$ crystals	1.515	1.010	N/A	$0.3 < a < 30$	$\geqslant 30$

After France, P. W., in *Fluoride Glass Optical Fibers,* P. W. France, Ed., CRC Press,
Boca Raton, FL, 1990.

For some limiting cases of x and m, there are simplified solutions for Q that provide
relevant cases for particles in fluoride fiber. The scattering efficiency has a characteristic wavelength dependence, either Rayleigh (λ^{-4}), Rayleigh–Gans (λ^{-2}), or wavelength-independent scattering (λ^0). Table 3.16 lists the results for five important
scattering defects found in fluoride fiber. Values for defect size vs. wavelength
dependence of the scattering are given.

Over the wavelength range 0.3 to 6 μm, the scattering for platinum particles
becomes wavelength independent for a particle radius as small as 0.1 μm. The
presence of platinum particles of this size in a fiber cannot be detected via optical
microscopy; however, the authors have plotted a theoretical scattering loss curve for
10^7 particles/km of Pt of 0.07 μm radius, and found good agreement with measured
scatter loss in a fiber. Other types of scattering centers were considered but did not
give a good fit. Using electron microscopy, Lu and Bradley[119] verified the presence
of 0.25 μm-radius platinum particles in fluoride glasses. Clearly, platinum incorporated in the melting process has severe deleterious effects on scattering.

Wavelength-independent scattering also becomes important for gas bubbles or
ZrO_2 crystals for radii $\geq$ 0.3 μm. For crystals of LaF_3 or $\beta BaZrF_6$, the scattering
contribution is less severe because of the similarity of their refractive indices to the
glass. Table 3.17 gives results for the calculated loss of various particles at 2.55 μm
which have sizes observed experimentally, along with the maximum tolerable density
to achieve low loss. Platinum is again the worst offender since even for submicron
particle size there must be less than 100/km to achieve low loss. Bubbles and LaF_3
crystals also cause severe scattering, and for bubbles the effective scattering may be
even higher since most will be drawn out into forward-scattering capillaries. In
summary, these results indicate that great care must be taken in the fiber fabrication
processes to preclude excess scattering loss.

TABLE 3.17
Extrinsic Defect Loss Calculated at 2.55 μm for Particles of Sizes
Observed in Fluoride Fiber

Particle	Radius (μm)	Loss (dB) per particle	Minimum Density (No./km) for $\alpha < 0.02$ dB/km
Platinum	0.25	2×10^{-4}	<100
Gas bubbles or LaF_3 crystals	4	0.06	0
$\beta BaZrF_6$ crystals	3	4×10^{-4}	<50

After France, P. W., in *Fluoride Glass Optical Fibers,* P. W. France, Ed., CRC Press, Boca Raton, FL, 1990.

3.8.3 FIBER LOSS RESULTS

3.8.3.1 Scattering Loss

A number of scattering measurements were made on short sections of various fluoride fiber compositions to determine the wavelength dependence of scattering. Hattori et al.[120,121] used an angular scattering system with HeNe and argon lasers to measure the scattering from 2-mm segments of fiber whose compositions were based on ZBGA as well as ZBLA with additions of either Na or Li. They obtained a good Rayleigh scattering fit of the angular dependence of the scattering in ZBGA, except for a slight forward-scattering component which they attributed to guided modes in the cladding. In addition, their measured Rayleigh scattering coefficient B for this fiber was only 0.6 $μm^4$ dB/km, resulting in a lower measured scattering loss than that of silica, as had been predicted. Their results for fibers of other compositions yielded values for B which were slightly higher, and there was evidence of back-scattering, which the authors attributed to phase separation.[121] The authors used electron microscopy to examine fracture surfaces of these fibers and confirmed the presence of phase separation. In other studies, micro-Raman scattering was used to detect evidence of microcrystallites in fluoride fibers, such as ZrO_2 and AlF_3.[122,123]

Evidence for near-intrinsic scattering levels in the IR in short sections of fluoride fiber was found by Busse et al.,[124] in which the well-established scattering sphere method was used.[125] Three different laser sources were launched into the fiber: an HeNe at 0.6328 μm; a fiber-pigtailed diode laser at 0.826 μm; and a color center laser at 2.55 μm. An indium antimonide detector for the IR and a silicon photodiode for the visible and near-IR were used to measure the light scattered from a 5-cm section of fiber threaded through an integrating sphere. Another detector monitored the output from the fiber end during the experiment. Both detectors were operated using the reference frequency from a chopper with a lock-in amplifier to preclude background contributions. The scattered loss was obtained from

---

TABLE 3.18

Scattering Losses in dB/km Measured in Short Lengths of Fluoride Fiber

Fiber Section	$\lambda = 0.63$ μm	$\lambda = 0.826$ μm	$\lambda = 2.55$ μm
Type A	7.5	—	0.025
Type B	8.26	5.13	0.10

After Busse, L. E. et al., *Opt. Lett.*, 15, 423, 1990.

$$\alpha\,(\text{dB/km}) = 4.34 \times 10^5 / L\left(P_{sc}/P_o\right) \qquad (3.28)$$

where P_{sc} is the scattered signal, P_o is the transmitted signal measured in the scattering sphere from the fiber end, and L is the diameter of the scattering sphere in centimeters. The cladding modes were stripped with index-matching oil just before and after the scattering sphere.

The scattering losses measured in 5-cm sections of a double-crucible-drawn fiber are given in Table 3.18. The fiber composition was a ZBLAN core and HZBLAN cladding. Two types of sections with different scattering characteristics were identified in the fiber: one that had very little scattering when visible light was launched into the fiber (type A); and another section that appeared to have high scattering (type B). The measured minimum loss at 2.55 μm for type A fiber was determined to be 0.025 ± 0.013 dB/km.

When the measured loss at 0.63 μm for the type A sections was extrapolated assuming a Rayleigh dependence to 2.55 μm, the extrapolated loss value was 0.028 dB/km, which was similar to the measured loss. Thus, these results indicated the presence of Rayleigh scattering behavior in these sections and that there was no measurable wavelength-independent scattering contribution. Losses measured on other sections of fiber from this same spool were higher, however, as shown by the values in Table 3.18 for type B sections indicating non-Rayleigh scattering contributions from defects.

3.8.3.2 Total Attenuation

The most important parameter to determine for fluoride fibers is the total attenuation, which is composed of absorption and scattering terms. The standard cut-back technique[125] was used to measure the total attenuation as a function of wavelength. Early results for minimum attenuation were reported to be 0.9[126] and 0.7 dB/km[100] on relatively short (<100 m) sections of fiber. As the purification and fiberization techniques improved, longer lengths were measured with low loss, as shown in Table 3.19.[107,127,128] Significant progress in lowering the optical loss was made over the 6-year period shown.

The separate contributions of scattering and absorption were obtained by deconvoluting the total attenuation, based on the known absorption bands for OH, transition

TABLE 3.19
Minimum Attenuation Measured on Long Lengths
of Fluoride Fibers

Minimum Attenuation (dB/km)	Wavelength (μm)	Fiber Length (m)	Ref.
9.5	2.55	400	127
3.9	2.55	100	127
0.65 ± 0.25	2.59	110	107
1.9 ± 0.1	2.3	320	128
0.45 ± 0.15	2.3	60	128

TABLE 3.20
Contributions of Various Impurity Ions to the
Absorption Loss for the Fluoride Fiber in Figure 3.29
after Deconvolution of the Total Attenuation

Impurity Ion	Wavelength of Absorption Peak(s) (μm)	Concentration[a] (ppb)
Ho^{3+}	0.64 and 1.95	80
Nd^{3+}	0.74 and 0.81	15
CO_2	2.68	—
Cu^{2+}	0.97	18
OH^-	2.87	4

[a] Based on known extinction coefficients.

After Carter, S. F. et al., *Electron. Lett.,* 26, 2115, 1990.

metal ions, and rare earth ions in the fluoride glass,[45] as well as the wavelength dependence of the scattering. Results for the fiber with minimum attenuation of 0.65 dB/km are shown in Figure 3.29, where the positions of the individual absorption peaks are labeled.[107] The concentrations of these various absorptive species are given in Table 3.20. The deconvolution of the minimum attenuation at 2.59 μm was shown to give a total of 0.33 dB/km of extrinsic absorption loss and 0.30 dB/km of wavelength-independent scattering loss. The absorption loss was mainly due to the tail of the Cu^{2+} absorption centered at 0.97 μm and the Nd^{3+} absorption centered at 2.52 μm. The wavelength-independent scattering was believed to be due to platinum submicron particles.[107] The OH band at 2.9 μm was large, with a peak height of about 20 dB/km. By fabricating glasses using SF_6 atmosphere, the OH band loss contribution can be reduced to less than 5 dB/km, as shown in Figure 3.30. For this fiber, a minimum loss of 2.6 dB/km was measured at 2.58 μm on a 110 m length.[91]

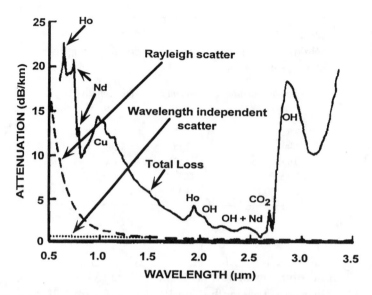

FIGURE 3.29   The total loss for a core-clad fluoride glass optical fiber. Several absorption bands are identified. The fiber was fabricated using the total-vacuum-casting process. (From Carter, S. F. et al., *Electron. Lett.,* 26, 2116, 1990. With permission.)

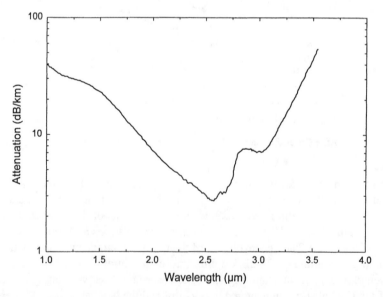

FIGURE 3.30   The total loss for a core-clad fluoride glass optical fiber made from glasses that were treated with SF_6 gas to lower the OH peak height at about 2.9 μm.

The lowest reported loss for fluoride fibers fabricated with the preform/casting technique was 0.45 dB/km at 2.3 μm, measured on 60 m of fiber.[128] Losses have also been measured on long lengths of fibers fabricated using the double-crucible

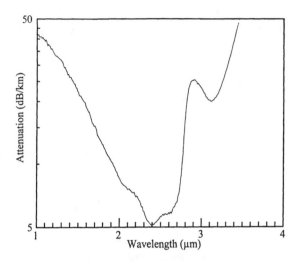

FIGURE 3.31 The total loss for a core-clad fiber fabricated using the double-crucible process. The fiber was 110 m in legnth.

process. The lowest loss for 125 m of this fiber was 5.1 dB/km at 2.39 μm.[25] The loss curve is shown in Figure 3.31.

3.9   STRENGTH OF FLUORIDE GLASS FIBERS

Several questions arise when talking about the strength of fluoride fibers. The first question which comes to mind is what is the ultimate theoretical strength of fluoride glass fibers. Second, what is the highest strength obtained to date and what are the typical strengths obtained in practice. Third, why are the practical strengths lower than the theoretically predicted value. In addition, values quoted in the literature cannot be compared directly since some parameters may be different. These include tensile strength and bending strength values, fiber gauge length, fiber diameter, fiber composition, load and strain rate, and testing atmosphere/environment.

Let us first consider the case for the theoretical strength. For silica, the theoretical strength has been estimated to be about $\frac{1}{5}$ the value of Young's modulus. The value of Young's modulus of ZBLAN is approximately 55 GPa, which gives a value of 11 GPa for the theoretical strength. In practice, the highest bending strength reported has been about 1.4 GPa for ZBLA and ZBLAN fibers.[129] Unfortunately, neither the load rate nor the strain rate, which are critical parameters, were specified. Nevertheless, these strengths were obtained by first chemically etching the preforms in HCl–ZrOCl$_2$ solution and then drawing the fibers in an N$_2$/NF$_3$ reactive atmosphere. Chemical etching removed any surface flaws, and the reactive atmosphere suppressed the hydrolysis reaction which led to the formation of ZrO$_2$ crystals on the surface. The etching action of HCl–ZrOCl$_2$ can be represented as follows:

$$\text{ZBLAN}_{(\text{glass})} + \text{water} \rightarrow \text{Ba}^2_{(\text{aq})} + \text{Zr}^{4+}_{(\text{aq})} + 6\text{F}^-_{(\text{aq})} \qquad (3.29)$$

In the presence of water and absence of a complexing agent, the components such as Ba and Zr precipitate out as fluoride crystals on the surface of the preform as follows:

$$Ba^{2+}_{(aq)} + 2F^{-}_{(aq)} \rightarrow BaF_{2(s)} \tag{3.30}$$

$$Zr^{4+}_{(aq)} + 4F^{-}_{(aq)} \rightarrow ZrF_{4(s)} \tag{3.31}$$

However, in the presence of HCl–ZrOCl$_2$ the fluoride ions are removed from solution and the cations remain soluble:

$$4F^{-}_{(aq)} + ZrOCl_{2(aq)} \rightarrow ZrOF^{2-}_{4(aq)} + 2Cl^{-}_{(aq)} \tag{3.32}$$

In addition to this, the formation of ZrO$_2$ and subsequent beneficial effect of N$_2$/NF$_3$ in maintaining a dry atmosphere during the fiberizing process can be represented as follows:

$$ZrF_4 + 2H_2O \rightarrow ZrO_2 + 4HF \tag{3.33}$$

$$2NF_3 + 3H_2O \rightarrow N_2 + 3/2O_2 + 6HF \tag{3.34}$$

The fiber bending strength was increased slightly to 1.6 GPa by etching the fibers with HCl–ZrOCl$_2$ solution. The major improvement was the shift in the low-strength population to higher values.

The tensile and bending strengths have been determined by other research groups, and some of these results are highlighted in Table 3.21. The tensile strengths are typically less than 50% of the bending values. This is primarily due to the longer gauge length that is under uniaxial strain in a tensile strength measurement compared with the small region that experiences maximum strain in a bending strength measurement. This is highlighted in Figure 3.32, which also demonstrates that the strength is dependent upon the type of treatment to which the preform is subjected.[133] The strength increases in the order of mechanical polishing < chemical etching < mechanical polishing/etching + plasma activated fluorine.

Sakaguchi et al.[132] used an alternative approach to reduce the surface crystallization tendency in the neck-down region of the preform. This was performed by extending the length of the furnace, thereby lowering the fiber draw temperature and hence reducing the crystal growth rates. The strengths were comparable to fibers obtained from chemically etched preforms.

Generally speaking, it is apparent that crystallization is a major problem affecting not only the optical properties of fibers but also their strength. Aside from this, the fluoride fibers exhibit zero stress aging due to corrosion by the environment, specifically water. This is easily seen by immersing fluoride fibers in water. Diffusion of water molecules into the glass causes break up of the glass structure. Since the glass components have low solubilities in water, they readily precipitate on the fiber

TABLE 3.21
Fiber Tensile and Bending Strengths

Glass Type	ZBLAN	ZBLA	ZBLYAL	ZBLAN	ZBLAN	ZBLA	ZBLA
Strength test	Tensile	Tensile	Tensile	Tensile	Bending	Bending	Bending
Gauge length (m)	0.4	0.25	0.2	1.0	—	—	—
Loading rate (mm/min)	20	1	10	1	1	NA	1
Preform treatment							
Polished	Yes	NA	Yes	Yes	Yes	NA	Yes
Etched	No	$ZrOCl_2$/HCl	No	H_3BO_3/HCl	H_3BO_3/HCl	$ZrOCl_2$/HCl	$ZrOCl_2$/HCl
Other	No	No	40 mm furnace	Ar/NF_3 plasma	Ar/NF_3 plasma	Fiber etched	No
Clad	FEP/ACC	FEP	No	ACC	ACC	NA	FEP
Drawing atmosphere	He	Ar	Ar	N_2	N_2	N_2/NF_3	N_2
Strength (MPa)	200–310	200–400	300–400	250–400	720–1200[a]	1400–1600	400–820
References	130	131	132	133	134	129	135

Note: NA = Not available; FEP = Teflon; ACC = acrylate.

[a] Corrected for geometric factor of 1.198.

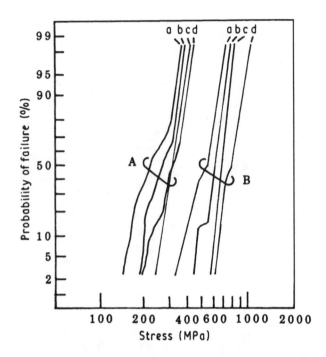

FIGURE 3.32 Weibull failure probability as a function of stress measured in air at 25°C and 60% relative humidity. Preform surface treatments: (a) mechanical polishing, (b) chemical etch, (c,d) mechanical polishing/etch plus plasma-activated fluorine. (A) Dynamic tensile loading; (B) bending strength measured under the same conditions. (From Pureza, P. C. et al., *J. Mater. Sci.,* 26, 5151, 1991. With permission.)

surface causing noticeable visible changes and producing extremely fragile fibers. In addition, work on bulk glasses has demonstrated that fluorides exhibit stress corrosion, and a mechanism has been postulated.[135] Therefore, it seems obvious to protect the fibers from moisture.

3.10 PROTECTIVE COATINGS

Polymer coatings such as Teflon are usually applied to the outside surface of fluoride glass fibers. Unfortunately, it has been shown that they are not impervious to moisture. For example, Nakata et al.[136] have demonstrated that the fiber strength degrades with time at 100% relative humidity. However, Moynihan et al.[137] have pointed out that only condensed water causes damage. It may be that temperature fluctuations (especially at night) can lead to condensation and, therefore, fiber damage. Nevertheless, water is a problem, and so the fiber needs to be protected from moisture. Since the chemical durability of chalcogenide glasses is well known, Nakata et al.[138] applied a chalcogenide glass coating to a fluoride glass preform by a dip coating process. Fibers drawn from this modified preform were subjected to 100% relative humidity for 30 days. The chalcogenide-coated fibers retained their strength for up to 30 days. For comparison, Teflon-coated fibers exhibited approximately 50% of the original strength after 30 days. Other attempts have been made to apply chalcogenide coatings using ion beam deposition prior to pulling fluoride fibers.[139] Unfortunately, there is no strength data available for fibers.

Various other vacuum-deposited coatings have been investigated. For instance, Simmons and Simmons[140] had favorable results using diamondlike carbon (DLC), although some pitting was evident in the coating. Schultz et al.[141] have demonstrated that MgO is an excellent protective coating since there was no damage to the glass even after 100 h exposure to water. More recently, it has been demonstrated that MgO coatings applied via cylindrical magnetron sputtering exhibit unflawed behavior for up to 1500 h on bulk samples and 1000 h on coated fibers stressed at 25 kpsi.[91] Using this technique, it should be possible to provide a minimum coating thickness of 0.03 μm required for hermeticity using a 2-m-long sputter coater and drawing fiber at a rate of 20 m/min.

3.11 APPLICATIONS FOR FLUORIDE OPTICAL FIBER

3.11.1 LOW-LOSS APPLICATIONS

Demonstrations of using low-loss fluoride single-mode fiber in a data transmission link at 2.55 μm have been reported.[142,143] In the first test, the setup consisted of a tunable color center laser with single mode output at 2.55 μm injected into a fluoride single-mode fiber, whose output was coupled into an external LiNbO$_3$ waveguide modulator and an InGaAs/InAsP photodiode.[142] A 400-m-long fiber with a core diameter of 11 μm and a clad diameter of 125 μm was used which possessed an attenuation of 40 dB/km at 2.55 μm. Up to 400 Mb/s transmission was demonstrated with this system. Improvements to the system resulted in a 14 dB increase in the minimum received power.[143]

Another important consideration for using fluoride fiber for communications applications is the ability to make low-loss splices in the fiber. Various splicing techniques have been investigated, including resistance-heating fusion splicing, mechanical splicing, and arc fusion splicing.[91] Fusion splicing of fluoride fiber is difficult to control and reproduce, leading to effects such as fiber bulging at the splice and an offset in core centers. Typical losses measured on splices showed large variation, with an average splice loss of 0.5 dB and a minimum loss of 0.1 dB. Mechanical splicing has not been successful because of the lack of availability of appropriate index-matching adhesives, and, as a result, splice losses are not available.

The lowest splice losses obtained to date were with an arc fusion splicer specifically designed for low-temperature splicing. To achieve lower temperatures, a Power Technologies, Inc., Model PTS-330 splicer was modified to operate at lower current levels and for shorter times and, using a He gas purge, to provide an even distribution of the heat.[144] The optical loss was calculated by measuring the fiber output intensity before and after breaking and splicing the fiber without disturbing the fiber-to-detector or source-to-fiber ends. The splice loss α in decibels is given by

$$\alpha = -10 \log(I_0/I_i) \qquad (3.35)$$

where I_0 and I_i are the measured output intensities before and after the splice was made, respectively. The average splice loss was improved to 0.08 dB/splice, and the minimum splice loss achieved was 0.024 dB/splice.[144] The improvements were attributed to two major factors, namely, reduction in the cleave angle of the fiber ends to <1° and reduction in the surface hydrolysis upon reheating.

Splice strengths were determined from bending tests, where the fiber surface was stressed to a level dependent upon the radius of curvature at rupture. This curvature was determined by using a micrometer driven at a strain rate of 0.1%/s and by assuming the displacement at fracture was equal to the diameter of a circle. The highest strength was obtained for splices made using a dry helium purge gas during fusion to reduce surface hydrolysis. Strengths up to approximately 600 MPa were achieved, which indicates that fluoride fiber systems would not be limited by the strength of splices.

3.11.2 MEDIUM-LOSS APPLICATIONS

There are a number of applications requiring short to medium lengths of fluoride fiber due to its good transparency beyond 2 μm, where silica fiber loss becomes high. These applications include chemical sensors, power delivery for mid-IR lasers, and, with appropriate rare earth dopants added to the glass composition, fiber lasers and amplifiers. Medium-loss fluoride fibers (loss ≤1 dB/m) are more easily fabricated than those with ultralow loss and are available from commercial vendors. Some of the applications for these fibers are described below.

3.11.2.1 Chemical Sensors

Many chemical species have absorption bands or overtones in the 1 to 4 μm region, such as OH or C–H. Spectroscopic instruments (such as a Fourier transform infrared, or FTIR, spectrometer) can be used to characterize the contamination of a soil sample with toxic chemicals, such as toluene or benzene. The use of fiber attached to the FTIR spectrometer allows for remote capability of such measurements whereby the soil sample can be tested *in situ*.[145] By coupling absorption cells via the fiber to an FTIR spectrometer, researchers have made web sensors to monitor polymer film thicknesses on-line in a production plant.[146] Fiber-optic-coupled gas or liquid sensing devices, which utilize evanescent spectroscopy or attenuated total reflection, have also been used with FTIR spectrometers, as described in Chapter 8. Recently, products utilizing AOTF (acousto-optic tunable filter) technology have also become available for chemical sensing using fluoride fiber optic coupling.[147] Another application using fluoride fibers involved the detection of water in jet fuels.[91]

Fluoride fibers have been tested in high-radiation environments (for example, irradiation with ^{60}Co sources) and have proved more resistant to radiation-induced losses in the IR than silica fibers.[148] This result makes them superior to silica for sensing applications in which high radiation is present.

3.11.2.2 Infrared Power Delivery

Er:YAG lasers have an output at 2.94 μm, which is near the fundamental OH absorption peak. Thus, these lasers are useful for surgery, and a flexible fiber to deliver this power would be beneficial. Itoh et al.[149] have demonstrated up to 970 mJ input from an Er:YAG laser launched into a fluorozircoaluminate with a core diameter of 440 μm, a clad diameter of 500 μm, and a numerical aperture of 0.3. The pulse repetition frequency (PRF) was 1 to 10 Hz, and pulse width was 200 μs. The total irradiation time was 15 min. The maximum power output from the fiber was 8.7 W, measured at the highest PRF. No damage was observed up to this power level, which was limited by the available laser output. Power transmission was tested when the fiber was bent into a 360° loop, since flexibility is an important issue for surgical applications. No change in output was observed for bend radii above 40 mm. The maximum power output reported by these authors was twice that previously obtained with a fluorozirconate composition,[150] which the authors attribute to the higher softening temperature for fluorozircoaluminate glasses. Studies of the damage threshold limitations using 2.94 μm laser power incident on both single and multimode ZBLAN fibers were made by Wuthrich et al.[151] They used a laser emitting pulses of 200 μs duration with 4 Hz PRF. No damage was observed in the fibers for transmitted intensities below 80 kW/cm². Above this value, there were limitations due to damage at the input or output end faces of the fibers, although they measured a maximum transmitted intensity of 2.7 MW/cm². By using a tightly focused beam waist, the maximum power intensity transmitted was 14 MW/cm² without damage.

3.11.2.3 Fluoride Fiber Lasers and Amplifiers

Much work has been devoted to the development of rare earth–doped fluoride glasses and fibers.[152,153] This is due to the fact that the IR transmission region of fluoride fibers extends beyond that of silica fibers, making the fluoride materials good rare earth hosts for generation of long-wavelength radiation. Both fluorescence and upconversion have been investigated experimentally in rare earth–doped fluoride fibers. In fluorescence, photon input energy excites electrons to higher energy states, whereby nonradiative decay occurs by lattice phonons, and thereafter decay by photon emission occurs (fluorescence) at an energy less than that of the incident photon. In the upconversion process, the excited electron absorbs another photon and is excited to a higher energy state, thereafter decaying back while emitting a photon of energy greater than that of the original excitation; thus IR pumping can lead to visible emission. For both of these processes, fibers have great advantages over bulk glass due to the much longer path length and the confinement of the pump and emitted beams. For this reason there has been emphasis on fabricating rare earth–doped fluoride fibers for laser and amplifier applications.

A thorough review of results achieved for laser oscillation and amplification using rare earth–doped fluoride fibers is beyond the scope of this chapter; however, highlights are presented in Tables 3.22 and 3.23, respectively. In the development of lasers, early work was driven by the need for near-IR sources for communications applications, such as for 0.8 to 0.9, 1.3, and 1.55 μm.[154-161] The 1.3 μm transition in Nd^{3+}-doped silica fibers is quenched by excited state absorption, but this effect is less favorable in Nd^{3+}-doped fluorozirconate fiber, making it a better host. Brierly and Hunt[159] demonstrated a slope efficiency of 57%, which corresponded to a resultant quantum efficiency (QE) of 97% for the 1346 nm transition. This is one

TABLE 3.22
Laser Oscillation Results Using Rare Earth–Doped Fluoride Optical Fibers[a,b]

Laser Wavelength (μm)	Dopant	Pump Wavelength(s) (nm)	Maximum Output Power (mW)	Laser Efficiency[c,d] (%)	Ref.
Visible					
0.455	Tm^{3+}	645 and 1064	3	1.5	171
0.480	Tm^{3+}	647.1 and 676.4	0.4	NA	172
0.492	Pr^{3+}	1017 and 835	22	7.5	173
0.521	Pr^{3+}/Yb^{3+}	833 and 1016	0.7	1.6	174
0.544	Er^{3+}	801	3	16	175
0.55	Ho^{3+}	647.1	10.0	20.0[d]	176
0.635	Pr^{3+}/Yb^{3+}	833 and 1016	6.2	8.7	174
Near IR					
0.75	Ho^{3+}	647.1	2.0	NA[d]	154
0.82	Tm^{3+}	676.4	0.5	1.6	155
0.85	Er^{3+}	801	NA	38	156
0.983	Er^{3+}	647.1	10	8.7	157
1.02	Yb^{3+}	911	100	56[d]	158
1.05	Nd^{3+}	795	55	70	159
1.346	Nd^{3+}	795	30	57	159
1.38	Ho^{3+}	48	NA	0.28	160
1.47	Tm^{3+}	1064	100	59	161
1.6–1.72	Er^{3+}	514, 488	NA	NA	162
1.82	Tm^{3+}	1640	37	84[d]	163
Mid IR					
2.024	Tm^{3+}/Ho^{3+}	820–830	250	60[d]	164
2.3	Tm^{3+}	790	0.95	10	165
2.715	Er^{3+}	802	2.1	8[d]	166
2.7	Er^{3+}/Pr^{3+}	650, 795, 980	30	>13	167
2.9	Ho^{3+}	640	12.6	2.9	168

[a] All results listed are for CW laser operation.

[b] NA = not available.

[c] Also called slope efficiency.

[d] Indicates single-mode fiber was used.

of the highest QE results achieved for a fiber laser. Recently, pumping of Tm^{3+}-doped fluoride fiber has resulted in laser oscillation at 1.47 μm (close to the 1.55 μm window) with a slope efficiency as high as 59% and a maximum output power of 100 mW.[161] Other fiber laser sources were developed containing Yb^{3+}, Tm^{3+}, and Ho^{3+} for output in the visible/near-IR region (0.7 to 1.8 μm). Er^{3+}-doped fluoride fiber exhibits many transitions throughout the IR, including the 1.6 to 1.7 μm transition,[162] important for chemical sensors to detect methane, which has an absorption band at 1.67 μm.

TABLE 3.23
Laser Amplification Results Using Rare Earth–Doped Fluoride Optical Fibers[a]

Amplifier Wavelength (μm)	Dopant	Pump Wavelength (nm)	Maximum Gain (dB)	Ref.
0.806	Tm^{3+}	780	26	179
1.33	Nd^{3+}	780	10.0	180
1.30	Pr^{3+}	1017	28.3	181
1.30	Pr^{3+}	1047	40.6	182
1.47	Tm^{3+}	1064	>10	161
1.55	Er^{3+}	1485	18.0	183
1.82	Tm^{3+}	1640	36.5	163
2.72	Er^{3+}	647	35.9	184

[a] All results are for single-mode fiber.

In the mid-IR, use of Tm^{3+}/Ho^{3+} codoping resulted in an output of 250 mW at 2.024 μm from a single-mode fiber laser.[164] Laser sources in the 2.3 to 2.9 μm region[165-168] are of interest because of their proximity to the fluoride fiber minimum loss, as well as for surgical applications. Recently, an Er^{3+}/Pr^{3+} codoped fluoride fiber gave a maximum output of 30 mW at 2.7 μm.[167] The first system demonstration using an Er^{3+}-doped multimode fluoride fiber as a source for sending data at 2.7 μm through a fluoride fiber resulted in transmission at 34 Mb/s with a bit error rate of 10^{-9} and receiver sensitivity equal to –26 dB/m.[169] If low-loss single-mode fluoride fiber had been available for such a system, spans of 1000 km would be possible without penalty due to dispersion. Allen and Esterowitz[165] reported laser oscillation in a Tm^{3+}-doped fluoride fiber, pumped by a 790 nm AlGaAs laser diode, which had a tunable range of 2.2 to 2.5 μm. At 2.9 μm, an Ho^{3+}-doped fluoride fiber laser produced over 12 mW output when pumped at 640 nm.[168] Recently, Schneider[170] has reported the first demonstration of superfluorescence at 3.9 μm in an Ho^{3+}-doped fluoride fiber when pumped at 640 nm and cooled to 173 K; this wavelength is important for applications in the 3 to 5 μm atmospheric transmission window.

During the last 6 years, much work has been done in developing upconversion fluoride fiber lasers,[171-176] driven in part by need for compact blue/green sources for optical disk recording and laser printers and displays. By pumping a Pr^{3+}-doped fluoride fiber with both 835 and 1017 nm laser diode sources, Zhao et al.[173] reported up to 22 mW output at 492 nm. Pr^{3+}/Yb^{3+}-codoped, as well as Tm^{3+}- and Ho^{3+}-doped, fluoride fibers have also been developed for visible upconversion lasers, with results shown in Table 3.22. In addition, upconversion blue/green fluorescence has been recently reported for doped fluoride compositions that are not ZrF_4 based, such as BIZYTGaZr glasses[177] and Al–Zr based fluoride glasses.[178]

Many of the fluoride fiber lasers developed were multimode, but, with improved fabrication techniques, single-mode fluoride fibers have become practical for amplifier applications, with results given in Table 3.23. The first Nd^{3+}-doped single-mode fluoride fiber used for 1.3 μm amplification had a maximum of 10 dB gain using a

780 nm pump source.[180] The amplifier tuning range was 1.31 to 1.38 µm. More recently, use of Pr^{3+}-doped fiber with near-IR pumping has resulted in 30 to 40 dB of gain at 1.3 µm.[181,182] Devices utilizing such amplifiers for 1.3 µm have been developed and tested.[185] For example, a single-mode Pr^{3+}-doped fluoride fiber amplifier (PDFFA) has been used to demonstrate 100 Mb/s digital transmission at 1.3 µm through silica fiber.[186] It is projected that the fiber length could be increased to more than 27 km with no degradation in transmission quality. In another experiment, 24 analog FM-SCM-TV channels were amplified by a single-mode PDFFA without loss of image quality.[187]

Other notable results for 0.8, 1.5, and 2.7 µm amplification using doped fluoride fibers are given in Table 3.23. The 1.55 µm Er^{3+}-doped fluoride fiber amplifier was tunable from 1.52 to 1.60 µm.[183] A system demonstration of an Er^{3+}-doped fluoride fiber amplifier has been reported by Yoshida et al.,[188] in which 10 Gb/s ten-channel transmission was successfully achieved over 600 km of dispersion-shifted fiber with 100 km repeater spacing (23 dB total fiber loss). As development and optimization continues, further system tests are anticipated with doped fluoride fiber amplifiers.

3.12   SUMMARY

The ultralow-loss applications for fluoride glass fibers have not been realized because of problems associated with microcrystallization, crucible contamination, and bubble formation during fiber fabrication. Nevertheless, fibers with losses greater than 10 dB/km are routinely obtained and can be used for chemical sensor applications as well as high-power UV, visible, and IR laser transmission. Their major application may be as glass hosts for rare earth ions for making fiber lasers and amplifiers, especially for 1.3 µm communication systems. Currently, a number of commercial vendors throughout the world produce rare earth–doped and –undoped fluoride glass fibers for many of the applications discussed. One of the remaining concerns with these fibers lies in their relatively poor chemical durability compared with silica fibers. For practical applications, this problem could potentially be overcome with appropriate hermetic coatings.

REFERENCES

1. H. Rawson, in *Inorganic Glass-Forming Systems*, J. P. Roberts and P. Popper, Eds., Academic Press, New York, 1967, 236.
2. W. H. Dumbaugh and D. W. Morgan, *J. Non-Cryst. Solids,* 211, 38 (1980).
3. A. G. Pincus, *J. Am. Opt. Soc.,* 35, 92 (1945).
4. W. H. Dumbaugh, D. W. Morgan, and B. D. Spivack, Annual Report, Contract No. EY-76-C-02-4079, U.S. Energy Research and Development Administration, C00-4079-2, 1977.
5. A. Glass, Lawrence Livermore Laboratory Laser Program Annual Report, UCRL 50021-74, 1974, 256.
6. M. J. Weber, C. F. Cline, W. L. Smith, D. Milam, D. Herman, and R. W. Hellworth, *Appl. Phys. Lett.,* 32, 403 (1978).
7. J. R. Bettis, A. H. Gunther, and R. A. House, *Opt. Lett.,* 4, 256 (1979).

8. C. M. Baldwin, R. M. Almeida, and J. D. Mackenzie, *J. Non-Cryst. Solids,* 43, 309 (1981).

9. M. Poulain, M. Poulain, J. Lucas, and P. Brun, *Mater. Res. Bull.,* 10, 243 (1975).

10. M. Poulain, Fluoride glass composition and processing, in *Fluoride Glass Fiber Optics,* I. D. Aggarwal and G. Lu, Eds., Academic Press, New York, 1991, 1.

11. M. Poulain, M. Chanthansinh, and J. Lucas, *Mater. Res. Bull.,* 12, 151 (1977).

12. A. Lecoq and M. Poulain, *J. Non-Cryst. Solids,* 34, 101 (1979).

13. A. Lecoq and M. Poulain, *J. Non-Cryst. Solids,* 41, 209 (1980).

14. D. C. Tran, R. J. Ginther, G. H. Sigel, Jr., *Mater. Res. Bull.,* 17, 1177 (1982).

15. S. Mitachi, Y. Terunuma, Y. Ohishi, and S. Takahashi, *Jpn. J. Appl. Phys.,* 22, L537 (1983).

16. K. Ohsawa, T. Shibata, K. Nakamura, and M. Kimura, in *Proc. 1st. Int. Symp. on Halide Glasses,* Cambridge, U.K. (1982).

17. K. Ohsawa and T. Shibata, *J. Lightwave Technol.,* LT2, 602 (1984).

18. C. F. Rapp, Development of Large Diameter Fluoride Glass Windows, Part II. Demonstration of 51 cm and 79 cm Diameter Windows, Final Report, Contract Number N00014-89-C-2459, Naval Research Laboratory, 1991.

19. T. Kanamori and S. Takahashi, *Jpn. J. Appl. Phys.,* 24, L758 (1985).

20. H. Iwasaki, SPIE 618, Infrared optical materials and fibers IV, 2 (1986).

21. H. Tokiwa, Y. Mimura, O. Shinbori, and T. Nakai, *J. Lightwave Technol.,* LT3 (3), 569 (1985).

22. Y. Mimura, H. Tokiwa, and O. Shinbori, *Electron. Lett.,* 20, 100 (1984).

23. H. Tokiwa, Y. Mimura, O. Shinbori, and T. Nakai, *J. Lightwave Technol.,* LT3 (3), 574 (1985).

24. H. Tokiwa, Y. Mimura, T. Nakai, and O. Shinbori, *Electron. Lett.,* 21, 1131 (1985).

25. M. L. Nice, C. F. Rapp, and R. Mossadegh, Ultra-Low Loss Fluoride Fiber Development at Owens-Corning Fiberglass, Part II. Evaluation of the Double and Triple Crucible Techniques, Final Report. Contract number N00014-91-C-2361, 1993.

26. K. H. Sun, U.S. Patent 2,466,509, 1949.

27. J. J. Videau, J. Portier, and B. Piriou, *Rev. Chem. Min.,* 16, 393 (1979).

28. T. Kanamori, K. Oikawa, S. Shibata, and T. Manabe, *Jpn. J. Appl. Phys.,* 20, L326 (1981).

29. M. Poulain, M. Poulain, and M. Matecki, *Mater. Res. Bull.,* 16, 555 (1981).

30. P. Tick, *Mater. Sci. Forum,* 3–33, 115 (1988).

31. T. Izumitani, T. Yamashita, M. Tokida, K. Miura, and H. Tajima, in *Halide Glasses for Infrared Fiberoptics,* R. M. Almeida, Ed., Martinus Nijhoff, Boston, 1987, 187.

32. K. Miura, I. Masuda, K. Itoh, and T. Yamashita, *Mater. Sci. Forum,* 67–68, 335 (1991).

33. J. Nishii, Y. Kaite, and T. Yamagishi, *Phys. Chem. Glasses,* 30, 55 (1989).

34. M. Poulain, *J. Non-Cryst. Solids,* 56, 1 (1983).

35. J. Lucas, H. Slim, and G. Fonteneau, *J. Non-Cryst. Solids,* 44, 31 (1981).

36. J. Guery, G. Courbin, C. Jacoboni, and R. De Pape, *Mater. Res. Bull.,* 19, 1437 (1984).

37. M. Poulain and M. Poulain, *Mater. Sci. Forum,* 67–68, 129 (1990).

38. S. Mitachi, Y. Terunuma, Y. Ohishi, and S. Takahashi, *Jpn. J. Appl. Phys.,* 22, L537 (1983).

39. R. C. Folweiler, *SPIE,* 799, 106 (1987).

40. S. Mitachi, Y. Terunuma, Y. Ohishi, and S. Takahashi, *J. Lightwave Technol.,* LT2, 587 (1984).

41. K. Kobayashi, *Bull. Chem. Soc. Jpn.,* 61, 2965 (1988).

42. K. J. Ewing and J. A. Sommers, Purification and analysis of heavy metal fluorides for use in heavy metal fluoride glasses, in *Fluoride Glass Fiber Optics,* I. Aggarwal and G. Lu, Eds., Academic Press, New York, 1991, 142.

43. G. Maze, V. Cardin, and M. Poulain, *Electron. Lett.,* 21, 884 (1984).

44. S. Carter, P. W. France, M. W. Moore, and E. A. Harris, *Phys. Chem. Glasses,* 28, 22 (1987).

45. P. W. France, Manufacture of infrared fibers, in *Fluoride Glass Optical Fibers*, P. W. France, Ed., CRC Press, Boca Raton, FL, 1990, 100.

46. S. Takahashi, T. Kanamori, Y. Ohishi, K. Fujiura, and Y. Terunuma, *Mater. Sci. Forum,* 32–33, 87 (1988).

47. J. S. Sanghera, P. Hart, M. G. Sachon, K. J. Ewing, and I. D. Aggarwal, *J. Am. Ceram. Soc.,* 73, 1339 (1990).

48. M. L. Nice and C. F. Rapp, Ultra-Low Loss Fluoride Fiber Development at Owens-Corning Fiberglass, Final Report Contract No. N00014-87-C-2165, 1989.

49. G. Lu and J. Bradley, *Mater. Sci. Forum,* 19–20, 545 (1987).

50. M. Robinson, R. C. Pastor, R. R. Turk, M. Braunstein, and R. Braunstein, *Mater. Res. Bull.,* 15, 735 (1980).

51. M. L. Robinson and G. L. Tangman, *Mater. Res. Bull.,* 23, 943 (1989).

52. D. C. Tran and C. Fisher, U.S. Patent 4,539,032, 1985.

53. T. Nakai, Y. Mimura, H. Tokiwa, and O. Shinbori, *J. Lightwave Technol.,* 4, 87 (1986).

54. J. S. Sanghera, Naval Research Laboratory. Private communication (1989).

55. J. S. Sanghera, P. Pureza, and I. D. Aggarwal, U.S. Patent 5,364,434, 1994.

56. A. Sarhangi and D. A. Thompson, *Mater. Sci. Forum,* 19–20, 259 (1987).

57. L. Busse, I. Aggarwal, K. Ewing, and B. Harbison, U.S. Patent 5,211,731, 1993.

58. K. Fujiura, Y. Nishida, H. Sato, S. Sugawara, K. Kobayashi, Y. Terunuma and S. Takahashi, *J. Non-Cryst. Solids,* 161, 14 (1993).

59. R. E. Riman, M. Dejneka, J. Eamsiri, E. Snitzer, A. Mailot, and A. Leaustic, *J. Sol-Gel Sci. Technol.,* 2, 849 (1994).

60. M. Saad and M. Poulain, *Am. Ceram. Soc. Bull.,* 74, 66 (1995).

61. M. G. Drexhage, Heavy-metal fluoride glasses, in *Treatise on Materials Science & Technology*, Vol. 26, *Glass IV*, M. Tomozawa and R. Doremus, Eds., Academic Press, New York, 1985, 151.

62. S. Shibata, M. Horiguchi, K. Jinguchi, S. Mitachi, T. Kanamori, and T. Manabe, *Electron. Lett.,* 17, 775 (1981).

63. T. Miyashita and T. Minabe, *IEEE J. Quantum. Electron,* QE-18, 1432 (1982).

64. Y. Ohishi, S. Mitachi, T. Kanamori, and T. Manabe, *Phys. Chem. Glasses,* 24, 135 (1983).

65. P. W. France, S. F. Carter, M. W. Moore, and J. R. Williams, *SPIE,* 618, 51 (1986).

66. S. F. Carter, P. W. France, and J. R. Williams, *Phys. Chem. Glasses,* 27, 46 (1986).

67. P. W. France, S. F. Carter, J. R. Williams, K. J. Beales, and J. M. Parker, *Electron. Lett.,* 20, 607 (1984).

68. P. W. France, S. F. Carter, and J. R. Williams, *J. Am. Ceram. Soc.,* 67, C243 (1984).

69. M. Poulain and M. Saad, *J. Lightwave Technol.,* 2, 599 (1984).

70. S. F. Carter, P. W. France, M. W. Moore, and E. A. Harris, *Phys. Chem. Glasses,* 28, 22 (1987).

71. H. Hu and J. D. Mackenzie, *J. Non-Cryst. Solids,* 80, 495 (1986).

72. J. S. Sanghera, B. B. Harbison, and I. D. Aggarwal, *J. Non-Cryst. Solids,* 140, 146 (1992).

73. G. Fonteneau, *Mater. Res. Bull.,* 19, 685 (1984).

74. S. Mitachi and T. Miyashita, *Electron. Lett.,* 18, 170 (1982).

75. A. M. Busswell, R. L. Maycock, and W. H. Rhodebush, *J. Chem. Phys.,* 8, 362 (1940).

76. W. C. Hasz, S. N. Crichton, and C. T. Moynihan, *Mater. Sci. Forum,* 32–33, 589 (1988).

77. A. J. Drehman, *Mater. Sci. Forum,* 19–20, 483 (1987).
78. P. Hart, G. Lu, and I. D. Aggarwal, *Mater. Sci. Forum,* 32–33, 179 (1988).
79. J. S. Sanghera, P. Hart, M. G. Sachon, K. J. Ewing, and I. D. Aggarwal, *Mater. Sci. Forum,* 67–68, 7 (1991).
80. I. D. Aggarwal, J. S. Sanghera, B. Harbison, L. Busse, and P. Pureza, *Mater. Sci Forum,* 67–68, 443 (1991).
81. G. Lu and J. P. Bradley, *J. Am. Ceram. Soc.,* 69, 585 (1986).
82. J. Parker, D. Seddon, and A. Clare, *Phys. Chem. Glasses,* 28, 4 (1988).
83. J. S. Sanghera, P. Hart, M. G. Sachon, L. Busse, and I. D. Aggarwal, *J. Am. Ceram. Soc.,* 73, 2677 (1990).
84. S. Sakaguchi, K. Fujiura, Y. Ohishi, Y. Terunuma, and T. Kanamori, *J. Non-Cryst. Solids,* 95–96, 617 (1987).
85. J. S. Sanghera, M. G. Sachon, P. Hart, and I. D. Aggarwal, *Mater. Lett.,* 10, 181 (1990).
86. G. Lu, C. Fisher, and J. P. Bradley, *J. Non-Cryst. Solids,* 94, 45 (1987).
87. G. Lu, I. Aggarwal, and J. P. Bradley, *J. Am. Ceram. Soc.,* 71, C156 (1988).
88. J. S. Sanghera and I. D. Aggarwal, *J. Am. Ceram. Soc.,* 76, 2341 (1993).
89. S. Mitachi, S. Sakaguchi, H. Shigematsu, and S. Takahashi, *Jpn. J. Appl. Phys.,* 24, L827 (1985).
90. S. Mitachi and P. Tick, *Mater. Sci. Forum,* 67–68, 169 (1991).
91. B. Harbison, L. Busse, P. Pureza, J. Sanghera, J. Jewell, K. Ewing, and I. Aggarwal, Ultra Low Loss Fiber Optic Program, Final Report, RL 2B RS34047, Naval Research Laboratory, Washington, D.C. (1994).
92. J. J. Videau and J. Portier, Fluoride glasses, in *Inorganic Solid Fluorides*, P. Hagenmuller, Ed., Academic Press, New York, 1985, 309.
93. C. J. Simmons, S. A. Azali, and J. H. Simmons, in *Proc. 2nd Int. Symp. on Halide Glasses,* New York, 1983.
94. J. J. Mecholsky, A. C. Gozales, C. G. Pantano, B. Bendow and S. W. Freimen, in *Proc. 3rd Int. Symp. Halide Glasses,* Rennes, France, June 24–28, 1985.
95. K. H. Levin, D. C. Tran, R. J. Ginther, and G. H. Sigel, Jr., *Glass Technol.,* 23, 143 (1983).
96. J. M. Jewell and I. D. Aggarwal, *J. Non-Cryst. Solids,* 142, 260 (1992).
97. J. S. Sanghera, T. Tsai, J. Jewell, and I. D. Aggarwal, *Mater. Lett.,* 13, 135 (1992).
98. S. Mitachi, S. Shibata, and T. Manabe, *Jpn. J. Appl. Phys.,* 20, L337 (1981).
99. S. Mitachi, T. Miyashita, and T. Kanamori, *Electron. Lett.,* 17, 591 (1981).
100. T. Kanamori and S. Sakaguchi, *Jpn. J. Appl. Phys.,* 25, L468 (1986).
101. S. Sakaguchi and S. Takahashi, *J. Lightwave Technol.,* LT5, 1219 (1987).
102. Y. Ohishi, S. Sakaguchi, and S. Takahashi, *Electron. Lett.,* 22, 1034 (1986).
103. D. C. Tran, C. F. Fisher, and G. H. Sigel, *Electron. Lett.,* 18, 657 (1982).
104. G. Lu and I. Aggarwal, *Mater. Sci. Forum,* 19-20, 375 (1987).
105. J. R. Williams, S. Carter, P. W. France, and M. W. Moore, *Mater. Sci. Forum,* 67–68, 317 (1991).
106. B. Harbison and I. D. Aggarwal, *J. Non-Cryst. Solids,* 161, 7 (1993).
107. S. F. Carter, M. W. Moore, D. Szebesta, J. R. Williams, D. Ranson, and P. W. France, *Electron. Lett.,* 26, 2115 (1990).
108. F. T. Geyling, K. L. Walker, and R. Csencsits, *J. Appl. Mech.,* 105, 303 (1983).
109. J. Kirchhof, *Phys. Status Solidi (A),* 60, K127 (1980).
110. J. A. Lewis, *J. Fluid Mech.,* 81, 129 (1977).
111. D. Tran, *Mater. Res. Bull.,* 17, 1177 (1982).
112. M. G. Drexhage, in *IOOC'82,* San Francisco, Paper M12 (1982).

113. J. L. Mansfield, Evaluation of Multicomponent Fluoride Glass Candidates for Double Crucible Drawn Optical Fiber, Contract No. RADC-TR-83-131, Final Technical Report, 1983.

114. H. Tokiwa, Y. Mimura, T. Nakai, and O. Shinbori, *Electron. Lett.,* 21, 1131 (1985).

115. D. Tran, M. J. Burk, G. H. Sigel, and K. H. Levin, Paper TUG2, in *Tech. Dig. Conf. on Optical Fiber Communication,* New Orleans, 1984.

116. J. S. Sanghera, P. C. Pureza, and I. D. Aggarwal, *Method of Forming Waveguides with Ion Exchange of Halogen Ions,* U.S. Patent 5,294,240 (1994).

117. L. E. Busse and I. D. Aggarwal, *J. Lightwave Technol.,* 9, 828 (1991).

118. H. Tokiwa and Y. Mimura, *J. Lightwave Technol.,* LT4, 1260 (1986).

119. G. Lu and J. Bradley, *Mater. Sci. Forum,* 19–20, 306 (1987).

120. H. Hattori, Y. Ohishi, T. Kanamori, and S. Takahashi, *Appl. Opt.,* 25, 3549 (1986).

121. H. Hattori, T. Kanamori, S. Sakaguchi, and Y. Ohishi, *Appl. Opt.,* 26, 650 (1987).

122. H. Hattori, S. Sakaguch, T. Kanamori, and Y. Terunama, *Appl. Opt.,* 26, 2683 (1987).

123. J. A. Freitas, Jr., J. S. Sanghera, P. Pureza, U. Strom, and I. D. Aggarwal, *J. Non-Cryst. Solids,* 140, 166 (1992).

124. L. E. Busse, G. H. McCabe, and I. D. Aggarwal, *Opt. Lett.,* 15, 423 (1990).

125. D. Marcuse, *Principles of Optical Fiber Measurement,* Academic Press, New York, 1981, 226.

126. D. C. Tran, K. H. Levin, M. J. Burk, C. F. Fisher, and D. Brower, *SPIE,* 618, 48 (1986).

127. P. W. France, S. F. Carter, M. W. Moore, and J. R. Williams, *Mater. Sci. Forum,* 19–20, 381 (1987).

128. D. Szebesta, S. T. Davey, J. R. Williams, and M. W. Moore, *J. Non-Cryst. Solids,* 161, 18 (1993).

129. H. W. Schneider, *Mater. Sci. Forum,* 32–33, 561 (1988).

130. T. Shibata, H. Takahashi, M. Kimura., T. Ijichi, K. Takahshi, Y. Sasaki, and S. Yoshida, *Mater. Sci. Forum,* 5, 379 (1985).

131. H. W. Schneider, A. Schoberth, A. Staudt, and C. Gerndt, *Electron. Lett.,* 22, 949 (1986).

132. S. Sakaguchi, Y. Terunuma, Y. Ohishi, and T. Kanamori, *J. Mater. Sci. Lett.,* 6, 1063 (1987).

133. P. C. Pureza, P. H. Klein, W. I. Roberts, and I. D. Aggarwal, *J. Mater. Sci.,* 26, 5149 (1991).

134. P. C. Pureza, D. T. Brower, and I. D. Aggarwal, *J. Am. Ceram. Soc.,* 72, 1980 (1989).

135. J. A. Wysocki, C. G. Pantano, and J. J. Mecholsky, *SPIE,* 843, 21 (1987).

136. A. M. Nakata, J. Lau, and J. D. Mackenzie, *J. Non-Cryst. Solids,* 74, 229 (1985).

137. C. T. Moynihan, S. R. Loehr, and B. Wilson, Paper presented at the 91st Annual American Ceramic Society Meeting, Indianapolis (1989).

138. A. Nakata, J. Lau, and J. D. Mackenzie, *Mater. Sci. Forum,* 6, 717 (1985).

139. P. W. France, European Patent Appl., EP 266,289.

140. C. J. Simmons and J. H. Simmons, *J. Am. Ceram. Soc.,* 69, 661 (1986).

141. P. C. Schultz, L. J. B. Vacha, C. T. Moynihan, B. B. Harbison, K. Cadien, and R. Mossadegh, *Mater. Sci. Forum,* 19–20, 343 (1987).

142. T. Komukai, Y. Miyajima, and Y. Katsuyama, *Electron. Lett.,* 26, 584 (1990).

143. T. Komukai and Y. Miyajima, *Electron. Lett.,* 27, 572 (1991).

144. B. B. Harbison, W. I. Roberts, and I. D. Aggarwal, *Electron Lett.,* 25, 1214 (1989).

145. G. Nau, F. Bucholtz, K. Ewing, S. Vohra, J. McVicker, J. Sanghera, I. Aggarwal, J. Adams, D. Eng, and T. King, Mid-IR fiber optic sensor system for detection of organic contaminants in soil, presented at Conference on Optical Remote Sensing for Environmental and Process Monitoring, Air & Waste Management Association, San Francisco (1995), *SPIE,* 2504, 291 (1995).

146. D. Moynihan and R. Driver, *Photon. Spectra,* 24, 107 (1990).

147. Aurora 2600/Prizma 2600M AOTF Spectrometer, Infrared Fiber Systems, Inc., Silver Spring, MD.

148. D. L. Griscom and E. J. Friebele, Effects of high energy radiation on halide glasses, in *Fluoride Glass Fiber Optics,* Ed. I. D. Aggarwal and G. Lu, Eds., Academic Press, San Diego, 1991, 307.

149. K. Itoh, K. Miura, I. Masuda, M. Iwakura, and T. Yamashita, *J. Non-Cryst. Solids,* 167, 112 (1994).

150. C. Whitehurst, M. R. Dickinson, A. Charlton, and T. A. King, *Proc. SPIE,* 1048, 141 (1989).

151. S. Wuthrich, W. Luthy, and H. P. Weber, *Appl. Opt.,* 31, 5833 (1992).

152. J. S. Sanghera and I. D. Aggarwal, Rare earth doped heavy-metal fluoride glass fibers, in *Rare Earth Doped Fiber Lasers and Amplifiers*, M. Digonnet, Ed., Marcel Dekker, New York, 1993, 423.

153. R. S. Quimby, Active phenomena in doped halide glasses, in *Fluoride Glass Fiber Optics,* I. D. Aggarwal and G. Lu, Eds., Academic Press, San Diego, 1991, 307.

154. J. Y. Allain, M. Monerie, and H. Poignant, *Electron. Lett.,* 26, 261 (1990).

155. J. Y. Allain, M. Monerie, and H. Poignant, *Electron. Lett.,* 25, 1660 (1989).

156. C. A. Millar, M. C. Brierley, M. H. Hunt, and S. F. Carter, *Electron. Lett.,* 26, 1871 (1990).

157. J. Y. Allain, M. Monerie, and H. Poignant, *Electron. Lett.,* 25, 1082 (1989).

158. J. Y. Allain, M. Monerie, H. Poignant, and T. Georges, *J. Non-Cryst. Solids,* 161, 270 (1993).

159. M. C. Brierley and M. H. Hunt, *SPIE,* 1171, 157 (1990).

160. M. C. Brierley, P. W. France, and C. A. Millar, *Electron. Lett.,* 24, 539 (1988).

161. T. Komukai, T. Yamamoto, T. Sugawa, and Y. Miyajima, *IEEE J. Quantum Electron.,* 31, 1880 (1995).

162. R. G. Smart, J. N. Carter, D. C. Hanna, and A. C. Tropper, *Electron. Lett.,* 26, 649 (1989).

163. R. M. Percival, D. Szebesta, C. P. Seltzer, S. D. Perrin, S. T. Davey, and M. Louka, *IEEE J. Quantum Electron.,* 31, 489 (1995).

164. R. M. Percival, D. Szebesta, S. T. Davey, N. A. Swain, and T. A. King, *Electron. Lett.,* 28, 2231 (1992).

165. R. Allen and L. Esterowitz, *Appl. Phys. Lett.,* 55, 721 (1989).

166. R. Allen, L. Esterowitz, and R. J. Ginther, *Appl. Phys. Lett.,* 56, 1635 (1990).

167. J. Schneider, D. Hauschild, C. Frerichs, and L. Wetenkamp, *Int. J. Infrared Millimeter Waves,* 15, 1907 (1994).

168. L. Wetenkamp, *Electron. Lett.,* 26, 883 (1990).

169. M. C. Brierley, P. W. France, R. A. Garnham, C. A. Millar, and W. A. Stallard, Long wavelength fluoride fiber system using a 2.7 µm fluoride fiber laser, in *Proc. Conf. on Optical Fiber Communications (OFC)*, Houston, TX (1989), paper PD14-1.

170. J. Schneider, *Electron. Lett.,* 15, 1250 (1995).

171. M. P. Leflohic, J. Y. Allain, G. M. Stephan, and G. Maze, *Opt. Lett.,* 19, 1982 (1994).

172. J. Y. Allain, M. Monerie, and H. Poignant, *Electron. Lett.,* 26, 166 (1990).

173. Y. X. Zhao, S. Fleming, and S. Poole, *Opt. Commun.,* 114, 285 (1995).

174. D. Piehler, D. Craven, N. Kwong, and H. Zarem, *Electron. Lett.,* 29, 1857 (1993).

175. J. F. Massicott, M. C. Brierley, R. Wyatt, S. T. Davey, and D. Szebesta, *Electron. Lett.,* 29, 2119 (1993).

176. J. Y. Allain, M. Monerie, and H. Poignant, *Electron. Lett.,* 26, 261 (1990).

177. B. Jacquier, C. Linares, R. Mahiou, J. L. Adam, E. Denoue, and J. Lucas, *J. Lumin.,* 60, 175 (1994).

178. A. Shikida, H. Yanagita, and H. Toratani, *J. Opt. Soc. Am.,* B11, 928 (1994).
179. R. M. Percival, D. Szebesta, J. R. Williams, R. D. T. Lauder, A. C. Tropper, and D. C. Hanna, *Electron. Lett.,* 30, 1598 (1994).
180. Y. Miyajima, T. Komukai, and T. Suguwa, *Electron. Lett.,* 26, 194 (1990).
181. M. Shimizu, T. Kanamori, J. Temmyo, M. Wada, M. Yamada, Y. Terunuma, Y. Ohishi, and S. Sudo, *IEEE Photon. Technol. Lett.,* 5, 654 (1993).
182. M. Yamada, M. Shimizu, T. Kanamori, Y. Ohishi, Y. Terunuma, K. Oikawa, H. Yoshinaga, K. Kikushima, K. Miyamoto, and S. Sudo, *IEEE Photon. Tech. Lett.,* 7, 869 (1995).
183. D. M. Spirit, G. R. Walker, P. W. France, S. F. Carter, and D. Szebesta, *Electron. Lett.,* 26, 1218 (1990).
184. T. Yamamoto, T. Komukai, and Y. Miyajima, *Jpn. J. Appl. Phys. Lett.,* 32, L62 (1993).
185. T. J. Whitley, *J. Lightwave Technol.,* 13, 744 (1995).
186. N. Tomita, K. Kimura, H. Suda, M. Shimizu, M. Yamada, and Y. Ohishi, *IEEE Photon. Tech. Lett.,* 6, 258 (1994).
187. K. Kikushima, T. Whitley, R. Cooke, K. Stalley, M. Fake, and E. Lawrence, *Electron. Lett.,* 30, 1431 (1994).
188. S. Yoshida, S. Kuwano, M. Yamada, T. Kanamori, N. Takachio, and K. Iwashita, *Electron. Lett.,* 31, 1678 (1995).

4 Chalcogenide Glass-Based Fibers

Junji Nishii and Toshiharu Yamashita

CONTENTS

4.1   INTRODUCTION

Chalcogenide compounds of some elements belonging to groups 4B and 5B in the periodic table exhibit excellent glass-forming ability; for example, stable glass formation of As_2S_3 was demonstrated more than a century ago.[1] Based on the wide infrared (IR) transmission range of As_2S_3 glass,[2,3] various glass compositions have been developed as optical component materials for the 3 to 5 μm and 8 to 12 μm bands. In the 1960s, chalcogenide glasses were researched as photosensitive electronic components in optical memory and optical switching devices. The rapid progress of fiber fabrication technology of oxide glasses in the 1970s has been followed by a considerable effort to develop chalcogenide glasses as candidate fiber materials because of their wide IR transmission range. In this chapter, the fabrication and characteristics of chalcogenide glass fibers and their applications to IR optics will be described.

4.2   GLASS COMPOSITIONS

Chalcogenide glasses can be classified into three groups: sulfide, selenide, and telluride. Glass formation can be achieved when the chalcogen elements are melted and quenched in an evacuated silica glass ampoule with one or more elements, such as As, Ge, P, Sb, Ga, Al, Si, etc. In order to change the electronic resistivity of the glass, other metals, such as Bi, Ag, Li, are occasionally added, but it is rare to use the elements for production of optical fibers. The properties of the glasses change drastically with glass compositions. For example, while some sulfide glasses are transparent in the visible wavelength region, the transmission ranges of selenide and telluride glasses shift to the IR region with increasing the contents of selenium and tellurium because of their increased mass. The mechanical strength and the thermal and chemical stabilities of chalcogenide glasses, which are typically lower than oxide glasses, are sufficient for practical fiber applications. The compositions of some representative glasses are summarized in Table 4.1.

4.2.1   SULFIDE GLASSES

Glass formation of binary As–S and Ge–S glasses were studied by Hruby.[4,5] As–S glasses reveal excellent chemical durability and thermal stability against crystallization.

TABLE 4.1
Representative Chalcogenide Glass-Forming Systems

System	Ref.	System	Ref.	System	Ref.
As–S	4	Ge–Sn–S	11	Pb–Ge–Se	19
Ge–S	5	As–Se	12	Ge–Se–Te	21
Ge–As–S	6	Ge–Se	13	Ge–Se–Te–Tl	23
Ge–S–P	7	Ge–As–Se	14	Ge–As–Se–Te	24
As–S–P	8	Ge–Sb–Se	15	Ge–Sb–Se–Te	15
As–Sb–S	9	As–Se–P	8	Si–As–Te	25, 26
Ge–Sb–S	10	Tl–Ge–Se	19	Si–Ge–As–Te	27

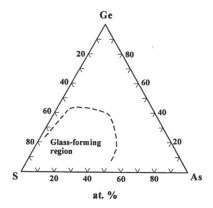

FIGURE 4.1 Glass-forming region of the Ge–As–S system. (From Feltz, A., *Amorphous Inorganic Materials and Glasses,* VCH, New York, 1993. With permission.)

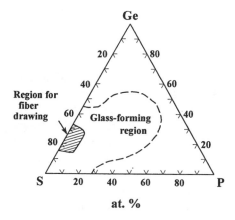

FIGURE 4.2 Glass-forming region of the Ge–S–P system. (From Shibata, S. et al., *Mater. Res. Bull.,* 16, 703, 1981. With permission.)

The glass transition temperature increases gradually with the content of As or Ge, and the Ge–S glasses exhibit higher glass transition temperatures than the As–S glasses. The preparation of As–S glasses is much easier than Ge–S glasses because As and S react in a relatively shorter melting time and lower melting temperature than Ge and S. As shown in Figure 4.1, a wide glass-forming region can be obtained in the ternary Ge–As–S system.[6] It is possible to obtain a glass with the total content of As and Ge exceeding 70 at. %. However, the stability of the glass toward moisture becomes low with increasing content of Ge.

The glass-forming ability of the Ge–S system is improved by the addition of small amounts of P (≤10 at. %).[7] The glass-forming region of this system is shown in Figure 4.2. Glass formation has been investigated in several ternary systems, such as As–S–P,[8] As–Sb–S,[9] Ge–Sb–S,[10] Ge–Sn–S,[11] but the fiber fabrication of these glasses has not been reported.

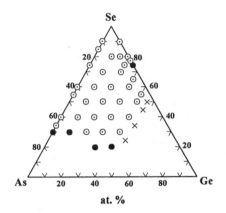

FIGURE 4.3 Glass-forming region of the Ge–As–Se system. (From Savage, J. A. and Nielsen, S., *Phys. Chem. Glasses,* 6, 90, 1965. With permission.)

4.2.2 SELENIDE GLASSES

Glass-forming tendencies of selenide systems are very similar to those of sulfide glasses. The preparation of selenide glasses is easier than sulfide glasses because of the low vapor pressure at the melting temperature (850 to 900°C) and rapid chemical reaction between Se and Ge or As. The range of glass formation in the As–Se system is 0 to 70 at. % of As, which is wider than the As–S system.[12] The glass-forming region of the binary Ge–Se glass is shown in Reference 13. These binary systems can incorporate several third components, e.g., S, P, Al, Si, Tl, Pb, Sn, Sb, Cl, Br, I. Ge–As–Se and Ge–Sb–Se glass systems possess excellent thermal stability for fiber drawing. Savage and Nielsen.[14] reported the formation and properties of the glasses in the Ge–As–Se system. They pointed out that the oxide impurities assist in the glass formation. The glass-forming region of this ternary system, which was determined using the purified raw elements, is shown in Figure 4.3. Glass formation was confirmed when the content of (Ge + As) = 70 at. %. The glasses in the Ge–Sb–Se system exhibit prominent thermal stability and chemical durability. The glass-forming region was investigated by Bordas et al.[15] Hilton[16] successfully prepared a large plate and prism by the casting method, and Klocek et al.[17,18] drew fibers from the glasses in this system.

The glass-forming region of the As–Se–P system, as reported by Blachnick and Hope,[8] is shown in Figure 4.4. Vitrification could be observed in the region P > 60 at. %. Although fiber fabrication has not been performed, the thermal stability against crystallization is sufficient for fiber drawing. Glass formation in the Tl–Ge–Se and Pb–G–Se systems was reported by Linke et al.[19] Phase separation was observed in the region Se > 70 at. %.

4.2.3 TELLURIDE GLASSES

Chalcogenide glasses containing Te have a wide transmission window in the mid-IR wavelength region. The vapor pressure of the telluride melts is generally much lower than those of sulfide and selenide melts, and the chemical reaction between tellurium

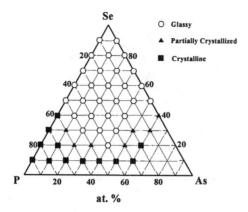

FIGURE 4.4 Glass-forming region of the As–Se–P system. (From Blachnik, R. and Hoppe, A., *J. Non-Cryst. Solids,* 34, 191, 1979. With permission.)

and the other components proceeds rapidly at relatively lower temperature. The systems Ge–Se–Te, Ge–As–Se–Te, and Ge–Se–Te–Sb are preferable candidates for fiber fabrication.

The glass formation in the Ge–Se–Te system has been investigated in detail.[20-22] As shown in Figure 4.5, two glass-forming regions, an Se-rich region and a Te-rich

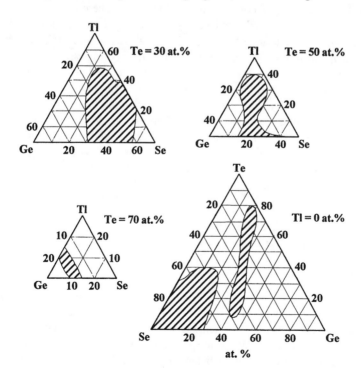

FIGURE 4.5 Glass-forming region of the Ge–Se–Te and Ge–Se–Tl systems. (From Nishii, J. et al., *Phys. Chem. Glasses,* 28, 55, 1987. With permission.)

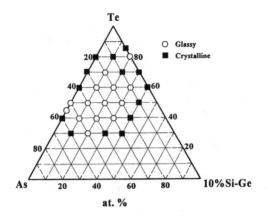

FIGURE 4.6 Glass-forming region of the Si–Ge–As–Te system. (From Savage, J. A., *J. Mater. Sci.*, 7, 64, 1972. With permission.)

region, are observed for the Ge–Se–Te system. The fiber can be drawn in each region. The addition of Tl is effective to suppress the crystallization tendency,[23] and the glass-forming region of the Ge–Se–Te–Tl system is also shown in Figure 4.5.

The Ge–As–Se–Te system exhibits excellent glass-forming ability. Inagawa et al.[24] investigated its glass-forming region over a wide range. No crystallization was confirmed until the Te content exceeded 60 at. %. There are several reports on the preparation of fiber from this quaternary system, and these will be described later.

The effect of the addition of Te to Ge–Sb–Se glasses was investigated by Bordas et al.[15] The upper limit of Te was 20 at. % when the cooling rate was 5°C/min. The IR absorption edge of the glasses in this system is located at shorter wavelengths than the Ge–As–Se–Te and Ge–Se–Te glasses due to the lower Te content.

The addition of Si was examined in several Te-based glasses in order to improve their thermal stabilities, which is very important for fabrication of fiber. Glass formation of Si–As–Te was reported by Feltz et al.[25] and Anthonis and Kreidl.[26] Savage[27] investigated the glass-forming region of the Si–Ge–As–Te system, and an example is shown in Figure 4.6. A glass transition temperature higher than 400°C was attained in these systems.

4.3 PHYSICAL PROPERTIES AND IMPURITY ABSORPTIONS

4.3.1 PHYSICAL PROPERTIES

Since the failure mechanisms and figures of merit of IR windows for high-power laser transmission have been discussed in detail for various crystalline and amorphous materials,[28] it follows that the physical properties of the window materials have also been investigated in detail. While most of the reported data have focused on the glass transition and crystallization behavior of several chalcogenide glasses, there is not sufficient optical and mechanical property data to design the optimum core

TABLE 4.2
Physical Properties Chalcogenide Glasses

Composition/at. %	$T_g/°C$	$n (\lambda/nm)$	$dn/dT/10^{-5}$	e/kpsi	ε	$\alpha/10^{-7} deg^{-1}$	Ref.
$As_{40}S_{60}$	204	2.410(5.0)	−1	2350	0.30	240	33
$Ge_{20}S_{80}$	170	2.401(5.0)	—	2200	0.27	—	31
$As_{40}Se_{60}$	178	2.784(8.0)	—	—	—	210	29
$Ge_{20}Se_{80}$	154	2.407(8.0)	—	1650	0.27	248	33
$Ge_{30}As_{20}Se_{50}$	361	2.569(8.0)	3	—	—	117	30
$Ge_{28}Sb_{12}Se_{60}$	300	2.619(8.0)	8	3100	0.24	141	17
$Ge_{27}Se_{18}Te_{55}$	205	3.045(5.0)	~14	2800	0.27	170	33
$Ge_{20}As_{30}Se_{30}Te_{20}$	236	2.866(5.0)	—	3500	0.27	154	34

Note: n: refractive index at λ (wavelength); dn/dT: temperature dependence of refractive index; E: Young's modulus; ε: Poisson's ratio, α: thermal expansion coefficient.

and cladding glass compositions. For example, Young's modulus, Poisson's ratio, and the refractive index are required for the estimation of scattering loss. Also, the temperature dependence of the refractive index is an important factor for high-power laser delivery.

The compositional dependence of the physical properties was investigated by Webber and Savage[29] for the Ge–As–Se system, Savage[30] for the Ge–As–Se–Te system, and Hilton and Hayes[31] for the Ge–As–S, Ge–Sb–Se, and Ge–As–Se–Te systems. Hilton[16] summarized the physical properties of several chalcogenide glasses compared with other IR transmitting crystalline materials. Hartouni et al.[32] reported on the physical properties of fiber candidates such as As–S, As–Se, As–Se–Te, Ge–As–Se, and Ge–As–Te. Table 4.2 lists the mechanical and optical properties for representative chalcogenide glasses, which have been or will be drawn into fibers. The Rayleigh scattering loss can be calculated using this data, as will be shown in Section 4.6.1.

4.3.2 IMPURITY ABSORPTIONS

The separation of intrinsic and extrinsic absorption at a specific wavelength is important to estimate the potential minimum loss. The common impurities contained in chalcogenide glasses are oxygen and hydrogen and/or water. These impurities form R–O or R–H bonds (R = chalcogen and 4B or 5B elements) in the glass matrix. The studies on the impurities in chalcogenide glasses have been reported since the 1960s. While one of the major purposes was to investigate the relationship between the electrical conductivity and the impurities, the majority of the reports described the influence of oxide and hydrogen impurities on the IR transmission characteristics. The assignments of the impurity absorption bands are summarized in Table 4.3.

There are only a few reports on the quantitative analysis of the absorption bands due to oxygen impurities. Nishii et al.[39] determined the relationship between the concentration of doped oxygen ([O] in ppm) and the absorption coefficients (α in

TABLE 4.3
Impurity Absorption Bands in Chalcogenide Glasses

Glass System	Wave Number (cm)	Wavelength (μm)	Assignment	Ref.
As–S	925	10.8	AsO–H	35
	1825	5.48	AsO–H$_{(overtone)}$	35
	3440	2.91	SO–H$_{(fundamental)}$	35
	4370	2.29	SO–H$_{(combination)}$	35
	5210	1.92	SO–H$_{(combination)}$	35
	6950	1.44	SO–H$_{(overtone)}$	35
	1580	6.32	H$_2$O$_{(molecule)}$	35
	3610	2.77	H$_2$O$_{(molecule)}$	35
	2480	4.03	S–H$_{(fundamental)}$	35
	2710	3.69	S–H$_{(combination)}$	35
	3215	3.11	S–H$_{(combination)}$	35
	3940	2.54	S–H$_{(combination)}$	35
	4880	2.05	S–H$_{(overtone)}$	35
	2025	4.94	Carbon	35
Ge–S	3570	2.80	SO–H	7
	2500	4.00	S–H	7
	2030	4.92	Ge–H	7
As–Se	3420	2.92	SeO–H	36
	3600	2.78	OH	36
	3520	2.84	OH	36
	1585	6.30	H$_2$O$_{(molecule)}$	36
	2830	3.53	Se–H	36
	2430	4.12	Se–H	36
	2190	4.57	Se–H	36
Ge–Se	1280	7.8	Ge–O$_{(combination)}$	37
	800	12.5	Ge–O$_{(stretching)}$	37
	500	20.0	Ge–O$_{(bending)}$	37
Ge–As–Se	3420	2.92	SeO–H	38
	2190	4.57	Se–H	38
	1270	7.90	Oxide	38
Ge–Se–Te	1230	8.13	Ge–O$_{(combination)}$	39
	765	13.07	Ge–O$_{(stretching)}$	39

cm^{-1}) of the Ge–O stretching vibration (α_1 at 765 cm^{-1}) and the combination vibration (α_2 at 1230 cm^{-1}) in a Ge–Se–Te glass. The spectra displaying these absorption bands are shown in Figure 4.7. The relationships between α and [O] are represented by the following equations, and these can be used to determine unknown [O] in glasses.

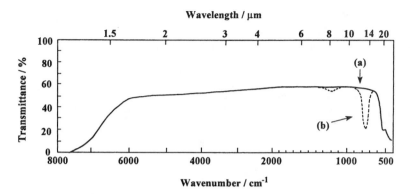

FIGURE 4.7 Transmission spectra of Ge–Se–Te glass containing oxygen impurities. Absorption bands at 765 and 1230 cm⁻¹ are due to Ge–O stretching and combination vibrations, respectively. (From Nishii, J. et al., *J. Mater. Sci.*, 24, 4293, 1989. With permission.)

$$\alpha_1 = 0.228[O] \tag{4.1}$$

$$\alpha_2 = 0.006\,[O] \tag{4.2}$$

For example, the oxygen content in a purified Ge–Se–Te glass was estimated to be approximately 0.04 ± 0.009 ppm. In a similar doping method, the extinction coefficients for S–H and Se–H in arsenic sulfide and arsenic selenide glasses have been determined, respectively. The values are 2.3 dB/m/ppm at 4.01 μm and 1.0 dB/m/ppm at 4.57 μm, respectively.[40] To date, the highest-quality glasses still contain 1 to 10 ppm of hydrogen impurities.

4.4 GLASS SYNTHESIS

4.4.1 PURIFICATION OF RAW ELEMENTS

Recently, high-purity raw elements (>6 nines purity) have became commercially available; for example, the purity of Ge ingots is higher than 10 nines as a result of the zone refining technique. However, the surface of each raw element is slightly oxidized or hydrated during handling. It is therefore necessary to purify each raw element before the preparation of the glasses. Representative methods for purification are summarized in Table 4.4.

Oxide layers due to chemical attack by oxygen gas are readily formed on Ge ingots and As chunks. A simple method to remove the oxide layer from the surface of a Ge ingot is chemical etching with an acidic solution, such as HNO_3.[35] However, a trace amount of oxide is apt to be formed again after the etching. One of the promising and easy methods is the reduction of Ge in an H_2 gas stream.[41] A Ge ingot is crushed into coarse grains, placed in a silica tube, and then high-purity H_2 gas is introduced into the vessel. The reduction is carried out at 900°C for more

TABLE 4.4
Purification Methods for Raw Elements

Elements	Purification Method	Ref.
Ge	Zone melting, chemical etching, H_2 reduction	35, 41
As	Sublimation in Ar stream or vacuum	35, 41
S	Distillation in vacuum or S_2Cl_2 stream	41, 43
Se	Distillation in vacuum or Se_2Cl_2 Stream	33, 44
Te	Zone melting, distillation in vacuum	24

than 2 h. After the treatment, the silica vessel is sealed and placed inside a glove box. The oxide layer on As chunks is removed by the sublimation method. The As chunks are placed in a silica vessel and heated up to 350°C under flowing Ar gas[35] or vacuum ($<10^{-5}$ torr).[41] The As oxide formed on the chunks can be easily removed by the sublimation method because of its high vapor pressure compared with elemental arsenic.

Chalcogen elements are easily vaporized under vacuum, and so the distillation method is employed for their purification. The commercially available Te ingot is usually purified by the zone melting method. However, distillation is preferable to remove oxide impurities formed on the surface. Dehydrated silica vessels are used for the distillation. The most difficult residue to remove is carbon or its compounds, which can be present in varying amounts. The most effective method to remove the carbon is to use quartz frits during distillation.[42] Almost all of the carbon impurities can be removed by the frits. The purified elements are sealed under vacuum and placed in a glove box immediately and without contact with air.

A unique method to minimize the IR absorption bands due to oxygen impurities is comelting with some metals, called *oxygen getters*, which have a free energy of formation (ΔG) of oxides lower than those of the chalcogen elements. The oxygen-gettering effect in chalcogenide glass melts was reported by Hilton et al.[42] Figure 4.8 shows the IR transmission spectra of TI-1173 (a Ge–Sb–Se glass) with and without 5 ppm added aluminum metal. The absorption band due to Ge–O stretching vibration around 12.5 μm is effectively suppressed for the glass containing Al metal. The upper limit of the content of oxygen getter is as high as 100 ppm. The excess amount of oxygen getter causes an attack on the silica ampoule, and the losses due to scattering or Si–O stretching absorptions are observed. Similar oxygen-gettering effects are confirmed by the addition of a trace amount of Mg.[24] In both cases, Al or Mg, it is preferable to distill the glass from the oxide residue. The distilled glass is then remelted for homogeneity.

Reactive atmosphere processing (RAP) is available to eliminate hydrogen impurities. Shibata et al.[43] reduced the absorption intensity due to S–H and O–H stretching vibrations by distillation of S under an S_2Cl_2 gas stream. The cause of these impurity absorption bands is H_2O, which can be eliminated by the following chemical reaction:

$$2S_2Cl_2 + 4H_2O \rightarrow 2SO_2 + 4HCl + 2H_2S \qquad (4.3)$$

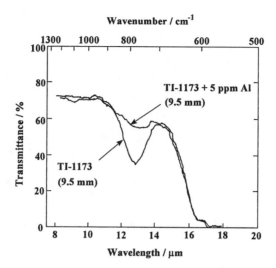

FIGURE 4.8   IR transmission spectra of TI-117 (Ge–Sb–Se) glasses with and without 5 ppm Al. The absorption band at 12.5 µm is due to the Ge–O stretching vibration. (From Hilton, A. R. et al., *J. Non-Cryst. Solids,* 17, 319, 1975. With permission.)

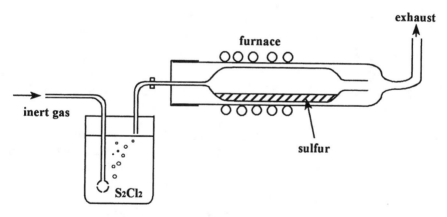

FIGURE 4.9   A setup for the distillation of S in an S_2Cl_2 gas stream. (From Shibata, S. et al., *J. Appl. Phys.,* 20, L13, 1981. With permission.)

The apparatus for the purification is shown in Figure 4.9. The investigators prepared Ge–S glass fibers using the purified S. The loss due to the S–H stretching vibration at 4 µm was decreased by one order of magnitude compared with the unpurified fiber. Katsuyama et al.[44] used the same method for the distillation of selenium under an Se_2Cl_2 atmosphere.

4.4.2   GLASS MELTING

The melting is carried out using high-purity and dry silica ampoules. The schematic procedure is represented in Figure 4.10. Before weighing, the inside of the ampoule

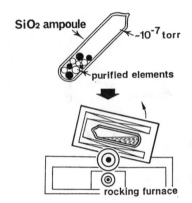

FIGURE 4.10   Melting procedure for chalcogenide glasses.

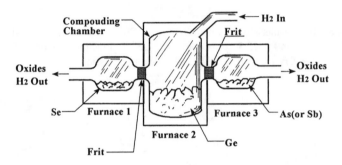

FIGURE 4.11   Advanced preparation method of Ge–Sb–Se glass. (From Hilton, A. R. et al., *J. Non-Cryst. Solids,* 17, 319, 1975. With permission.)

is chemically etched using HF solution, rinsed with distilled water, and then baked. The purified elements are weighed in a glove box containing recirculated inert gas atmosphere. The weighed raw elements are put into the ampoule and sealed under a vacuum lower than 1.3 Pa. The ampoule is heated in a rocking furnace for longer than 10 h between 700 and 900°C.

A more-advanced melting method was proposed by Hilton et al.[42] and is shown in Figure 4.11. The Se and As (or Sb) in the side chamber are distilled to the center chamber and then melted in the rocking furnace. A Ge–Sb–Se glass of about 4 to 5 kg mass was successfully prepared by this method. The glass block was polished to an optical finish and used as an IR window. This method was used for the preparation of optical fiber preforms. Inagawa et al.[24] suggested the distillation and melting technique shown in Figure 4.12. Purified raw elements As, Se, and Te, which exhibit relatively high vapor pressure, are placed in the right side of the chamber. The Ge chunks reduced in an H_2 gas stream are located in the left side. In the first step, the elements having high vapor pressure are melted with Mg metal (100 ppm) and then distilled to the left side. The right side is pulled off without losing the vacuum. The melting process of the glass is the same as that described above. A similar method was reported by Lezal et al.[45]

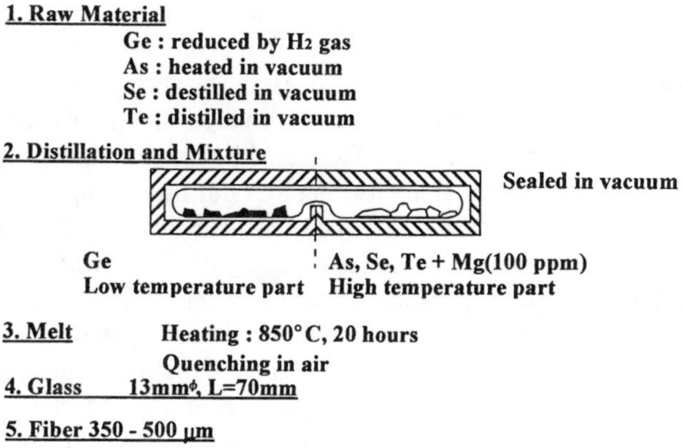

1. Raw Material
          Ge : reduced by H₂ gas
          As : heated in vacuum
          Se : destilled in vacuum
          Te : distilled in vacuum

2. Distillation and Mixture

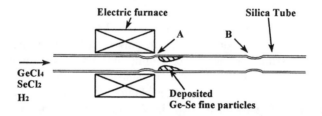

Sealed in vacuum

  Ge   **As, Se, Te + Mg(100 ppm)**
  Low temperature part   **High temperature part**

3. Melt   **Heating : 850°C, 20 hours**
              Quenching in air
4. Glass   **13mmᵠ, L=70mm**

5. Fiber 350 - 500 µm

FIGURE 4.12 Schematic diagram of glass preparation process of Ge–As–Se–Te glasses. (From Inagawa, I. et al., *J. Non-Cryst. Solids,* 95–96, 810, 1987. With permission.)

Electric furnace   **Silica Tube**

              A   B

GeCl₄
SeCl₂
H₂   **Deposited**
               Ge-Se fine particles

FIGURE 4.13 Setup for CVD. A and B are neck parts used for sealing off and separating. (From Katsuyama, T. et al., *J. Appl. Phys.,* 59, 1446, 1986. With permission.)

Katsuyama et al.[46] fabricated Ge–Se glasses by the chemical vapor deposition (CVD) method. The preparation apparatus is shown in Figure 4.13. Starting materials were $GeCl_4$ and $SeCl_2$ gas. The $SeCl_2$ gas was prepared by the thermal decomposition of Se_2Cl_2 liquid. The H_2 gas was used in order to improve the reaction efficiency. These gases were introduced into the silica tube with Ar as a transport gas and chemically decomposed into Ge–Se fine particles at 800°C. After deposition, the silica tube was sealed under evacuation at the neck parts (A, B) and then heated to the melting temperature of the Ge–Se compounds. While some residual oxygen and hydrogen impurities were observed in the transmission spectrum, this was the first successful preparation of chalcogenide bulk glass by the CVD process. Sergent[47] discussed the feasibility of preparation of chalcogenide glasses using a CVD method. Melling[48] reported the preparation of an amorphous Ge–S compound by the sol–gel method. There was unfortunately no report on the IR transmission characteristics of the glasses prepared by this method.

4.4.3 PREPARATION OF FIBER PREFORMS

After melting, the ampoule is quenched in air and annealed immediately in another furnace to room temperature. The bubbles are apt to be formed in the glass when the cooling of the melt is too rapid. In order to inhibit the formation of bubbles, the quenching must be done from the lowest possible temperature, but one higher than the precipitation temperature of crystalline phases. The surface of the glass rod must be polished to an optical finish to remove exfoliation or microcracks that formed during cooling. The cladding tubes are fabricated by rotation of the ampoule.[41] The ampoule is rotated horizontally along its long axis during cooling, and the thickness of the tube can be adjusted by the amount of batch. The inside of the tube is optically flat and smooth, which is required for the cladding of the optical fiber.

Other methods for the preparation of core rods and cladding tubes include the extrusion and casting processes, respectively. In the case of glasses exhibiting high thermal stability against crystallization and relatively low vapor pressure, casting in an inert atmosphere can be used for the preparation of glass blocks. Klocek et al.[17] successfully prepared cylindrical and square-shaped rods and tubes of Ge–Sb–Se glasses. Sergent[47] reported the preparation of a rod and tube of Ge–As–Se–Te glasses, while the precise preparation process was not described.

4.5   FIBER FABRICATION

Since the vapor pressure of chalcogen elements and chalcogenide glasses is relatively higher than those of oxide glasses, fiber drawing must be carried out at as low a temperature as possible. The precise control of atmosphere is required around the hot zone to prevent oxidization of glass by ambient oxygen or water. Therefore, some additional techniques are required in addition to the traditional drawing technology used for oxide glass fibers. Two kinds of drawing methods are reported, i.e., preform drawing and crucible drawing.

4.5.1   PREFORM DRAWING METHOD

In the case of fibers with no glass cladding or only a polymer cladding, such as Teflon FEP, silicone, etc., the fiber can be drawn using a conventional preform drawing method, such as that used for the manufacture of SiO_2 fiber. The preform drawing is carried out using a tube furnace with a narrow heat zone under an inert gas atmosphere. A typical drawing apparatus is shown in Figure 4.14, which was used for the drawing of Ge–S–P glass fibers.[7] The temperature gradient in the vicinity of the center portion of the furnace was about 50°C/cm. Fibers 150 to 200 µm in diameter can be drawn at a speed of 5 to 10 m/min. Fiber with a Teflon FEP cladding can be prepared using a similar apparatus. The glass rod is inserted into a heat-shrinkable Teflon FEP tube and heated in vacuum so that the tube contracts onto the glass rod tightly. This method is applicable only for glasses that can be drawn into fiber in the temperature region of 250 to 350°C. If the drawing temperature exceeds 350°C, the thermal decomposition of Teflon FEP leads to a deterioration in the mechanical strength of the fiber.

There are some reports on the fabrication of fibers with a chalcogenide glass cladding by the preform drawing method. The drawing apparatus is similar to that shown in Figure 4.14. The core rod and cladding tube, polished to an optical finish, are placed in an SiO_2 muffle furnace, zonally heated, and drawn into fiber. It is

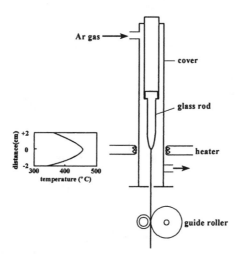

FIGURE 4.14 Apparatus used for fiber drawing of Ge–S–P glasses. (From Shibata, S. et al., *Mater. Res. Bull.,* 16, 703, 1981. With permission.)

preferable to evacuate the gas at the core/cladding interface in order to prevent oxidization of the surfaces of the preform and the formation of structural imperfections (bubbles, cracks, etc.). Klocek et al.[17] prepared Ge–Sb–Se fibers with core/cladding structure by the preform drawing method. Over 1 km length of fiber was successfully drawn. The fabrication of fibers with a glass cladding by the preform drawing method has also been reported by others.[47,49]

4.5.2 CRUCIBLE DRAWING METHOD

Because chalcogenide glasses (or chalcogen elements) possess high vapor pressure above their softening temperatures, it would appear that fiber drawing using the crucible process would be rather disadvantageous compared with the preform drawing method. Despite this, Kanamori et al.[38] have prepared As–S glass fibers with a core/cladding structure using a double-crucible technique, a process that has been conventionally used for the production of multicomponent oxide glass fibers. The schematic diagram of the double-crucible setup is shown in Figure 4.15. The crucibles were prepared from Pyrex glass. Fibers of 100 to 300 µm in outer diameter and 300 to 1000 m in length were drawn at 300 to 400°C under an Ar gas atmosphere. A similar crucible method was used for the drawing of unclad Ge–Se and Ge–Se–Sb fibers by Katsuyama et al.[44] In order to minimize the drawing temperature, they pressurized the inside of the crucible with inert gas between 0.5×10^5 and 1.5×10^5 Pa. Bornstein et al.[50] reported a similar drawing method, but the inside of the crucible was not pressurized. A novel drawing process for fibers with a glass cladding was demonstrated by Nishii et al.,[41] which incorporated the advantages of both the crucible drawing and preform drawing methods. A schematic of the setup is shown in Figure 4.16. The core rod and cladding tube were placed in an SiO_2 crucible with a nozzle at its lower end. The atmosphere in the crucible was purged with Ar gas, and only the vicinity of the nozzle was heated to the softening temperature of the

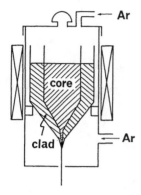

FIGURE 4.15 Double-crucible method for drawing of As–S glass fiber. (From Kanamori, T. et al., *J. Lightwave Technol.,* LT2(5), 507, 1984. With permission.)

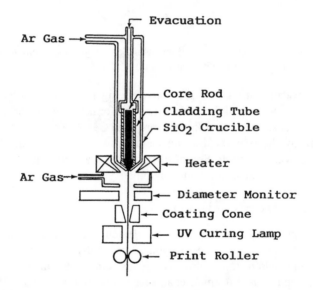

FIGURE 4.16 Modified crucible drawing method fo chalcogenide glass fibers with glass cladding. (From Nishii, J. et al., *Appl. Opt.,* 28, 5122, 1989. With permission.)

preform. The inside of the crucible was pressurized with Ar gas to above 2×10^5 Pa after the cladding tube uniformly adhered to the circumference of the nozzle. The space between the core and cladding tube was maintained at a vacuum lower than 1.3 Pa. The formation of structural imperfections, such as bubbles and microcracks at the boundary between the core and cladding, can be suppressed by this method.

4.6 TRANSMISSION LOSS

4.6.1 THEORETICAL ESTIMATION

The ultimate limiting loss of optical fibers can be expressed by the sum of the electronic absorption loss (Urbach tail), Rayleigh scattering loss (fluctuations of

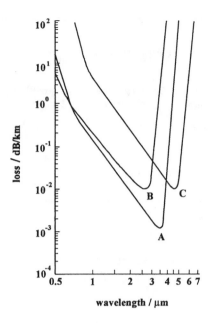

FIGURE 4.17 Calculated loss spectra of fluoride and chalcogenide glass fibers. A: Ba–Gd–Zr–F; B: Ca–Ba–Y–Al–F; C: GeS$_3$. (From Shibata, S. et al., *Electron. Lett.,* 17, 775, 1981. With permission.)

density and refractive index), and multiphonon absorption loss. The losses due to electronic and multiphonon absorptions are determined experimentally, while the Rayleigh scattering coefficient (A) can be calculated from the following equations:

$$A \approx (8/3)\pi^3 n^8 p^2 \beta k T_g$$
$$\approx (8/3)(n^2 - 1)^2 \beta k T_g$$

(4.4)

where n is the refractive index of glass, p is the average photoelastic constant, β is the isothermal compressibility, Tg is the glass transition temperature, and k is Bolzmann's constant. Figure 4.17 shows the calculated transmission loss of GeS$_3$ glass.[52] The Rayleigh scattering losses of several chalcogenide glasses estimated using this method are listed in Table 4.5. In spite of the relatively low glass transition temperature, the Rayleigh scattering loss of chalcogenide glasses is more than one order of magnitude higher than that of oxide glasses, due primarily to their high refractive indices.

Another estimation of Rayleigh scattering losses was reported by Van Uitert and Wemple[53] and Klocek and Colombo.[18] The $n^8 p^2$ factor in Equation 4.4 can be substituted as follows:

$$n^8 p^2 \approx B^2 Z_a^2 / E_o^4$$

(4.5)

TABLE 4.5
Calculated Rayleigh Scattering Loss — 1

Glass Composition (at. %)	Refractive Index at 5 μm	Isothermal Compressibility (10^{-11} Pa^{-1})	Glass Transition Temp. (K)	Scattering Loss (dB/km)	Ref.
SiO_2	1.458[a]	2.71	1480	$0.25/\lambda^4$	52
$Ge_{25}S_{75}$	2.132[a]	8.54	710	$3.97/\lambda^4$	52
$Ge_{20}Se_{80}$	2.401	8.850	443	$4.41/\lambda^4$	b
$Ge_{20}As_{30}Se_{30}Te_{20}$	2.866	6.579	508	$8.63/\lambda^4$	b
$Ge_{27}Se_{18}Te_{55}$	3.045	7.143	478	$11.58/\lambda^4$	b

[a] Refractive index at 589 nm.
[b] Non-Oxide Glass Co., Ltd. Technical Report. Yamashita, T. et al., to be published.

where Z_a is the formal chemical valence of the anion, E_0 is the average electronic excitation energy, and B is a dimensionless structure factor represented by the following equation:

$$B = N_a d \tag{4.6}$$

and where N_a is the volume density of the anion and d is first neighbor bond length. The final form of the modified function of the Rayleigh scattering coefficient is given by

$$A \approx \left(B^2 Z_a^2 / E_o^4\right)\left(T_g \beta / \lambda^4\right) \tag{4.7}$$

The Rayleigh scattering losses estimated by this method are summarized in Table 4.6. The predicted losses of Ge–Se and Ge–Se–Sb glasses are shown in

TABLE 4.6
Calculated Rayleigh Scattering Loss — 2

Glass Composition (at. %)	Scattering Loss (dB/km)
$Ge_{25}Se_{75}$	$65/\lambda^4$
$Ge_{40}Se_{60}$	$114/\lambda^4$
$Ge_{32}Sb_8Se_{60}$	$106/\lambda^4$
$Ge_{28}Sb_{12}Se_{60}$	$105/\lambda^4$
$Ge_{22}Sb_{18}Se_{60}$	$159/\lambda^4$
SiO_2	$0.8/\lambda^4$

After Klocek, P. and Colombo, L., *J. Non-Cryst. Solids,* 93, 1, 1987.

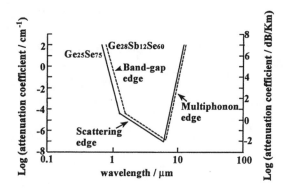

FIGURE 4.18   Calculated optical losses of (1) $Ge_{25}Se_{75}$ and (2) $Ge_{28}Sb_{12}Se_{60}$ glasses. (From Klocek, P. and Colombo, L., *J. Non-C ryst. Solids,* 93, 1, 1987. With permission.)

Figure 4.18. Although the loss values are one or more orders of magnitude higher than those obtained using Equation 4.4, the proposed minimum loss is still lower than that of silica glass fiber.

4.6.2   LOSS MEASUREMENT

Transmission loss of chalcogenide glass fiber is measured by the conventional cutback method. The light sources used are either an Ni–Cr heater[54] or a platinum lamp in the wavelength region $\lambda > 1.8$ μm and a halogen lamp below 1.8 μm.[38] The output light is separated into its spectral components by diffraction gratings and detected by a vacuum thermocouple or InSb and HgCdTe detectors. More recently, it was demonstrated that a FTIR spectrometer can be used to obtain fiber attenuation data.[55] The microcrack and unevenness of the fiber ends lead to significant experimental error. Bornstein et al.[56] reported that the fiber ends produced by breakage have usually three distinct regions: mirror region, knife penetration, and break deviation area from its initial propagation plane. Therefore, the fiber ends should be polished.

4.6.3   SULFIDE GLASS FIBERS

Fiber drawing of As–S glasses was first reported by Kapany and Simms.[57] While the transmission loss was relatively high (~20 dB/m), some applications of the fibers were described in detail. In 1978, Van Uitert and Wemple[53] predicted that IR transmitting materials are a potentially useful medium for ultralow-loss optical fibers. Since then, many reports on the preparation and loss characteristics of chalcogenide glass fibers have been presented. Typical sulfide glass fiber compositions and their losses are listed in Table 4.7. The drawing of Ge–P–S glass fibers for potential long-distance application was first reported by Shibata et al.[7] The transmission loss of 0.38 dB/m was attained at 2.4 μm for a 0.5 m length of unclad $Ge_{26}S_{65}P_9$ glass fiber. They predicted that the minimum loss of 10^{-1} to 10^{-2} dB/km would be attained in the wavelength region of 5 to 6 μm if the OH and SH impurities were lowered by four to five orders of magnitude. In 1981, an unclad GeS_3 glass fiber was prepared

TABLE 4.7
Preparation Methods and Losses of Representative Sulfide Glass Fibers

Core (at. %)	Cladding	Drawing Method	Minimum Loss/dBm (wavelength/μm)	Ref.
$Ge_{26}S_{65}P_9$	Air	Preform	0.38 (2.4)	7
$Ge_{25}S_{75}$	Air	Preform	0.36 (2.4)	43
$Ge_{25}S_{80}$	Air	Preform	0.56 (2.4)	38
$As_{40}S_{60}$	Air	Preform	0.078 (2.4)	58
$As_{40}S_{60}$	Air	Preform	0.035 (2.4)	38
$As_{40}S_{60}$	Teflon	Preform	0.15 (4.8)	60
$As_{40}S_{60}$	Teflon	Preform	0.1 (2.5)	33
$As_{38}S_{62}$	$As_{35}S_{65}$	Double crucible	0.2 (2.4)	38
As–S–Se	As–S	Rod-in-tube in crucible	0.2 (2.9)	33
$As_{35}S_{65}$	$As_{32}S_{68}$	Double crucible	0.023 (2.3)	61

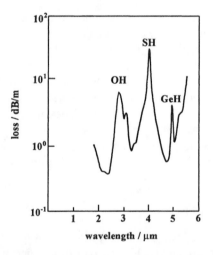

FIGURE 4.19 Loss spectrum of unclad GeS_3 glass fiber using S purified by distillation in an S_2Cl_2 gas stream. (From Shibata, S. et al., *J. Appl. Phys.*, 20, L13, 1981. With permission.)

using S purified by distillation under an S_2Cl_2/Ar gas stream.[43] The transmission loss is shown in Figure 4.19. The minimum losses of 0.36 and 0.56 dB/m were attained at 2.4 and 5 μm, respectively.

Since 1982, there have been many reports describing fiber drawing and loss characteristics of As–S glass fibers. Miyashita and Terunuma[58] and Kanamori et al.[38] attained minimum losses of 78 and 35 dB/km for unclad As_2S_3 glass fibers, respectively. The assignments of the impurity absorption bands are summarized in Table 4.3.

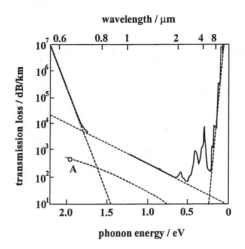

FIGURE 4.20 Relationship between photon energy and transmission loss of As_2S_3 glass fiber. (From Kanamori, T. et al., *J. Non-Cryst. Solids,* 69, 231, 1985. With permission.)

The effect of the weak absorption tail on the transmission loss was studied by Kanamori et al.[35] The weak absorption tail is observed below the exponential part of the absorption edge (Urbach tail) and is due to transitions from localized states deep in the band gap to the extended states.[59] Figure 4.20 shows the plot of transmission loss of an unclad As_2S_3 glass fiber against photon energy.[35] Curve A is the scattering loss estimated by measuring Rayleigh's ratio R_{90} at 633 nm (He–Ne laser). The loss between 0.7 and 6.0 µm is assigned to the weak absorption tail. Consequently, the expected minimum loss of this fiber is estimated as 23 dB/km at 4.6 µm. Wood and Tauc[59] suggested that the deep potential states causing the weak absorption tail are associated with disorder, defects, and impurities. The concentration of states giving rise to the weak absorption tail is approximately 10^{16} to 10^{17}/cm³. There is a possibility that the transmission loss characteristics of As–S glass can be considerably improved if the concentration of defects and impurities can be minimized.

An As–S glass fiber with a glass cladding was successfully prepared by Kanamori et al.[38] The fibers of 100 to 300 µm in cladding diameter and 1.4 to 2 in cladding/core diameter ratio were drawn at 300 to 400°C under an Ar gas atmosphere. A numerical aperture (NA) of 0.5 was attained by choosing the glass compositions of $As_{38}S_{62}$ for core and $As_{35}S_{65}$ for cladding. Unfortunately, the loss was approximately one order of magnitude higher than that of the unclad As–S glass fiber. The increase in loss was attributed to the precipitation of microcrystalline particles.

An As–S glass fiber with a glass cladding was drawn by the modified rod-in-tube method[41] (see Figure 4.16). About 3 at. % of selenium was added to the core in order to obtain an NA of 0.4. The transmission loss of the fiber is shown in Figure 4.21. The absorption bands due to OH and SH impurities at 2.9 and 4.1 µm could be minimized by the distillation of sulfur under an Se_2Cl_2 stream, and a minimum loss below 0.2 dB/m was attained near 2.6 and 4.6 µm.

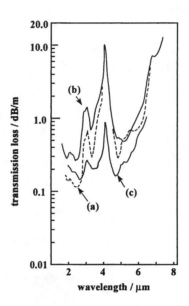

FIGURE 4.21 Loss spectra of As–S glass fibers: (a) Teflon cladding, (b) glass cladding, (c) glass cladding (using sulfur purified under Se$_2$Cl$_2$ gas stream). (From Nishii, J. et al., *J. Non-Cryst. Solids,* 140, 199, 1992. With permission.)

A Teflon FEP cladding can be applied to As–S glass fiber.[60] An example of the loss of the fiber with a Teflon cladding is also shown in Figure 4.21. Recently, a sulfide fiber with a glass cladding was drawn using the double crucible process and with a minimum loss of 23 dB/km (±8 dB/km) at 2.3 µm.[61] The core and cladding glass compositions were As$_{35}$S$_{65}$ and As$_{32}$S$_{68}$, respectively. The fiber NA was 0.49 at 2.94 µm.

4.6.4 SELENIDE GLASS FIBERS

Selenide glass fiber drawing has been reported since 1983. Representative fibers and their losses are summarized in Table 4.8. Physical and chemical properties of binary glasses, such as glass transition temperature, mechanical strength, chemical durability, etc., are improved by the addition of certain third elements. The glasses of the Ge–As–Se system possess candidate properties for optical fiber preparation. The fabrication of low-loss unclad fibers in this ternary system was reported by Kanamori et al.[35] A loss curve is shown in Figure 4.22. The minimum loss of 0.3 dB/m was attained at 3.9 µm. Four intense absorption bands are observed at 2.92, 4.57, 6.32, and 7.9 µm, which are due to hydrogen and oxygen impurities (see Table 4.2). The extrinsic absorptions in the wavelength region beyond 7 µm are mainly due to oxygen impurities associated with Ge or Se. Katsuyama et al.[44] successfully eliminated the oxygen impurities in Ge–Sb–Se glasses by the addition of 100 ppm Al metal as the oxygen getter, and prepared the unclad fiber by the crucible drawing method.

There are some reports on the preparation of selenide glass fibers with a resin cladding and glass cladding. Parant et al.[62] prepared As–Ge–Se glass fibers with UV

TABLE 4.8
Representative Selenide Glass Fibers and Their Losses

Core (at. %)	Cladding	Drawing Method	Minimum Loss/dB/m (wavelength/μm)	Ref.
$As_{40}Se_{60}$	Air	Crucible	5.8 (10.6)	32
$Ge_{20}Se_{80}$	Air	Crucible	0.2 (5.5)	44
$Ge_{15}As_{10}Se_{75}$	Air	Crucible	4.5 (10.6)	32
$As_{38}Ge_5Se_{57}$	Air	Preform	0.3 (3.9)	35
$Ge_{10}Sb_{26}Se_{64}$	Air	Crucible	0.2 (7.0)	44
$Ge_{28}Sb_{12}Se_{60}$	Air	Crucible	10.0 (10.6)	56
$As_{15}Ge_{30}Se_{55}$	Resin	Crucible	5.0 (7.0)	62
Ge–As–Se	Resin	Preform	0.2 (6.0)	33
$Ge_{28}Sb_{12}Se_{60}$	$Ge_{32}Sb_8Se_{60}$	Rod-in-tube	—	17
Ge–As–Se	Ge–As–Se	Rod-in-tube in crucible	0.2 (6.0)	33

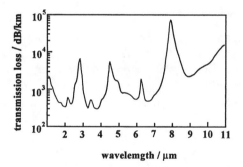

FIGURE 4.22 Loss spectra of an unclad $As_{38}Ge_5Se_{57}$ glass fiber. (From Kanamori, T. et al., *J. Non-Cryst. Solids,* 69, 231, 1985. With permission.)

curable epoxy-acrylate and polyolefin coatings. The minimum loss was 5 dB/m at 7 μm. As–Ge–Se glass fibers with a Teflon FEP cladding and a Ge–As–Se glass cladding were prepared by Nishii et al.[33] The transmission losses of these fibers are shown in Figure 4.23. The absorption bands between 7.5 and 11 μm are due to the Teflon cladding. Therefore, the Teflon cladding can not be used for fibers transmitting beyond 7.5 μm. On the contrary, the fiber with the glass cladding does not exhibit such excess loss. The preparation of Ge–Sb–Se glass fibers with a glass cladding using a rod-in-tube method was reported by Klocek et al.[17] An NA higher than 0.45 was achieved by changing the atomic ratio of Ge/Sb in the core and cladding glasses. While there is no comment on the minimum loss in the literature, these fibers are applicable for short-distance applications, such as temperature sensors, image bundles, etc.

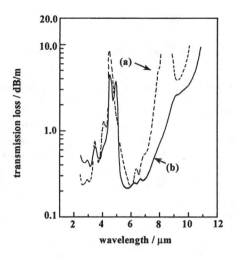

FIGURE 4.23   Loss spectra of Ge–As–Se glass fibers with (a) Teflon cladding and (b) glass cladding. (From Nishii, J. et al., *J. Non-Cryst. Solids,* 140, 199, 1992. With permission.)

4.6.5   TELLURIDE GLASS FIBERS

The addition of tellurium to selenide glasses is effective in shifting the multiphonon absorption tail to longer wavelengths. Telluride glasses generally, but not always, tend to be less stable against crystallization with increasing tellurium content. Therefore, more-precise temperature control is required for fiber drawing compared with the drawing of sulfide and selenide glass fibers. Some examples of telluride glass fibers are listed in Table 4.9.

The Ge–As–Se–Te system exhibits a broad glass-forming region with reasonable thermal and chemical durabilities. The drawing of unclad fibers from this system was reported by Parant et al.[62] One of the applications of tellurium-based glass fibers

TABLE 4.9
Preparation Methods and Losses of Several Telluride Glass Fibers

Core (at. %)	Cladding	Drawing Method	Minimum Loss/dB/m (wavelength/μm)	Ref.
$Ge_{30}As_{13}Se_{27}Te_{30}$	Air	Preform	4 (7), 7.5 (10.6)	62
$Ge_{22}Se_{20}Te_{58}$	Air	Crucible	1.5 (10.6)	63
$Ge_{15}As_{20}Se_{15}Te_{50}$	Air	Crucible	0.9 (9.5), 1.5 (10.6)	24
$Ge_{27}Se_{18}Te_{55}$	Air	Crucible	0.6 (8.2), 1.5 (10.6)	54
$As_{40}Se_{40}Te_{20}$	Air	Crucible	5 (8.7), 7 (10.6)	66
As–Se–Te	Air	Preform	3.8–6 (10.6)	67
$Ge_{30}As_{10}Se_{30}Te_{30}$	Air	Preform	0.1 (6.6), 1.9 (10.6)	64
Ge–Se–Te (Te = 45)	Ge–As–Se–Te	Rod-in-tube in crucible	0.4 (8.2), 2 (10.6)	65

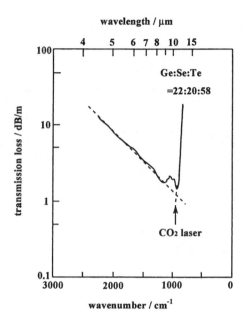

FIGURE 4.24 Loss spectrum of an unclad $Ge_{22}Se_{20}Te_{58}$ glass fiber. (From Katsuyama, T. and Matsumura, H., *Appl. Phys. Lett.*, 49, 22, 1986. With permission.)

is in the delivery of CO_2 laser power (emission wavelength = 10.6 μm). In order to minimize the loss at 10.6 μm, it is preferable to maximize the Te content. Glasses in the Ge–Se–Te system can be obtained in the region where the tellurium content exceeds 60 at. %. Katsuyama and Matsumura[63] succeeded in drawing unclad $Ge_{22}Se_{20}Te_{58}$ fiber using a crucible drawing method. The fiber was drawn very rapidly (1 to 5 m/s) from the crucible which was pressurized with an inert gas in order to inhibit the crystallization of the glass. The transmission loss of the fiber is shown in Figure 4.24. The minimum transmission loss of 1.5 dB/m was attained at 10.6 μm. The loss between 4 and 8 μm changes linearly with $1/\lambda^3$, and could possibly be attributed to scattering from crystalline particles.

Figure 4.25 shows the loss spectra of the unclad $Ge_{20}Se_{80}$ and $Ge_{27}Se_{18}Te_{55}$ glass fibers.[54] The fibers were prepared by the crucible drawing method. The pressure inside the crucible was 3×10^5 Pa, and the drawing speed was 1 cm/min, which is two orders of magnitude slower than that used by Katsuyama and Matsumura.[63] The loss of the former fiber in the wavelength region between 1 and 8 μm decreased linearly against photon energy, which implies that the weak absorption tail is responsible for the loss. A similar linear section is not observed for the telluride fiber. This nonlinear change in loss can be represented by the equation loss $\alpha\lambda^{-4}$ and can, therefore, be attributed to Rayleigh scattering, which is much higher than the calculated value (see Table 4.4). The origin of this high scattering loss remains unknown. Transmission loss characteristics of unclad Ge–As–Se–Te glass fibers were investigated by Inagawa et al.[24] and Sanghera et al.[64]

By taking the contribution from the evanescent absorption loss into account, it might be difficult for resin-clad fibers to attain low loss in the wavelength region

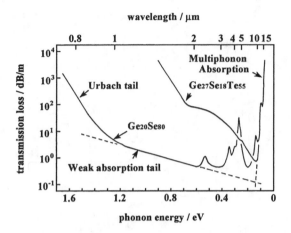

FIGURE 4.25   Loss vs. photon energy for unclad $Ge_{20}Se_{80}$ and $Ge_{27}Se_{18}Te_{55}$ glass fibers. (From Nishii, J. et al., *J. Non-Cryst. Solids,* 95–96, 541, 1987. With permission.)

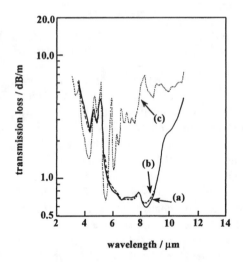

FIGURE 4.26   Loss spectra of Ge–As–Se–Te glass fibers: (a) $Ge_{25}As_{20}Se_{25}Te_{30}$ core and $Ge_{20}As_{30}Se_{30}Te_{20}$ cladding, (b) $Ge_{25}As_{20}Se_{25}Te_{30}$ unclad, and (c) resin-clad fibers. (From Nishii, J. et al., *Appl. Phys. Lett.,* 53, 553, 1988. With permission.)

beyond 7.5 μm. However, Nishii et al.[34] have drawn a Ge–As–Se–Te glass fiber with a glass cladding using the modified rod-in-tube method using a crucible (see Figure 4.16). The calculated NA of the fiber was 0.67 and the diameters of the core and cladding were 300 and 420 μm, respectively. Figure 4.26 shows the transmission loss of the fiber. The losses of the unclad and UV curable resin-clad fibers are also shown for comparison. The minimum loss of 0.6 dB/m was achieved at 8.5 μm for both the unclad and glass-clad fibers.

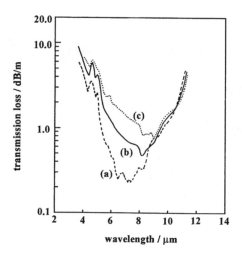

FIGURE 4.27 Loss spectra of Ge–Se–Te core and Ge–As–Se–Te clad glass fibers with Te content in core of (a) Te = 48 at. %, (b) Te = 52 at. %, and (c) Te = 55 at. %. (From Nishii, J. et al., *Appl. Opt.*, 28, 5122, 1989. With permission.)

Ge–Se–Te glass fibers with a Ge–As–Se–Te glass cladding have been prepared by the same drawing method.[41] The lowest loss attained at 10.6 μm was 1.8 dB/m. The loss spectra are shown in Figure 4.27.

4.7 MECHANICAL STRENGTH OF FIBER

It is needless to mention that chalcogenide glass fibers are required with sufficient flexibility and mechanical strength for practical applications. In this section, the tensile and bending strength characteristics of the fibers are summarized. The tensile strength (σ_T) and bending strength (σ_B) are obtained from the following equations:[68]

$$\sigma_T = P \cdot E_t / \left[\left(D_t^2 \cdot E_f \right) + \left(D_a^2 - D_f^2 \right) \cdot E_c \right] \tag{4.8}$$

$$\sigma_B = 1.198 \cdot E_f \cdot \left(D_f / D_b \right) \tag{4.9}$$

where P is the breaking load, E_f and E_c are Young's moduli of fiber and coating, D_f and D_c are the diameters of fiber and coating, D_b is the bending diameter. Table 4.10 lists the strength values of the Ge–As–Se, As–Se, Ge–Sb–Se, and Ge–As–Se–Te glass fibers. For comparison, a tensile strength of 126 kpsi was reported for a ZBLAN glass fiber (ZrF$_4$-based glass).[70] While the values for Young's modulus of chalcogenide glasses are approximately one half of those of fluoride glasses,[71] it is expected that the mechanical strength of chalcogenide glass fibers will become higher than the values shown in Table 4.10 by performing chemical etching of the preform or improvements in the fiber drawing technique.

TABLE 4.10
Average Strength (Bending and Tensile) of Several Chalcogenide Glass Fibers

Composition	Cladding	Fiber Diameter (μm)	Strength (kpsi) Bending/Tensile
Ge–As–Se(AMTIR® 1c)	Glass and resin	150	30.7/16.9
As–Se(AMTIR® 2)	Resin	300	63.5/12.0
Ge–Sb–Se(AMTIR® 3c)	Glass and resin	150	30.7/14.2
Ge–Sb–Se(AMTIR® 3c)	Glass and resin	300	21.6/12.2
$As_{40}Se_{60}$	Resin	300	23.2/9.3
$Ge_{28}Sb_{12}Se_{60}$	Resin	300	34.8/15.8
$Ge_{25}Sb_{15}Se_{60}$	Resin	300	20.4/10.7
$Ge_{25}Sb_{15}Se_{60}$	Glass and resin	300	26.7/9.2
$Ge_{33}As_{12}Se_{55}$	Resin	300	21.6/9.0
$Ge_{33}As_{12}Se_{49}Te_6$	Resin	300	20.3/8.7
$Ge_{33}As_{12}Se_{49}Te_6$	Glass and resin	300	30.7/14.7
$Ge_{33}As_{12}Se_{45}Te_{10}$	Resin	300	17.6/5.2

After McEnroe, D. J. et al., *Proc. SPIE,* 799, 39, 1987; 843, 62, 1987.

TABLE 4.11
Mechanical Strength of Some Chalcogenide Glass Fibers before and after Treatment at 80°C, 85% RH for 24 h

Fiber	Cladding	Fiber Diameter (μm)	Before Treatment (kpsi) Bending/Tensile	After Treatment (kpsi) Bending/Tensile
As–S	Teflon	260/300/—	—/15.9	—/9.6
As–S	Glass and UV resin	205/265/403	90.1/14.7	84.1/11.6
Ge–As–Se	Teflon	220/240/—	—/16.8	—/16.3
Ge–As–Se	Glass and UV resin	260/325/430	51.4/14.0	40.6/10.0
Ge–Se–Te	Glass and UV resin	260/335/430	50.1/16.1	52.1/12.6

Nishii et al.[33] have investigated the effect of heat treatment under humid conditions on the mechanical strength of some representative fibers. The results are listed in Table 4.11. The mechanical strength generally decreased after the treatment, and, therefore, appropriate protection of the fibers from attack by moisture is required before their practical applications can be realized.

4.8 APPLICATIONS OF CHALCOGENIDE GLASS FIBERS

The transmission losses of chalcogenide glass fibers are still high for long-distance applications. However, chalcogenide glass fibers have various potential uses in the short-distance area such as remote temperature monitoring, thermal imaging, chemical sensing, and power delivery of intense IR laser emission. In this section, some recent results on these applications are described.

4.8.1 TEMPERATURE MONITORING AND THERMAL IMAGING

Chalcogenide glass fiber radiometry enables one to monitor the temperature and to perform thermal imaging in a narrow region without any interference from microwave or electromagnetic waves, etc. Ueda et al.[72] measured the grinding temperature of the surface layer of ceramic plates using an As–S glass fiber with a Teflon cladding. This gives important information on the grinding mechanism, lifetime of the wheel, deterioration in dimensional accuracy, etc. The experimental setup is shown in Figure 4.28a. The fiber was inserted into a pinhole in the silicon carbide just below the grinding surface. The IR radiation power from the top surface of the pinhole is delivered through the fiber to an InSb detector and converted to temperature. Figure 4.28b shows the change in temperature with time of the grinding surface of silicon carbide when the distance between the fiber end and the grinding surface is 20 μm. Rapid changes in temperature of around 50 to 100°C can be detected. Figure 4.28c represents the plots of measured temperature against the distance between the fiber end and grinding surface. The temperature on the grinding surface of silicon carbide is estimated as 200°C. A similar experiment was carried out for silicon nitride using a fluoride glass fiber.

Preparation of a coherent fiber bundle of As_2S_3 glass was suggested by Kapany and Simms.[57] The optical characteristics of a rigid fiber bundle were investigated by Saito et al.[73] Up to 1000 fibers of As_2S_3 glass with a Teflon cladding were bundled in a heat-shrinkable Teflon tube and then drawn into the fiber bundle. The fibers adhered to each other during drawing. The transmission loss of the bundle was estimated as low as 0.6 dB/m at 3.25 and 3.8 μm. Although a fluctuation of ±20 K was observed in the indicated temperature at 773 K, the thermal image of a hot electric iron was successfully delivered with good coherency through the 1.3-m-long bundle.

Klocek et al.[17] reported the preparation of a Ge–Se–Sb glass fiber bundle. In order to maintain flexibility of the bundle, only the fibers in the end areas were bundled by fusion and the length of the fused area was minimized to inhibit cross talk. A thermal image of a test pattern could be delivered by their bundle.

An As_2S_3 fiber bundle having practical flexibility and coherency was developed by Nishii et al.[74] The bundle was designed to be applicable to the IR television camera system, AVIO TVS-2100. The detector used was an InSb crystal with an element size of 83×83 μm and an active wavelength range of 3 to 5.4 μm. Three kinds of fibers with a Teflon cladding were prepared by the preform drawing method, i.e., As_2S_3, $Ge_{10}As_{30}S_{40}Se_{20}$, and $Ge_{15}As_{20}Se_{65}$. The relationship between the setting temperature of the black body (Tb) and the temperature detected by the TVS-2100 through each fiber (Tm) was investigated before the preparation of the bundle. The results are shown in Figure 4.29A. If the fiber has no transmission loss and no Fresnel loss, then Tm should be identical to Tb. It is apparent from the figure that the As–S glass fiber exhibits the largest Tm/Tb because of its relatively lower transmission

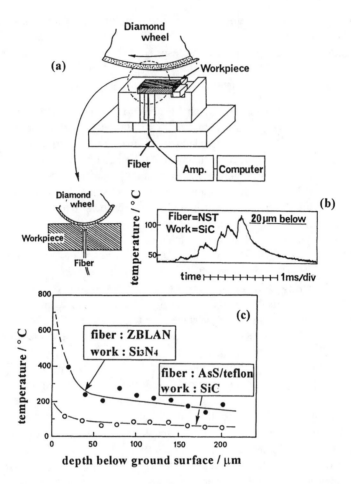

FIGURE 4.28   (a) Experimental setup for the measurement of temperature on the grinding surface. (b) Time dependence of the temperature change when the distance between the fiber end and grinding surface is 20 μm. (c) Plots of the measured temperature and the distance between the fiber end and grinding surface. (From Ueda, T. et al., *J. Eng. Ind.,* 114, 317, 1992. With permission.)

loss and Fresnel loss when compared with the other fibers. Two kinds of bundles of As_2S_3 glass fibers with Teflon cladding, 1550 and 8400 cores, were prepared by a modified-layer-winding method. The bundle made by this method was flexible because both its ends only were cemented together. Figure 4.29B and C show the IR imaging system and the end cross section of the bundle, respectively. Precise control of the fiber diameter to less than ±2.5% is required to attain the regular array of the fiber. The thermal images of a human body and operating LSI packages were detected using this system. This coherent IR image bundle would make possible the remote location of the camera and realize the full 360° field of view.

Some physical properties of Ge–Se–Te glasses were investigated for applications in the 8 to 12 μm thermal imaging systems.[30]

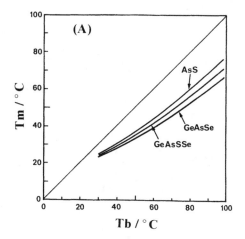

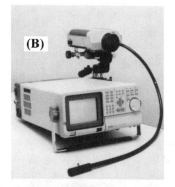

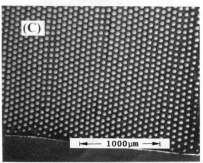

FIGURE 4.29 (A) Relationship between the setting temperature of black body (Tb) and the temperature measured by TVS-2100 via fibers (Tm). (B) Photograph of IR imaging system. (C) End cross section of bundle. (From Nishii, J. et al., *Appl. Phys. Lett.,* 59, 2639, 1991. With permission.)

4.8.2 CHEMICAL SENSING

The IR absorptions of fundamental vibrational modes of gases, liquids, and solids can be detected remotely using the chalcogenide glass fibers and an FTIR spectrometer.

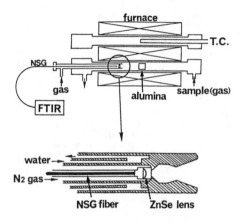

FIGURE 4.30 Schametic of experimental apparatus for the measurement of gas species at high temperature. (From Maeda, M., in *Chemical Sensor Technology,* Vol. 3, N. Yamazoe, Ed., Kodansha, Ltd., Tokyo, 1991, 185. With permssion.)

Particular chemical species can be detected from their intrinsic vibrational absorptions by appropriate optical filtering or chopping. This remote monitoring technique is applicable to *in situ* and continuous pollution monitoring, process control, etc.

Heo et al.[75] used an unclad $Ge_{27}Se_{18}Te_{55}$ fiber of 380 μm diameter for the evanescent wave spectroscopy of liquids. They quantitatively measured the presence of small amounts of organic and acid species in aqueous solutions.

In situ quantitative analysis of gas species at high temperature using IR optical fibers was reported by Maeda.[76] A schematic diagram of the experimental setup is shown in Figure 4.30. Alumina balls (5 mm diameter) were placed at the inlet end of the gas in order to preheat the gas before the measurement. The IR optical fiber and IR lens (ZnSe) were installed in a metal probe cooled by water. The IR emissions from the gaseous species were delivered via the fiber to the FTIR spectrometer. Measurements were performed on several gaseous species with concentrations as low as 100 ppm in the furnace at high temperature (>1500°C).

4.8.3 POWER DELIVERY OF CO_2 AND CO LASERS

Optical fibers transmitting in the mid-IR wavelength region are applicable for the delivery of high power from CO_2 and CO lasers. The oscillating wavelength of these tunable lasers are located at 9.2 to 10.8 μm and 5.3 to 5.9 μm, respectively. There have been many reports on the fabrication of fibers for use in surgical and industrial IR laser systems. Chalcogenide glass is a candidate fiber material because of its superior physical and chemical properties and unique optical properties.

4.8.3.1 CO_2 Laser Power Transmission

The power transmission characteristics of a CO_2 laser through Ge–Se–Te (GST-1, GST-2) and Ge–As–Se (GAS) glass fibers were investigated by Nishii et al.[65] The characteristics of the fibers are summarized in Table 4.12. The relationship between

TABLE 4.12
CO₂ and CO Laser Power Transmission Characteristics of Chalcogenide Glass Fibers

Laser	Fiber[a]	Diameter of Core (μm)	Length (cm)	Loss (dB/m) 5.4/10.6 μm	Cooling	AR Coating	Laser Power (W) Input/Output	Efficiency (%)
CO_2	NTEG-1	450	100	1.0/21	No	No	24.8/5.9	23.8
		450	100	1.0/2.1	Gas	PbF_2	21.8/9.5	43.6
		450	100	1.0/2.1	Water	PbF_2	19.4/10.7	55.2
	NTEG-2	450	100	1.9/2.0	No	No	10.6/3.1	29.2
		450	100	1.9/2.0	Water	No	11.6/3.6	31.0
	NTEG-1	450	100	0.5/5.5	No	No	18.0/2.2	12.2
		450	100	0.5/5.5	Water	No	19.4/3.3	17.0
CO	NST	900	100	0.6/—	No	No	111/59	53.2
	NSG	780	100	0.6/—	No	No	169/85	50.3
	NSEG-1	700	20	0.5/5.5	No	No	13/8.5	65.4
	NSEG-2	800	52	1.9/2.0	No	No	17/5	29.4

[a] NTEG-1: Ge–Se–Te core (Te = 45 at. %) with Ge–As–Se–Te cladding; NTEG-2: Ge–Se–Te core (Te = 51.5 at. %) with Ge–As–Se–Te cladding; NST: As–S core with Teflon cladding; NSG: As–S–Se core with As–S cladding; NSEG-1: Ge–As–Se core with Ge–As–Se cladding; NSEG-2: Ge–As–Se–Te core with Ge–As–Se cladding.

After Nishii, J. et al., *J. Non-Cryst. Solids*, 140, 199, 1992.

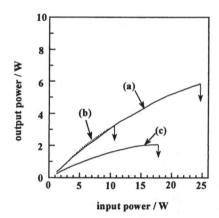

FIGURE 4.31 Relationship between input and output powers of CO_2 laser for (a) Ge–Se–Te (Te = 45 at. %), (b) Ge–Se–Te (Te = 51.5 at. %), and (c) Ge–As–Se glass fibers. (From Nishii, J. et al., *Proc. SPIE,* 1228, 224, 1990. With permission.)

the input and output powers was investigated under several conditions. The continuous wave (CW) CO_2 laser beam was focused using a ZnSe lens and launched into the fiber. The fiber ends were polished using Al_2O_3 or CeO_2 powders with water to an optical finish. Cooling of the fiber ends by blowing N_2 gas was effective in suppressing the thermal damage to the fiber ends. The optical flatness of the fiber ends is also important to inhibit damage by the laser beam. Figure 4.31 represents the relationship between input and output powers of a CO_2 laser through three different fibers of 100 cm length. The output powers increased linearly until the input power exceeded 5 W. The slope in this power region agrees with the calculated value from the Fresnel loss and transmission loss of the fiber. Figure 4.32A shows the surface temperature of the fiber when the input power was 7 W. The apparent rise in temperature in the region between 20 and 50 cm from the input end was confirmed in each fiber. Figure 4.32B represents the temperature dependence of the losses of each fiber. The fiber was placed in a tube heater with inner diameter of 10 mm and heated uniformly. The apparent leveling off of the output power in Figure 4.31 was caused by the increase in loss with temperature of the fiber shown in Figure 4.32B.

When the fiber ends were cooled by flowing gas, the core section in the region between 20 and 50 cm from the input end was fused. Figure 4.33 shows the damaged cross sections of Ge–Se–Te fibers. The origin of this damage was attributed to the focusing of the laser beam caused by the formation of a thermally induced lens in the core. This arises from the large positive temperature dependence of both the refractive index of the core glass and the transmission loss of the fiber.

Power transmission efficiency is improved by using antireflection coatings on both fiber end faces and also by cooling the fiber with water. The Fresnel losses can be suppressed to less than 2% by deposition of PbF_2 thin films by a thermal evaporation method. The maximum output power of 10.7 W was attained by these methods. Figure 4.34 shows a fiber cable with a ZnSe lens ($f = 30$ mm) at the output end. The fiber was cooled by flowing gas from the input connector to the hand piece.

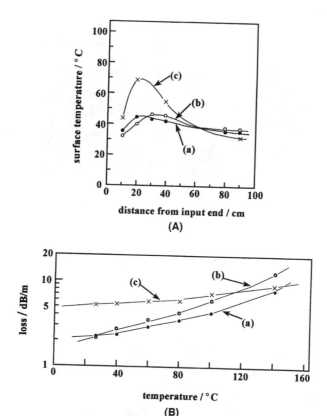

FIGURE 4.32 (A) Surface tempreatures of the fibers during CO_2 laser power transmission (launched power = 7 W) and (B) temperature dependence of the transmission losses for (a) Ge–Se–Te (Te = 45 at. %), (b) Ge–Se–Te (Te = 51.5 at. %), and (c) Ge–As–Se glass fibers. (From Nishii, J. et al., *Proc. SPIE*, 1228, 224, 1990. With permission.)

The output beam was focused into a spot of approximately the same size as the core diameter. There was no change in the beam quality and output power when the cable was bent to less than 10 cm in radius.

4.8.3.2 CO Laser Power Transmission

In 1984, Dianov et al.[49] reported the power transmission from a CO laser through As–S, As–Se and Ge–As–Se glass fibers with a Teflon cladding. Approximately 6 to 7 W of power was delivered for longer than 7 to 8 h.

A feasibility study on the medical use of CO laser power delivery through As_2S_3 fiber with a Teflon cladding was reported by Arai et al.[77] The maximum output power of 13.5 W was attained using a 400 μm core diameter fiber of 43.7 cm length. The upper limit of the transmitted power was restricted as a result of melting of the core at the output end of the fiber. This fiber is applicable for laser angioplasty and laser endoscopy.

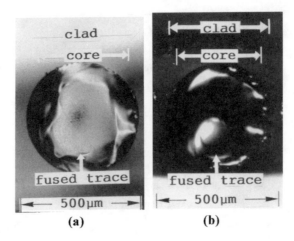

FIGURE 4.33 Fiber cross sections of Ge–Se–Te glass fibers — (a) Te = 45 at. % and (b) Te = 51.5 at. % — damaged by CO_2 laser beam. (From Nishii, J. et al., *Proc. SPIE,* 1228, 224, 1990. With permission.)

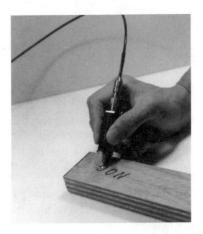

FIGURE 4.34 Ge–Se–Te glass fiber cable with hand piece. (From Nishii, J. et al., *J. Non-Cryst. Solids,* 140, 199, 1992. With permission.)

Currently, the maximum output power available from a CO laser exceeds 1 kW. Such high emission power is useful for industrial laser welding, cutting, melting, etc. High power delivery of CO laser energy via a chalcogenide glass fiber was reported by Hattori et al.[78] and Nishii et al.[33] Power transmission characteristics are summarized in Table 4.12. The maximum output power, which is one order of magnitude higher than that of Ge–Se–Te glass fibers, was restricted by the thermal damage to the fiber near the fiber ends. The output power should be increased by appropriate cooling of the fiber and by reducing the loss of the fiber.

TABLE 4.13
Representative Chalcohalide Glasses

Glass-Forming System	Glass Transition (T_g) or Softening (T_s) temp. (°C)	Ref.
Sb–S–Br	$T_s = 110$–155	81
Sn–Sb–S–I	$T_g = 154$–168	82
Pb–Sb–S–I	$T_g = 136$–158	82
As–Se–I	$T_g = 38$–62	83
Se–S–Br	$T_g = 53$–72	84
Cd–As–Br	$T_g = 180$	85
Se–Te–I	$T_g = 54$–81	86
Ge–S–Br	$T_g = 200$–430	87
Se–Te–Br	$T_g = 58$–78	88

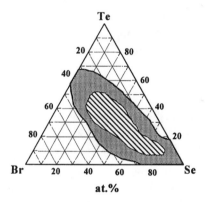

FIGURE 4.35 Glass-forming region of the Se–Te–Br system. The shaded region represents the glass-forming region. The diagonal lines indicate the region where no crystallization peak was observed in the DSC curve. (From Lucas, J. et al., *Proc. SPIE,* 1228, 109, 1990. With permission.)

4.9 OTHER GLASS SYSTEMS AND THEIR APPLICATIONS

4.9.1 CHALCOHALIDE GLASSES AND FIBERS

Halogen elements (Cl, Br, I) can be incorporated in chalcogenide glasses to shift the multiphonon absorption edge to longer wavelengths. As summarized in Table 4.13, various glass-forming systems ahve been developed. The glass-forming region of Se–Te–Br is shown in Figure 4.35. The fiber drawing of chalcohalide glasses was reported by Lucas et al.[79] who attained a minimum loss of 5 dB/m in the vicinity of 10.6 μm. Inagawa et al.[80] prepared unclad $Se_{25}Te_{30}I_{45}$ glass fiber, and

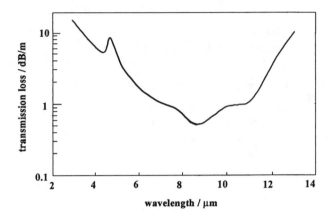

FIGURE 4.36 Transmission loss of unclad $Se_{25}Te_{50}I_{45}$ glass fiber. (From Inagawa, I. et al., *Jpn. J. Appl. Phys.,* 30, 2846, 1991. With permission.)

TABLE 4.14
Chalcogenide Glasses Containing Rare Earth Ions

System	Ref.
L_2S_3–Ga_2S_3 (L = Y, La, Er, Eu)	92
La–Ga–S–Ho	93
La–Ga–S–Er	94
Eu–As–S	95
Eu–Ga–Ge–S	96

the transmission loss curve is shown in Figure 4.36. A loss of 0.9 dB/m was achieved at 10.6 μm. However, the glass transition temperature of this glass is 48°C, which is relatively lower than the other chalcohalide glasses. Some additives are required in order to improve the thermal characteristics of the glass.

4.9.2 CHALCOGENIDE GLASSES CONTAINING RARE EARTH METALS

Rare earth chalcogenide compounds form glasses when they are mixed with other chalcogenide materials having high glass-forming ability. Some glass systems are listed in Table 4.14. Recently, rare earth–doped chalcogenide glasses have received considerable attention as fiber material for optical amplifiers. La–Ga–S glasses have been proposed as a host material for Pr^{3+}-doped fibers.[89] The emission band fo the Pr^{3+} ion (1.3 μm: $^1G_4$ to $^3H_5$) can be used for amplification of optical signals at 1.3 μm, which is an important wavelength used in optical commuincation networks. Calculated quantum efficiencies of this emission band increased from ~5 to 26% by changing the host material from fluoride glasses to Ga–La–S glasses.[90] The fiber drawing of the glasses in this system has not been achieved because of their thermal instability.[91]

4.9.3 OPTICAL SWITCHING AND FIBER GRATINGS

Ultrafast all-optical switching devices are required for the distributed optical fiber network. Optical switching using an As_2S_3 waveguide was reported by Tanaka and Odajima.[97] Switching of a red light by the illumination with blue light, which is due to the photostructural transofrmations of the glass, was carried out in the switching time of 10 s. Recently, the formation of Bragg reflectors, similar to Hill gratings,[98] was confirmed in As_2S_3 glass by Shiramine et al.[99] The reflectivity of 10 to 20% was attained by irradiation with an He–Ne laser operating at 0.6328 μm.

The third-order optical nonlinear susceptibilities ($\chi^{(3)}$) of chalcogenide glasses are much larger than that of SiO_2 glass.[100] Chalcogenide glasses are, therefore, applicable for optical switching devices. Asobe et al.[101] confirmed the n_2 value of 4.2×10^{-14} cm²/W in a single-mode As–S glass fiber of 48 cm length and transmission loss of 6.6 dB/m at 1.319 μm. The NA and the effective core area were 0.22 and 24.6 μm², respectively. While an incident power of 14 W was required for switching, the efficiency can be improved by optimization of the fiber dimensions and reduction in the fiber loss.

The quality of chalcogenide glass fiber has been considerably improved. Recently, several products, such as single-mode fiber, fiber image bundles, etc., have become commercially available. In the future, chalcogenide glass fibers will be employed in various IR optical functional devices.

REFERENCES

1. C. S. Sellack, *Ann. Phys.*, 139, 182 (1870).
2. R. Frerichs, *J. Opt. Soc. Am.*, 43, 1153 (1953).
3. W. A. Fraser and J. Jerger, Jr., *J. Opt. Soc. Am.*, 43, 322A (1955).
4. A. Hruby, *J. Non-Cryst. Solids*, 28, 139 (1978).
5. A. Hruby, *Czech. J. Phys.*, B23, 1263 (1973).
6. A. Feltz, Ed., *Amorphous Inorganic Materials and Glasses*, VCH, New York, 1993, 257.
7. S. Shibata, Y. Terumuma, and T. Manabe, *Mater. Res. Bull.*, 16, 703 (1981).
8. R. Blachnik and A. Hoppe, *J. Non-Cryst. Solids*, 34, 191 (1979).
9. Y. Kawamoto and S. Tsuchihashi, *Yogyo Kyokai Shi*, 77, 328 (1969).
10. D. Linke and I. Böckel, *Z. Anorg. Allg. Chem.*, 419, 97 (1976).
11. V. A. Feltz, E. Schlenzig, and D. Arnold, *Z. Anorg. Allg. Chem.*, 403, 243 (1974).
12. M. B. Myers and E. J. Felty, *Mater. Res. Bull.*, 2, 535 (1967).
13. G. Parthasarathy, G. M. Naik, and S. Asokan, *J. Mater. Sci. Lett.*, 6, 214 (1987).
14. J. A. Savage and S. Nielsen, *Phys. Chem. Glasses*, 6, 90 (1965).
15. S. Bordas, J. Casas-Vazquez, N. Clavaguera, and M. T. Clavaguera-Mora, *Thermochim. Acta*, 28, 387 (1979).
16. A. R. Hilton, *Appl. Opt.*, 5, 1877 (1966).
17. P. Klocek, M. Roth, and R. D. Rock, *Opt. Eng.*, 26, 88 (1987).
18. P. Klocek and L. Colombo, *J. Non-Cryst. Solids*, 93, 1 (1987).
19. D. Linke, M. Gitter, and F. Krug, *Z. Anorg. Allg. Chem.*, 444, 217 (1978).
20. J. A. Muir and R. J. Cashman, *J. Opt. Soc. Am.*, 57, 1 (1976).

21. S. Bordas, N. Clavaguera, M. Geli, J. Casas-Vazquez, and M. T. Clavaguera-Mora, in *Proc. 5th Int. Thermal Analysis,* p. 14 (1977).
22. J. Wieder and S. Aronson, *J. Non-Cryst. Solids,* 33, 405 (1979).
23. J. Nishii, Y. Kaite, and T. Yamagishi, *Phys. Chem. Glasses,* 28, 55 (1987).
24. I. Inagawa, R. Iizuka, T. Yamagishi, and R. Yokota, *J. Non-Cryst. Solids,* 95–96, 810 (1987).
25. A. Feltz, H. J. Büttner, F. J. Lippmann, and W. Maul, *J. Non-Cryst. Solids,* 8–10, 64 (1972).
26. H. E. Anthonis and N. J. Kreidl, *J. Non-Cryst. Solids,* 13, 13 (1973/74).
27. J. A. Savage, *J. Mater. Sci.,* 7, 64 (1972).
28. T. F. Deutsch, *J. Electron. Mater.,* 4, 663 (1975).
29. P. J. Webber and J. A. Savage, *J. Non-Cryst. Solids,* 20, 271 (1976).
30. J. A. Savage, *NATO ASI Ser. E.,* 123, 367 (1987).
31. A. R. Hilton and D. J. Hayes, *J. Non-Cryst. Solids,* 17, 339 (1975).
32. E. Hartouni, F. Hulderman, and T. Guiton, *Proc. SPIE,* 505, 131 (1984).
33. J. Nishii, I. Inagawa, S. Morimoto, R. Iizuka, T. Yamashita, and T. Yamagishi, *J. Non-Cryst. Solids,* 140, 199 (1992).
34. J. Nishii, T. Yamashita, and T. Yamagishi, *Appl. Phys. Lett.,* 53, 553 (1988).
35. T. Kanamori, Y. Terunuma, S. Takahashi, and T. Miyashita, *J. Non-Cryst. Solids,* 69, 231 (1985).
36. C. T. Moynihan, P. B. Macedo, M. S. Maklad, R. K. Mohr, and R. E. Howard, *J. Non-Cryst. Solids,* 17, 369 (1976).
37. M. Vlcek, L. Tichy, J. Klikorka, and A. Triska, *J. Mater. Sci.,* 22, 2119 (1987).
38. T. Kanamori, Y. Terunuma, S. Takahashi, and T. Miyashita, *J. Lightwave Technol.,* LT2(5), 507 (1984).
39. J. Nishii, T. Yamashita, and T. Yamagishi, *J. Mater. Sci.,* 24, 4293 (1989).
40. M. F. Churbanov, *J. Non-Cryst. Solids,* 140, 324 (1992).
41. J. Nishii, T. Yamashita, and T. Yamagishi, *Appl. Opt.,* 28, 5122 (1989).
42. A. R. Hilton, D. J. Hayes, and M. D. Rechtin, *J. Non-Cryst. Solids,* 17, 319 (1975).
43. S. Shibata, T. Manabe, and M. Horiguchi, *J. Appl. Phys.,* 20, L13 (1981).
44. T. Katsuyama, K. Ishida, S. Satoh, and H. Matsumura, *Appl. Phys. Lett.,* 45, 925 (1984).
45. D. Lezal, B. Petrovska, G. Kuncova, M. Pospisilova, and J. Götz, *Proc. SPIE,* 799, 44 (1987)
46. T. Katsuyama, S. Satoh, and H. Matsumura, *J. Appl. Phys.,* 59, 1446 (1986).
47. C. L. Sergent, *Proc. SPIE,* 799, 18 (1987).
48. P. J. Melling, *Ceram. Bull.,* 63, 1427 (1984).
49. E. M. Dianov, V. J. Masychev, V. G. Plotnichenko, V. K. Sysoev, P. J. Baikalov, G. G. Devjatykh, A. S. Konov, J. V. Schipchev, and M. F. Churbanov, *Electron. Lett.,* 20, 129 (1984).
50. A. Bornstein, N. Croittoru, and E. Marom, *Proc. SPIE,* 320, 102 (1982).
51. R. Olshansky, *Rev. Mod. Phys.,* 51, 341 (1979).
52. S. Shibata, M. Horiguchi, K. Jinguji, S. Mitachi, T. Kanamori, and T. Manabe, *Electron. Lett.,* 17, 775 (1981).
53. L. G. Van Uitert and S. H. Wemple, *Appl. Phys. Lett.,* 33, 57 (1978).
54. J. Nishii, S. Morimoto, R. Yokota, and T. Yamagishi, *J. Non-Cryst. Solids,* 95–96, 541 (1987).
55. R. D. Driver, G. M. Leskwitz, L. E. Curtis, D. E. Moynihan, and L. B. Vacha, *Proc. Mater. Res. Soc. Symp.,* 172, 169 (1990).
56. A. Bornstein, N. Croitoru, and E. Marom, *Proc. SPIE,* 484, 99 (1984).

57. N. S. Kapany and R. J. Simms, *Infrared Phys.,* 5, 69 (1965).
58. T. Miyashita and Y. Terunuma, *J. Appl. Phys.,* 21, L759 (1982).
59. D. L. Wood and J. Tauc, *Phys. Rev.,* 5, 3144 (1972).
60. M. Saito and M. Takizawa, *J. Appl. Phys.,* 59, 1450 (1986).
61. A. V. Vasilev, G. G. Devyatykh, E. M. Dianov, A. N. Guryanov, A. Yu. Lapter, V. G. Plotnichenko, Yu. N. Pyrkov, G. E. Snopatin, I. V. Skripachev, M. F. Churbanov, and V. A. Shipunov, *Quantum. Electron.,* 23, 89 (1993).
62. J. P. Parant, C. Sergent, D. Guignot, and C. Brehm, *Glass Technol.,* 24, 161 (1983).
63. T. Katsuyama and H. Matsumura, *Appl. Phys. Lett.,* 49, 22 (1986).
64. J. S. Sanghera, V. Q. Nguyen, P. C. Pureza, F. H. Hung, R. Miklos, and I. D. Aggarwal, *J. Lightwave Technol.,* 12, 737 (1994).
65. J. Nishii, I. Inagawa, S. Morimoto, R. Iizuka, T. Yamashita, and T. Yamagishi, *Proc. SPIE,* 1228, 224 (1990).
66. N. J. Pitt, G. S. Sapsford, T. Y. Clapp, R. Worthington, and M. G. Scott, *Proc. SPIE,* 618, 124 (1986).
67. A. R. Hilton, Sr., A. R. Hilton, Jr., and J. McCord, *SPIE,* 1048, 85 (1989).
68. D. J. McEnroe, M. J. Finney, P. H. Prideaux, and P. C. Schultz, *Proc. SPIE,* 799, 39 (1987).
69. D. J. McEnroe, M. J. Finney, P. H. Prideaux, and P. C. Schultz, *Proc. SPIE,* 843, 62 (1987).
70. J. R. Williams, S. F. Carter, P. W. France, and M. W. Moore, *Proc. 6th Int. Symp. on Halide Glasses,* Clausthal, FRG, October 1–5 (1989).
71. M. G. Drexhage, *Proc. SPIE,* 128, 2 (1990).
72. T. Ueda, K. Yamada, and T. Sugita, *J. Eng. Ind.,* 114, 317 (1992).
73. M. Saito, M. Takizawa, S. Sakuragi, and F. Tanei, *Appl. Opt.,* 24(5), 2304 (1985).
74. J. Nishii, T. Yamashita, T. Yamagishi, C. Tanaka, and H. Sone, *Appl. Phys. Lett.,* 59, 2639 (1991).
75. J. Heo, M. Rodorigues, S. J. Saggese, and G. H. Sigel, Jr., *Appl. Opt.,* 30, 3944 (1991).
76. M. Maeda, in *Chemical Sensor Technology,* Vol. 3, N. Yamazoe, Ed., Kodansha Ltd., Tokyo, 1991, 185.
77. T. Arai, M. Kikuchi, S. Sakuragi, M. Saito, and M. Takizawa, *Proc. SPIE,* 576, 24 (1985).
78. T. Hattori, S. Sato, T. Fujioka, S. Takahashi, and T. Kanamori, *Electron. Lett.,* 20, 811 (1984).
79. J. Lucas, X. H. Zhang, H. L. Ma, and G. Fonteneau, *Proc. SPIE,* 1228, 109 (1990).
80. I. Inagawa, T. Yamagishi, and T. Yamashita, *Jpn. J. Appl. Phys.,* 30, 2846 (1991).
81. I. D. Turyanitsa and B. M. Koperles, *Inorg. Mater. Transl.,* 9, 38 (1973).
82. I. D. Turyanitsa, I. M. Migolinets, B. M. Koperles, and I. F. Kopinets, *Inorg. Mater. Transl.,* 10, 1234 (1974).
83. D. Herrmann, *Z. Anorg. Allg. Chem.,* 438, 75 (1978).
84. D. Hermann, *Z. Anorg. Allg. Chem.,* 438, 83 (1978).
85. S. F. Marenkin, B. Khuseinov, R. A. Karieva, P. Sherov, M. Ulugkhodzhaeva, and Sh. Mavlonov, *Izv. Akad. Nauk SSSR, Neorg. Mater.,* 18, 686 (1982).
86. X. H. Zhang, G. Fonteneau, and J. Lucas, *J. Non-Cryst. Solids,* 104, 38 (1988).
87. J. Heo and J. D. Mackenzie, *J. Non-Cryst. Solids,* 111, 29 (1989).
88. I. Chiaruttini, G. Fonteneau, X. H. Zhang, and J. Lucas, *J. Non-Cryst. Solids,* 111, 77 (1989).
89. P. C. Becker, M. M. Broer, V. G. Lambrecht, A. J. Bruce, and G. Nykolak, Post deadline paper PDP5, in *Tech. Dig. of Topical Meeting Optical Amplifiers and Their Applications,* OSA, Washington, D.C., 1992, 20–23.

90. D. W. Hewak, R. S. Deol, J. Wang, G. Wylangowski, J. A. Mederios Neto, B. N. Samson, R. I. Laming, W. S. Brocklesby, D. N. Payne, A. Jha, M. Poulain, S. Otero, S. Surinach, and M. D. Baro, *Electron. Lett.,* 29 237 (1993).

91. A. J. Bruce, V. G. Lambrecht, L. R. Copeland, W. A. Reed, G. Nykolak, and D. Tran, *Ex. Abst. of 1994 Fall Meeting of the Glass and Optical Materials Division,* Nov. 9–11, Columbus, OH, 1994, 89–90.

92. A. M. Loireau-Lozac'h and M. G. Jean Flahau, *Mater. Res. Bull.,* 11, 1489 (1976).

93. R. Reisfeld, A. Bornstein, J. Flahaut, M. Guittard, and A. M. Loireau-Lozac'h, *Chem. Phys. Lett.,* 47, 408 (1977).

94. R. Reisfeld and A. Bornstein, *J. Non-Cryst. Solids,* 27, 143 (1978).

95. P. S. Bornier, M. Guittard, and J. Flahaut, *Mater. Res. Bull.,* 14, 973 (1979).

96. P. S. Bornier, M. Guittard, and J. Flahaut, *Mater. Res. Bull.,* 15, 689 (1980).

97. K. Tanaka and A. Odajima, *Appl. Phys. Lett.,* 38, 481 (1981).

98. K. O. Hill, Y. Fujii, D. C. Johnson, and B. S. Kawasaki, *Appl. Phys. Lett.,* 32, 647 (1978).

99. K. Shiramine, H. Hisakuni, and K. Tanaka, *Appl. Phys. Lett.,* 64, 1771 (1994).

100. H. Nasu, K. Kubodera, M. Kobayashi, M. Nakamura, and K. Kamiya, *J. Am. Ceram. Soc.,* 73, 1794 (1990).

101. M. Asobe, T. Kanamori, and K. Kubodera, *IEEE Photon. Tech. Lett.,* 4(4), 362 (1992).

5 Single-Crystal Fibers

Robert S. F. Chang and Nicholas Djeu

CONTENTS

5.1 INTRODUCTION

Single-crystal fibers have been the subject of many studies because of the potential of making fiber-optic devices, both active and passive, based on the unique physical properties found in crystalline materials. The simplest single-crystal fiber device is a passive one in which the fiber acts as a light guide. The subject of this chapter is the development of single-crystal fibers for infrared (IR) transmission. Because of their inherent slow growth rate, crystalline fibers cannot compete with silica glass fibers for the same applications. Instead, they are regarded as specialty fibers for applications that do not require long length but do require fiber performance that noncrystalline fibers cannot provide.

This chapter focuses on the research and development of single-crystal, solid-core fibers for IR transmission in the last 20 years. It describes the various growth techniques used to fabricate single-crystal fibers, as well as their limitations. Some techniques are more adaptable for growing different materials than others considering that crystal melting point temperatures can vary from a few hundred to over 2000°C. Fiber materials that have been studied are mostly well-known IR optical materials. However, in recent years the only crystal fiber material still studied for IR transmission is sapphire. The excellent physical properties of this optical material are the reason that there is a sustained interest in sapphire fibers. These fibers are currently the most-developed single-crystal fibers. Therefore, a significant portion of this chapter is devoted to the growth and characterization of sapphire fibers.

5.2   MOTIVATIONS

Unlike glass fibers, which are pulled at a high speed from a heated preform, single-crystal fibers have to be grown at a much slower rate from a melt. Long-distance transmission using crystal fibers is therefore not practical. Instead, the early development of crystalline IR transmitting fibers was driven primarily by the interest in fibers with good transmission at the 10.6 μm wavelength of the CO_2 laser. Such fibers would deliver laser power to target for surgery, machining, welding, and heat treatment. Excellent IR optical materials, such as the halides of alkali metals, silver, and thallium, were considered as promising candidates for fiber development.

More recently, solid-state lasers with output wavelengths near 3 μm have emerged as excellent medical lasers because of the strong water absorption at these wavelengths in human tissues. Currently, silica-based fibers do not have good transmission at 3 μm. Although there are efforts to improve the IR performance of these fibers, it is not clear whether or not the absorption edge can be extended much beyond 2 μm. Fluoride glass fibers, on the other hand, have excellent transmission in the 2 to 3 μm region, but their chemical stability in a wet environment is a problem. Therefore, single-crystal fibers that are free of the above constraints and that can handle high laser fluence are sought for the new medical lasers, and sapphire fiber is a prime candidate for fiber delivery of Er:YAG laser operating at 2.94 μm.[1]

Besides applications in optical fiber beam delivery systems, single-crystal fibers also find potential usage in fiber-based sensors. In applications where sensors must operate in harsh environments, the optical property of fiber material is not the only consideration. High melting temperature, chemical inertness, and mechanical strength often dictate the choice of fiber materials. Sapphire is one example of a single-crystal material that possesses an unusual combination of these properties. Applications involving high-temperature fiber-optic-based sensors will be discussed later in this chapter.

5.3   MATERIALS

About 80 crystals are listed as IR optical materials.[2] Many of these materials have similar IR transmission characteristics, but may possess very different physical properties. Materials selected for IR transmitting fiber growth are generally the same materials that have proved their performance in low-loss IR windows. Table 5.1 lists

TABLE 5.1
Optical Properties of Crystalline Materials
for IR Transmitting Fibers

Crystal	Structure	Transmission Range[a] (μm)	Refractive Index[b]
Si	Cubic	1.2–7.0	3.426 (5 μm)
Ge	Cubic	1.8–23	4.00
Al_2O_3	Trigonal	0.15–6.5	1.70 (3.3 μm)
BaF_2	Cubic	0.14–15	1.450 (5 μm)
CaF_2	Cubic	0.13–10	1.400 (5 μm)
CsBr	Cubic	0.21–50	1.662
CsI	Cubic	0.25–60	1.739
CuCl	Cubic	0.19–30	1.88
MgF_2	Tetragonal	0.11–9	1.337 (5 μm)
KCl	Cubic	0.2–24	1.457
AgBr	Cubic	0.45–35	2.00 (5 μm)
KRS-5	Cubic	0.6–40	2.36
TlCl	Cubic	0.4–30	2.193
ZrO_2–Y_2O_3	Cubic	0.35–7	2.009 (5 μm)

[a] Taken from References 2 and 3, $\geq$10% external transmission for a 1-mm sample thickness.
[b] Taken from Reference 2 and *Optovac Handbook 82*. Unless specified otherwise, the value of refractive index at 10 μm is given.

the optical properties of these materials. While low transmission loss is usually a prime consideration, mechanical strength of the material in fiber form cannot be overlooked. Fibers that tend to be brittle will be difficult to bend and, therefore, lose much of their attractiveness as optical fibers. The physical properties of IR transmitting crystalline materials are given in Table 5.2. In the following sections, we describe the different approaches to single-crystal fiber growth that in many cases are material specific.

5.4 FIBER FABRICATION TECHNIQUES

The principal techniques that have demonstrated fabrication of single-crystal fibers with usable lengths for IR transmission are the melt growth methods. The latter include (1) capillary-fed growth, pressurized or drawing, (2) pulling through a shaper, (3) edge-defined growth, and (4) zone melting. The source of heat can vary from conventional resistive and RF induction heating to laser melting. While most methods require the use of crucible and die, zone melting does not. The choice will greatly depend on the material, and the following gives a general description of the special features of each melt growth method.

TABLE 5.2
Physical Properties of Crystalline Materials for IR Transmitting Fibers

Crystal	Density (g/cm^3)	Solubility ($g/100$ g H_2O)	Young's Modulus (GPa)	Hardness (Knoop)[a]	Thermal Expansion Coef. (10^{-6}/K at T K)	Thermal Conductivity (W/m·K at T K)
Si	2.33	Insoluble	130.91	1100	2.6(293)	163(313)
Ge	5.35	Insoluble	102.66	800	6.1(298)	59(293)
Al_2O_3	3.98	Insoluble	344.5	1370	5.6(293)∥ 5.0(293)⊥	35.1(300)∥ 33.0(300)⊥
BaF_2	4.89	0.17	53.05	82(500)	19.9(300)	11.7(286)
CaF_2	3.18	0.0017	75.79	160–178(500)	18.9(300)	10(273)
CsBr	4.44	124.3	15.85	19.5(200)	47(273)	0.94(273)
CsI	4.51	44	5.30	—	48.3(293)	1.1(298)
CuCl	4.14	0.0061	—	2–2.5(Mohs)	10(313)	—
MgF_2	3.18	Insoluble	138.5	415	14(310)∥ 8.9(310)⊥	21[b](300)
KCl	1.984	34.7	29.63	7.2(200)	36.6(300)	6.7(315)
AgBr	6.473	Insoluble	31.97	7	33.8(300)	1.21(273)
KRS-5	7.371	0.05	15.85	40.2(200)	58(293)	0.544(293)
TlCl	7.004	0.32	31.69	12.8(500)	53(293)	0.75(311)
ZrO_2–Y_2O_3	5.64	Insoluble	—	7.5–8.5(Mohs)	8.8(293)	10.5(260)

[a] Load in g given in parenthesis.
[b] Crystal orientation not specified.

Data taken from Browder, J. S. et al., in *Handbook of Infrared Optical Materials*, P. Klocek, Ed., Marcel Dekker, New York, 1991.

5.4.1 Pressurized Pulling Method

Bridges et al.[4] used nitrogen gas to pressurize one arm of a fused-quartz U-tube containing a liquid charge of AgBr kept molten by an oven. The other end of the tube was capped by a shaped tip whose temperature was independently controlled. The gas pressure determined the rate of feed of liquid to the tip. The melt–solid interface was positioned close to the exit of the tip.

5.4.2 Capillary-Fed Growth Method

Pastor[5] reported the growth of single-crystal $NaNO_2$ fiber from the melt using a capillary-fed growth method. Capillary action through a tapered capillary tube filled the die with melt and a thin-walled platinum capillary tubing was used as a nucleating tip that dipped into the melt-filled die orifice. Control of the melt–solid interface position in the die was accomplished by adjusting the opening of the shutter above the crucible and die assembly.

5.4.3 Modified Pulling Method

Mimura and his co-workers[6-10] have grown several alkali halide single-crystal fibers using this technique. Continuous fiber growth is accomplished by "pulling" the fiber from the melt through a shaper nozzle. The shaper is attached to the bottom of the crucible as shown in Figure 5.1. The Pt–Rh crucible and nozzle are heated by two separate RF induction units. A Pt–Rh needle placed in the nozzle is used to control the amount of molten liquid falling through the nozzle. A Pt–Rh wire seed is then inserted into the nozzle to contact a hanging drop of liquid to initiate the growth. The molten zone was found to be stable over a wide range of ratio of the radius of the nozzle and of the crystal fiber, but fiber quality was best when this ratio was slightly greater than unity.

Halide fibers that have been successfully grown include KCl, KBr, KBr–KCl, KRS-5,[6] CsBr,[7,8] and CsI.[10] The potassium halide fibers grown in the preferred $\langle 100 \rangle$ axis were brittle and tended to fracture along the (100) plane.[8]

5.4.4 Zone Melting Method

The starting materials for this method are polycrystalline (pc) fibers usually formed by extrusion through a heated die. The pc fiber is converted to a single-crystalline (sc) fiber by moving the pc fiber through a region heated by a heater coil, in which a small stationary molten zone is established. Fiber diameter is preserved, but slight diameter reductions may be possible by changing the speeds of the drive wheels that move the pc and sc fibers. Harrington et al.[11] have fabricated single-crystal fibers of TlBr/TlI(KRS-5), TlBr, CuCl, AgCl, and AgBr using this method. In general, materials suitable for extrusion are those with high ductility and tensile strength. Single-crystal silicon fibers have also been grown from pc silicon using float zoning.[12]

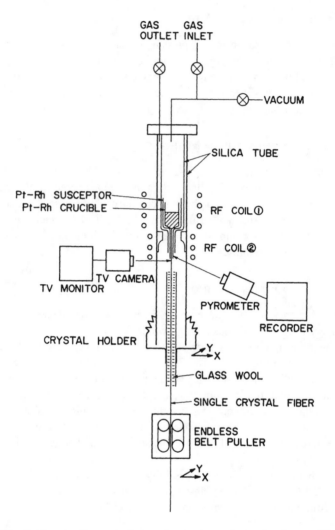

FIGURE 5.1 Schematic diagram of crystal growth apparatus for the modified pulling method. (From Mimura, Y. et al., *J. Appl. Phys.,* 53, 5491, 1982. Reprinted with permission from the American Institute of Physics.)

5.4.5 Edge-Defined Film-Fed Growth (EFG) Method

This method, developed by LaBelle and his co-workers,[13-16] relies on capillary action to create a pool of liquid on the top surface of a die from which the crystal is grown. If the angle between the top surface of the die and the vertical edge of the capillary hole exceeds a certain angle, the liquid will spread to the edge of the die and the shape of the crystal withdrawn is defined by the edge of the die rather than by the hole. By properly designing the die, any cross-sectional shape can be grown. The EFG method has been so far used exclusively to study the growth of sapphire fibers. Molybdenum is the die material of choice that is wetted by the liquid and yet does

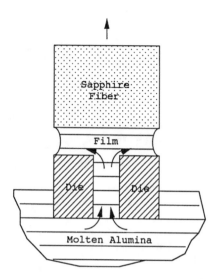

FIGURE 5.2 A planar die for EFG of sapphire fibers. (Adapted from LaBelle, H. E., Jr. and Mlavsky, A. I., *Mater. Res. Bull.*, 6, 571, 1971.)

not contaminate the melt. Figure 5.2 shows an EFG die with a 90° angle between the flat top of the die and the edge of the hole. Up to 30 fibers can be grown from a single die, and that greatly reduces the production cost of single-crystal sapphire fibers.

Saphikon, Inc., of Milford, NH is currently using the EFG method to produce kilometers of structural-grade sapphire fibers for use in aerospace lightweight, high-temperature composites. These fibers are not suitable for optical applications because of their high scattering loss. Recently, Saphikon has made some effort toward producing optical-quality sapphire fibers.[17]

5.4.6 LASER-HEATED PEDESTAL GROWTH (LHPG) METHOD

This method is well suited for growing crystalline fibers in that the laser provides tight focusing for directed and instantaneous heating. The CO_2 laser is most often used because its 10.6 μm radiation is readily absorbed by oxide and fluoride materials. The starting material can be either crystalline, polycrystalline, or ceramic. For fabrication of IR transmitting fibers, crystalline feeds are easily available. As illustrated in Figure 5.3, the laser is focused on the tip of the feed rod to create a melt and an oriented seed crystal is lowered into the melt to initiate the growth. The molten zone is held together by surface tension. The pedestal growth is similar to the float zone method except the fiber diameter is smaller than the feed diameter. The diameter reduction ratio is determined by the speeds of the feed and fiber transport mechanisms. For stable growth, this ratio will depend on the material but is generally ≤3.

The technique itself has gone through several stages of improvement since Haggerty[18] first used LHPG to produce fibers. Fejer et al.[19] implemented a mirror

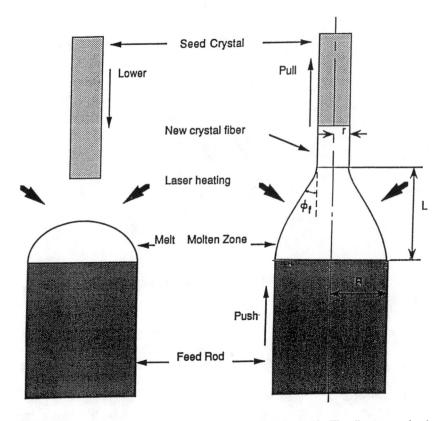

FIGURE 5.3 Schematic diagram of laser-heated pedestal growth. The diameter reduction ratio is given by R/r, and the molten zone is defined by the zone length L and the meniscus angle Φ_f.

system that enabled axially-symmetric heating of the molten zone, thereby eliminating the need for multiple beams and rotation of the feed. The optical arrangement of their design is shown in Figure 5.4. The reflaxicon changes the profile of the laser

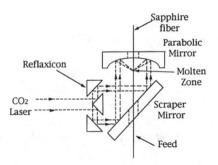

FIGURE 5.4 Cross-sectional diagram of laser beam-focusing optics used for LHPG. (Adapted from Fejer, M. M. et al., *Rev. Sci. Instrum.*, 55, 1791, 1984.)

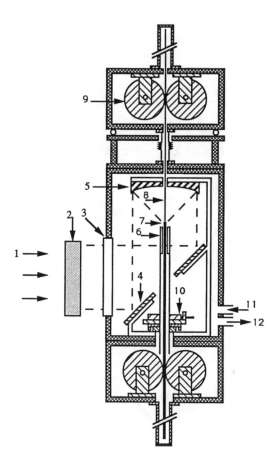

FIGURE 5.5 Schematic of the LHPG apparatus at USF. 1: incident CO_2 laser beam; 2: Gaussian reflector; 3: ZnSe window; 4: turning mirror; 5: paraboloidal focusing mirror; 6: capillary guide; 7: feed rod; 8: grown fiber; 9: motor-driven roller pair; 10: translational stage assembly; 11: gas inlet; 12: gas outlet. (From Phomsakha, V. et al., *Rev. Sci. Instrum.*, 65, 3860, 1994. Reprinted with permission from the American Institute of Physics.)

beam from Gaussian to one with a hole in the center. This beam is redirected to the top parabolic mirror for tight focusing onto the molten zone. For fiber diameter uniformity, a high-speed fiber diameter measurement system was developed to provide feedback to the drive mechanisms that moved the feed and fiber.[20] More recently, simultaneous growth of five sapphire fibers was demonstrated by scanning the laser beam across the molten zones and modulating the beam intensity to deliver a linear laser energy distribution to the zones.[21] This supposedly reduced overheating and stabilized the shape of the molten zone. Optical characterization of these fibers has not been reported.

The latest modification to LHPG was implemented by the group at the University of South Florida (USF).[22,23] In their fiber-drawing apparatus, shown in Figure 5.5, the expanded and collimated CO_2 laser beam is incident on a mirror with a graded reflectivity profile. The transmitted beam is transformed into a top-hat-shaped beam. This provides a softer focusing of the laser radiation on the shoulder of the molten zone. Unlike the reflaxicon approach, which places a ring-shaped pattern far from the center of the parabolic mirror, the beam in this approach fills the mirror almost uniformly. As a result, part of the laser radiation impinges on a short length of the fiber above the freezing interface. The roller transport systems are far removed from the molten zone. The feed is guided through a precision capillary positioned just below the molten zone. The flexible seed needs only to contact the melt, and surface tension automatically centers the fiber with respect to the zone. Thus, the fiber being grown is much less susceptible to vibrations caused by the roller motors. Sapphire fibers grown under a helium atmosphere in this apparatus can be produced with a growth rate of 2 cm/min without sacrificing optical quality. This is almost seven times faster than the previous maximum LHPG growth speed of 3 mm/min.

5.5   FIBER CHARACTERISTICS

There have been a number of attempts (listed in Table 5.3 in chronological order) to develop single-crystal fibers for IR transmission. The early attempts were well documented in review articles of crystalline fibers,[24-27] but, with the exception of sapphire fibers and perhaps CsBr, none of the other crystal fibers has been well characterized. Efforts seemed to end with the demonstration of feasibility of fiber growth for these fibers. In contrast, interest in sapphire fibers has not waned over the years, and recent advances in sapphire fiber technology have significantly moved these fibers toward the commercial market. The following subsections will describe the characteristics of sapphire fibers grown in different laboratories.

5.5.1   TRANSMISSION

Fiber loss measurements in the IR spectral region have been made with either a Fourier-transform IR (FTIR) spectrometer or the usual broadband lamp source/monochromator combination. The cutback method used so frequently with silica glass fibers to determine fiber loss is less appropriate for the much shorter single-crystal fibers. The method also assumes uniformly distributed losses, which may not be the case for crystal fibers. Instead, the optical power incident on the fiber entrance face and that transmitted by the fiber were measured to determine the external transmittance. Fiber propagation loss was then determined after correcting for Fresnel loss at both ends of the fiber. (Coupling loss can generally be made negligible.) Table 5.4 shows the measured loss at the Er:YAG laser wavelength of 2.94 μm for sapphire fibers reported by various groups.

Figure 5.6 shows the attenuation of a sapphire fiber grown at USF between 700 nm and 3.7 μm.[23] The 100-μm fiber was grown in dry air at 3 mm/min, and was annealed in a furnace at 1200°C for 24 h. The loss shown in Figure 5.6 represents what can be routinely achieved. The best fibers from USF have measured losses as low as 0.1 to 0.2 dB/m in the near IR. For fiber attenuation loss measurement, a tungsten lamp after being filtered by a monochromator was used as the source, and a pinhole slightly smaller than the fiber diameter served to define the input beam

TABLE 5.3
Summary of Past Works on SC Fibers (Excluding Sapphire Fibers)

Material	Melting Point (°C)	Refractive Index[a]	Diameter (mm)	Length (m)	Method of Growth	Pull Rate (mm/min)	Attenuation	Ref.
Ge	937	4.00	—	—	Pulling	—	—	28
KCl	776	1.457	0.075–0.100 square	0.10	Vapor-solution method	—	0.5 cm^{-1} at 10.6 µm	29
AgBr	432	2.00 (5 µm)	0.35–0.75	2	Pressurized capillary-fed growth	20	4 W cw CO_2 laser limited by laser	4
KRS-5	440 480	2.36	0.6–1.0	2	Modified floating zone	0.5–3	—	6
CsI	621	1.739	0.7	1.5	Modified pulling	5–6	33 dB/m at 10.6 µm 13 dB/m at 663 nm	10
CsBr	636	1.662	0.7–2	1.5	Modified pulling	5–10	5 dB/m at 10.6 µm 8 dB/m at 633 nm	7
TlCl	430	2.193	0.75	>1	Capillary-fed	10	3 dB/m at 10.6 µm	30
CsBr	636	1.662	0.7–1.2	1–3	Modified pulling	6–10	2–5 dB/m at 10.6 µm	8
CsBr	636	1.662	1–2	—	Modified pulling	4–5	0.3 dB/m at 10.6 µm 47W cw CO_2 laser	9
NaNO$_2$	271	—	0.5	0.6	Tapered capillary-fed Czochralski	0.25	—	5
AgBr	432	2.00 (5 µm)	0.62	0.8	Traveling zone melting	10	6.6 dB/m at 10.6 µm	11
CaF$_2$	1360	1.400 (5 µm)	0.6	—	LHPG	1	Oxygen contamination	31
BaF$_2$	1280	1.450 (5 µm)	0.2–0.6	—	LHPG	1	Oxygen contamination	31
Si	1410	3.42	1–2	0.3	Float zoning	2–5	—	12
ZrO$_2$–Y$_2$O$_3$	2690	2.01 (5 µm)	0.3–0.5	10–25	LHPG	1–20	10 dB/m at 850 nm	32

[a] Values given for 10.6 µm, unless specified otherwise.

TABLE 5.4
Sapphire Fiber Loss at Er:YAG Laser Wavelength (2.94 μm)

Source	Method	Diameter (μm)	Length (cm)	Loss (dB/m)	Ref.
Rutgers	LHPG	340	15	1.6	33
		—	—	0.7	35
Saphikon	EFG	300	100	2.6–9.9	17
Stanford	LHPG	110	30	1.7	34
		110	30	0.9[a]	34
USF	LHPG	100	88	0.7	23

[a] Fiber grown in pure oxygen.

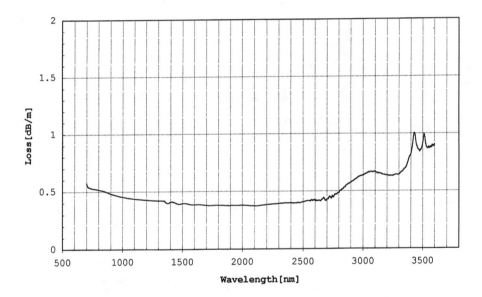

FIGURE 5.6 IR loss of a 88-cm-long LHPG sapphire fiber with 100 μm diameter. This level of fiber performance can be routinely achieved with fibers from USF. (From Chang, R. S. F. et al., *SPIE,* 2396, 48, 1995. Reprinted with permission from SPIE.)

launched into the fiber. The signals with and without the fiber gave the relative wavelength-dependent transmission. The curve was then put on an absolute scale with laser measurements at 840 nm and 2.94 μm serving as anchors. The attenuation constant was inferred after correction for Fresnel losses at the ends.

Attenuation of this sapphire fiber is below 0.5 dB/m from 900 to 2800 nm. At the Er:YAG laser wavelength, the loss is about 0.7 dB/m. This level of attenuation is typical for LHPG sapphire fibers (see Table 5.4), but not for EFG sapphire fibers. Their loss at this wavelength is typically 3 to 4 dB/m. The predicted loss at 2940 nm based on a multiphonon absorption model developed by Thomas et al.[36] is slightly higher than 0.1 dB/m.

Absorption peaks near 3400 and 3500 nm in Figure 5.6 are likely due to CH and they have been observed in fibers grown elsewhere.[33,34] Absent in this plot is the OH absorption. The latter is dominated by an intense peak at 3095 nm followed by weaker peaks at 3139 and 3053 nm. The loss at 3095 nm for sapphire fibers grown in room air can be as high as 3 to 4 dB/m. It is not clear whether or not the broad maximum centered at 3100 nm in Figure 5.6 is still OH related.

The transmission of sapphire fibers grown at high speeds depends strongly on the growth atmosphere used. It is well known that LHPG sapphire fibers grown faster than 3 to 4 mm/min in air, nitrogen, or argon have increasingly poorer transmission. The increase in scattering loss is believed to be due to microvoids formed by the entrapment of the ambient gas in the molten zone. Helium as a growth atmosphere for sapphire fibers was recently studied at USF,[22] and it was found that high-optical-quality fibers could be grown at speeds as high as 20 mm/min. The difference is attributed in part to the fact that the entrapped helium can diffuse out to the surface of the molten zone faster than other gases. The study further showed a dependence of fiber quality on the pressure of helium used. The attenuation measured at four laser wavelengths for sapphire fibers grown at different helium pressures is shown in Figure 5.7. The optimum helium pressure for a growth rate

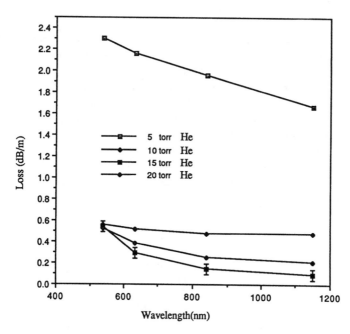

FIGURE 5.7 Attenuation at four laser wavelengths for fibers grown at different He pressures. All fibers were 1 m long, 100 μm diameter, and grown at 20 mm/min. The data for 15 torr He are averaged values for three fibers. All other data are for single fibers. (From Phomsakha, V. et al., *Rev. Sci. Instrum.*, 65, 3860, 1994. Reprinted with permission from the American Institute of Physics.)

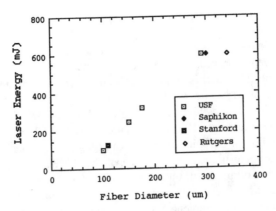

FIGURE 5.8 Er:YAG laser energy transmitted without damage as a function of sapphire fiber diameter. USF;[23] Saphikon;[17] Stanford;[34] Rutgers.[33] (From Chang, R. S. F. et al., *SPIE*, 2396, 48, 1995. Reprinted with permission from SPIE.)

of 20 mm/min appears to be near 15 torr. At higher pressures, gas entrapment becomes a problem, while at lower pressures nonstoichiometric material loss through evaporation is suspected to lead to the degradation.

5.5.2 LASER DAMAGE THRESHOLD

In a laser damage study of window materials, sapphire was found to have the highest bulk damage threshold at 2.7 μm (HF laser wavelength) of 100 GW/cm^2.[37] Like other optical materials, sapphire is limited by laser damage at defect sites on the surface. The surface damage threshold for sapphire is 3.8 GW/cm^2, much lower than the bulk value. Since the damage is defect related, careful surface preparation can be expected to increase the damage threshold of this hard material.

In fiber delivery of laser power, an extremely high fluence level is also encountered. Laser damage usually occurs at the entrance face of the fiber, and the primary concern is the maximum laser power or energy that the fiber can safely handle. Laser damage thresholds of sapphire fibers have been studied by several groups using the pulsed Er:YAG laser.[17,23,33,34] Both EFG and LHPG sapphire fibers have been tested. In the USF study the damage thresholds for fibers of four different diameters were compared.[23] The results are shown in Figure 5.8. The laser beam used, though not diffraction limited, had a smooth spatial profile. The focused laser spot was the same for all the fibers, and was comparable to the diameter of the largest fiber. Thus, while launching efficiency was quite good for the largest fiber, this was not the case for the smaller ones. The damage thresholds for the smaller fibers would be reduced under optimal launching conditions. For the 100-μm fiber, the 100-mJ threshold translates to a damage fluence of about 1.3 kJ/cm^2. This is to be compared with the surface damage fluence of 0.65 kJ/cm^2 for bulk sapphire at 2.7 μm.[37] In making this comparison, however, one must keep in mind that the latter is for an HF laser pulse of 175 ns while the fiber data were taken using an Er:YAG laser with pulses lasting for approximately 100 μs.

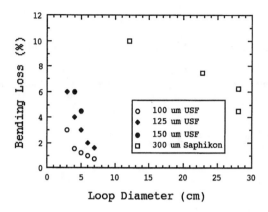

FIGURE 5.9 Bending loss of sapphire fibers as a function of loop diameter. The USF data were measured at 840 nm and the Saphikon[17] data were measured at 2.94 µm. (From Chang, R. S. F. et al., *SPIE,* 2396, 48, 1995. Reprinted with permission from SPIE.)

For fiber delivery of high laser energies, it is desirable to have both the higher damage thresholds found in larger diameter fibers and the flexibility of the thinner fibers. This can be accomplished in a tapered fiber with larger ends, since the subsurface damage threshold is much higher. A 1-m long, 100-µm-diameter sapphire fiber with 150 µm diameter at both ends has been successfully grown at USF.[23] The computer-controlled tapering at each end occurred over a length of 2 cm. Larger tapering may be possible, but the CO_2 laser power delivered to the molten zone will need to be computer controlled as well.

5.5.3 BENDING LOSS

Sapphire fibers are stiffer than glass fibers, but they are highly flexible when the fiber diameter is below 200 µm. A c-axis 150-µm fiber can be bent to form a loop of 0.8 cm in diameter without breaking.[38] The effect of bending on the fiber transmission will depend on the loop diameter, the fiber diameter, and the launching conditions of the input beam.[37] Figure 5.9 shows the bending loss measured at 840 nm as a function of loop diameter for sapphire fibers with diameters of 100, 125, and 150 µm grown at USF.[23] It is seen that bending will have a negligible effect on the throughput of these fibers if the loop diameter is 10 cm or more. For a given fiber diameter, the bending loss increases slowly at first with reduction in loop diameter and then sharply. The loop diameter at which this steep increase in bending loss occurs is different depending on the fiber diameter, 4 cm for the 100-µm fiber and 6 to 7 cm for the 125-µm fiber. For the same loop diameter, the thicker fiber has higher bending loss. These variations in propagation loss caused by bending have also been reported for large-core multimode silica fibers.[39] Also shown in Figure 5.9 are the bending loss measurements for the 300-µm EFG fibers measured at 2.94 µm.[17]

TABLE 5.5
Mechanical Characterization of Single-Crystal Sapphire Fibers

Fiber Source	Diameter (μm)	T (°C)	Strength (MPa)	Strain at Failure (%)	Strain Rate (min⁻¹)	Gauge Length (mm)	Ref.
Rutgers (LHPG)	38	20	—	0.43[a]	—	—	33
	550	–196	—	1.01[a]	—	—	33
Saphikon (EFG)	180	20	—	0.39[a]	—	—	33
	280	20	—	0.39[a]	—	—	33
	280	–196	—	0.75[a]	—	—	33
Saphikon (EFG)	160	20	2500	—	0.048	208	42
	160	800	1800	—	0.048	208	42
	160	1000	1500	—	0.048	208	42
	160	1200	1000	—	0.048	208	42
	160	1400	650	—	0.048	208	42
	160	1500	600	—	0.048	208	42
NASA Lewis	100	25	5385	—	—	—	21
(LHPG)	100	1000	1517	—	—	—	21
	100	1200	1043	—	—	—	21
Saphikon (EFG)	100	25	2434	—	—	—	21
	100	1000	724	—	—	—	21
	100	1200	606	—	—	—	21
Stanford (LHPG)	100	21	2540	1.20	0.1	20	41
	100	21	3060	1.85	0.1	20	41
USF (LHPG)	100	22	2190–4131	—	0.05	25.4	43
	200	22	1404–1980	—	0.05	25.4	43
	100	1093	1205–1628	—	0.05	12.7	43
	200	1093	488–936	—	0.05	12.7	43

[a]Mean value.

5.5.4 MECHANICAL STRENGTH

Of all the single-crystal fibers grown, only sapphire (see Table 5.5) and CsBr[8] fibers have been characterized mechanically. Interest in sapphire fiber strength stems from its potential application in reinforcement of metal and ceramic-matrix composites. These fibers have demonstrated tensile strength above 2000 MPa which is much higher than 500 to 700 MPa measured in bulk sapphire materials.[40] This is presumably due to greater structural perfection found in fibers. Both LHPG and EFG sapphire fibers have been tested in tension[21,41-44] and by four-point bending.[33] The tested fibers were nominally grown with the c-axis orientation along the fiber axis. The mechanical strengths of these fibers are dependent on fiber diameter, strain rate, and temperature. In general, smaller-diameter fibers have higher strengths than larger fibers. As the strain rate is decreased, the fiber strength also decreases.[42]

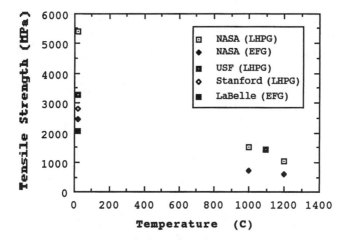

FIGURE 5.10   Tensile strength of sapphire fibers with 100 µm diameter as a function of temperature. NASA (LHPG);[21] NASA (EFG);[21] USF (LHPG);[43] Stanford (LHPG);[41] LaBelle (EFG).[44]

Table 5.5 summarizes the results of fiber strength measurements carried out by the different groups. The strain rates, the gauge lengths, and the diameters of fibers used in these studies are noted in the table. It appears that LHPG fibers exhibited superior strengths compared with the EFG fibers due to fewer defects in the higher-quality LHPG fibers. The tensile strengths of 100-µm sapphire fibers as a function of temperature are shown in Figure 5.10. There is about a factor of two difference in tensile strengths measured by different groups at room temperature for the LHPG fibers. Within each group, the fiber strength was lower at higher temperatures. Similar degradation was also observed in larger-diameter sapphire fibers, and slow crack growth in these fibers has been confirmed as the cause for degradation at elevated temperatures.[42]

5.5.5   HIGH-TEMPERATURE TRANSMISSION

The IR transmission property of sapphire as a function of temperature has been studied because of the frequent use of sapphire as windows and domes under high temperature conditions. Transmission of sapphire in the IR is limited by the onset of multiphonon absorption. For thin material at room temperature, usable transmission out to 6 µm is possible. At elevated temperatures, IR transmittance is lowered in general, but much more so at the long-wavelength limit.[45,46] Therefore, the useful IR range of sapphire at high temperatures is limited effectively to below 4 µm.

Transmission data measured using thin crystal samples may not be accurate enough to predict losses in fibers. Fibers offer much longer path lengths, and hence a more accurate measurement is possible. The high-temperature transmission measurement was carried out at a near-IR wavelength using a tube furnace which provided a uniformly heated zone of approximately 40 cm.[47] The 840-nm output

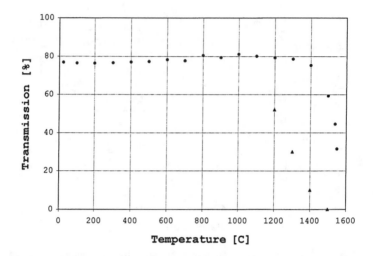

FIGURE 5.11 External transmission of a 88-cm-long LHPG sapphire fiber measured at 840 nm (solid circles) as a function of temperature.[47] The tube furnace provided a uniformly heated section of approximately 40 cm. The predicted transmission based on bulk loss measurement[46] is given by solid triangles.

from a laser diode was launched into the fiber, and the transmitted light was monitored as a function of the furnace temperature. Emission from the fiber at high temperatures was accounted for by measuring the background signal with the laser blocked off. Figure 5.11 shows the external transmittance of the sapphire fiber from room temperature to 1530°C, the highest attainable temperature in the furnace. The fiber transmission at 840 nm was essentially flat up to 1400°C. Above 1400°C the fiber was very lossy and its transmittance quickly dropped to 20% at 1530°C. This observation is significantly different from the temperature dependence reported for bulk sapphire at this wavelength.[46] By using the absorption coefficients measured in bulk sapphire samples from 1200 to 1600°C, the external transmission through a sample path length of 40 cm is calculated and compared with the measured fiber transmission in Figure 5.11. The latter is much better than predicted, and the sharp roll-off in fiber transmission did not occur until the temperature was above 1500°C. The discrepancy is too large to be explained by any uncertainty in temperature. The small sample thickness used in the bulk study might have been the difference. At longer wavelengths in the IR, the lower limit on the high-temperature transmission of sapphire fibers can be estimated using the bulk data shown in Figure 5.12.[46]

5.6   FIBER CLADDING DEVELOPMENT

All single-crystal fibers have been grown as bare fibers without cladding. Hygroscopic fibers such as the cesium halide fibers were placed in loose Teflon tubes to protect them from the environment.[7,10] These fibers were still essentially air clad.

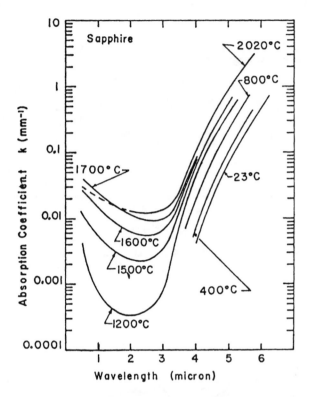

FIGURE 5.12 Absorption coefficient of bulk-grown sapphire as a function of temperature. (From Gryvnak, D. A. and Burch, D. E., *J. Opt. Soc. Am.,* 55, 628, 1965. Reprinted with permission from the Optical Society of America.)

Cladding of sapphire fibers was investigated by Harrington and his co-workers[33,35] at Rutgers University. Teflon FEP 100 and Teflon AF 1600, both with refractive index values around 1.3, have been used as cladding materials for sapphire fibers. The Teflon FEP was melt extruded onto the fiber with thickness of 100 to 300 μm, while the Teflon AF applied in a dip-coating process was only 5 μm thick. As shown in Figures 5.13 and 5.14, both claddings, though very different in thickness, were able to prevent light from leaking from the core through the cladding to the surrounding medium. Even though these polymers exhibit different IR transmission characteristics, the clad sapphire fibers showed no noticeable absorption features due to the Teflons. The IR absorption spectrum of bulk Teflon FEP with 1 mm thickness showed significant features beyond 3.7 μm.[33] Teflon AF has absorption features at wavelengths longer than 2.5 μm.[48]

For high-temperature cladding, alumina appears to be a natural choice because in PC form its thermal mechanical properties are very similar to those of sapphire. El-Sherif et al.[49] have developed a chemical deposition technique to coat sapphire fibers with a thin layer of PC alumina. Fine alumina powder (20 nm particle size)

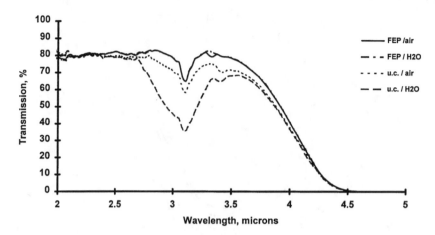

FIGURE 5.13 External transmission of an unclad (u.c.) sapphire fiber and a FEP-coated sapphire fiber placed in air and water. (From Nubling, R. et al., *SPIE,* 2131, 56, 1994. Reprinted with permission from SPIE.)

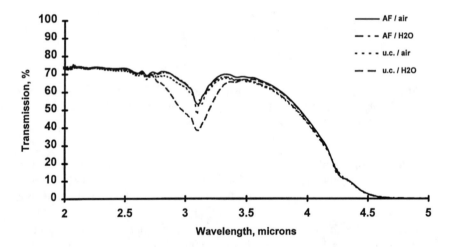

FIGURE 5.14 External transmission of a sapphire fiber placed in air and water before and after it was coated by Teflon AF. (From Nubling, R. et al., *SPIE,* 2131, 56, 1994. Reprinted with permission from SPIE.)

was added to a suspension made of a polymerizable monomer carrier. Fibers coated with this solution would polymerize at room temperature, and the binder was later removed by heating the fibers to 600°C. Further sintering the fibers at 1400°C produced the PC alumina cladding. No transmission measurement has been reported on these alumina-clad sapphire fibers. Another way to put alumina cladding on sapphire fibers is the sol–gel method.[35] After drying and firing at high temperature, the thin alumina gel film left a transparent coating of alumina on the fiber. Because of the high porosity of this film, the refractive index was only about 1.35 compared with 1.6 to 1.7 for PC alumina. Early results indicated the presence of water or OH in the sol–gel coating.

A process developed for coating ceramic fibers with precious metal or metal oxide for use in composite materials may also be suitable for cladding sapphire fibers.[50] Sapphire fibers pulled from a molten mixture of quanidine soap and a platinum precursor were sintered at a high temperature to remove the organics leaving no residues but a thin coat of platinum on the fibers. Zirconium oxide has also been deposited on sapphire fibers using a slurry of the oxide particles with the molten quanidine soap. This same technique may lend itself to coating sapphire fibers with alumina.

5.7 SUMMARY AND FUTURE TRENDS

The development of single-crystal fibers for waveguiding in the IR has thus far been limited to a few demonstrations in research laboratories. The intrinsically slow growth rates of crystalline fibers make it hard for them to compete with other types of fibers which are much more advanced in their fabrication technologies. Extruded PC silver halide fibers are now commercially available for delivering CO_2 laser power. Silica, fluoride glass, and chalcogenide glass fibers are commonly used in IR fiber-optic applications. Finding a need or niche for IR transmitting single-crystal fibers is therefore a task of matching the unique properties of crystalline materials to applications that absolutely demand them. These properties include high laser damage threshold, high melting point, high tensile strength, wide transmission range, and exceptional chemical stability. Such features are seldom found in glass or PC fibers.

It is no surprise then that the only current activity in single-crystal fiber research centers around sapphire fibers. With an extraordinary combination of physical and chemical properties, sapphire fibers offer an edge over other fibers. The recent advance[22] made in rapid drawing of these LHPG fibers without sacrificing optical quality represents a major breakthrough in sapphire fiber technology. This advance will significantly lower the cost of optical-grade sapphire fibers, which will in turn make viable the use of these fibers for specialty applications. Teflon-clad sapphire fibers will be developed for medical laser power delivery. Alumina-clad sapphire fibers, if proved satisfactory, will be the first high-temperature optical fibers for operation above 1000°C. There is a great demand for such fibers. Sapphire fiber-optic temperature sensors can provide much more reliable temperature measurements than thermocouples for industrial process control at high temperatures.[51] The fiber-optic temperature sensors will also find usage in monitoring the hot gas temperature of a turbine engine.[52] Sapphire fibers embedded in ceramic composites have been used as stress and displacement sensors in a high-temperature environment.[49] Fiber-optic imaging for *in situ* observations at elevated temperatures is possible with a coherent bundle of sapphire fibers. Even bare sapphire fibers can be used as fiber evanescent sensors to monitor composites during cure in an autoclave[53] or to determine concentration of highly absorbing materials in hostile environments. These are just some of the applications that demand the use of sapphire fibers. Many more applications for sapphire (and other high-temperature) fibers will arise as the cost of these fibers is lowered by making further advances in crystal fiber technology.

REFERENCES

1. G. N. Merberg, *Lasers Surg. Med.,* 13, 572 (1993).
2. J. S. Browder, S. S. Ballard, and P. Klocek, Physical property comparisons of infrared optical materials, in *Handbook of Infrared Optical Materials,* P. Klocek, Ed., Marcel Dekker, New York, 1991, 141.
3. *Handbook of Laser Science and Technology, Vol. IV,* M. J. Weber, Ed., CRC Press, Boca Raton, FL, 1986.
4. T. J. Bridges, J. S. Hasiak, and A. R. Strnad, *Opt. Lett.,* 5, 85 (1980).
5. A. C. Pastor, *J. Cryst. Growth,* 70, 295 (1984).
6. Y. Mimura, Y. Okamura, Y. Komazawa, and C. Ota, *Jpn. J. Appl. Phys.,* 19, L269 (1980).
7. Y. Mimura, Y. Okamura, Y. Komazawa, and C. Ota, *Jpn. J. Appl. Phys.,* 20, L17 (1981).
8. Y. Mimura, Y. Okamura, and C. Ota, *J. Appl. Phys.,* 53, 5491 (1982).
9. Y. Mimura and C. Ota, *Appl. Phys. Lett.,* 40, 773 (1982).
10. Y. Okamura, Y. Mimura, Y. Komazawa, and C. Ota, *Jpn. J. Appl. Phys.,* 19, L649 (1980).
11. J. A. Harrington, A. G. Standlee, A. C. Pastor, and L. G. DeShazer, *Proc. SPIE,* 484, 124 (1984).
12. A. R. Hilton, Sr. and J. McCord, *Proc. SPIE,* 1048, 32 (1989).
13. H. E. LaBelle, Jr. and A. I. Mlavsky, *Mater. Res. Bull.,* 6, 571 (1971).
14. H. E. LaBelle, Jr., *Mater. Res. Bull.,* 6, 681 (1971).
15. B. Chalmers, H. E. LaBelle, Jr., and A. I. Mlavsky, *Mater. Res. Bull.,* 6, 571 (1971).
16. H. E. LaBelle, Jr., *J. Cryst. Growth,* 50, 8 (1980).
17. J. J. Fitzgibbon, H. E. Bates, A. P. Pryshlak, and M. J. Philbrick, *Proc. SPIE,* 2131, Biomedical Fiber Optic Instrumentation, 50 (1994).
18. J. S. Haggerty, Production of Fibers by a Floating Zone Fiber Drawing Technique, Final Report NASA-CR-120948 (May 1972).
19. M. M. Fejer, J. L. Nightingale, G. A. Magel, and R. L. Byer, *Rev. Sci. Instrum.,* 55, 1791 (1984).
20. M. M. Fejer, G. A. Magel, and R. L. Byer, *Appl. Opt.,* 24, 2362 (1985).
21. L. Westfall, A. Sayir, and W. Penn, *NASA Tech. Briefs,* Aug., 77 (1994).
22. V. Phomsakha, R. S. F. Chang, and N. Djeu, *Rev. Sci. Instrum.,* 65, 3860 (1994).
23. R. S. F. Chang, V. Phomsakha, and N. Djeu, *Proc. SPIE,* 2396, Biomedical Opto-electronic Instruments, 48 (1995).
24. R. S. Feigelson, Growth of fiber crystals in *Crystal Growth of Electronic Materials,* E. Kaldis, Ed., Elsevier Science Publishers, North-Holland, Amsterdam, 1985, 127.
25. R. S. Feigelson, *J. Cryst. Growth,* 79, 669 (1986).
26. T. Katsuyama and H. Matsumura, *Infrared Optical Fibers,* Adam Hilger, Philadelphia, 1989, 146.
27. P. A. Tick and P. L. Bocko, Optical fiber materials, in *Optical Materials,* Vol. 1, S. Musikant, Ed., Marcel Dekker New York, 1990, 289.
28. T. Nakagawa, Inst. Electr. Eng. Jpn., 84, 1614 (1964)
29. G. Tangonan, A. C. Pastor, and R. C. Pastor, *Appl. Opt.,* 12, 1110 (1973).
30. A. V. Vasil'ev, E. M. Dianov, L. N. Dmitruk, V. G. Plotnichenko, and V. K. Sysoev, *Sov. J. Quantum Electron.,* 11, 834 (1981).
31. R. S. Feigelson, W. L. Kway, and R. K. Route, *Opt. Engr.,* 24, 1102 (1985).
32. L. Tong, Y. Wang, and Z. Ding, *Proc. SPIE,* 2292, Fiber Optic and Laser Sensors XII, 429 (1994).

33. G. N. Merberg and J. A. Harrington, *Appl. Opt.,* 32, 3201 (1993).
34. D. H. Jundt, M. M. Fejer, and R. L. Byer, *Appl. Phys. Lett.,* 55, 2170 (1989).
35. R. Nubling, R. Kozodoy, and J. A. Harrington, *Proc. SPIE,* 2131, Biomedical Fiber Optic Instrumentation, 56 (1994).
36. M. E. Thomas, R. I. Joseph, and W. J. Tropf, *Appl. Opt.,* 27, 239 (1988).
37. E. W. Van Stryland, M. Bass, M. J. Soileau, and C. C. Tang, NBS Spec. Publ. 541, 118 (1978).
38. G. A. Magel, D. H. Jundt, M. M. Fejer, and R. L. Byer, *Proc. SPIE,* 618, 89 (1986).
39. A. A. P. Boechat, D. Su, D. R. Hall, and J. D. C. Jones, *Appl. Opt.,* 30, 321 (1991).
40. J. B. Wachtman, Jr. and L. H. Maxwell, *J. Am. Ceram. Soc.,* 42, 432 (1959).
41. H. F. Wu, A. J. Perrotta, and R. S. Feigelson, *Light Metal Age,* April, 97 (1991).
42. S. A. Newcomb and R. E. Tressler, *J. Am. Ceram. Soc.,* 76, 2505 (1993).
43. J. Dobbs and R. S. F. Chang, unpublished (1994).
44. H. E. LaBelle and A. I. Mlavsky, *Nature,* 216, 574 (1967).
45. U. P. Oppenheim and U. Even, *J. Opt. Soc. Am.,* 52, 1078 (1962).
46. D. A. Gryvnak and D. E. Burch, *J. Opt. Soc. Am.,* 55, 625 (1965).
47. C. Cheong and R. S. F. Chang, unpublished (1994).
48. Dupont Publication H-16577-1.
49. M. A. El-Sherif, S. Hu, J. Rahdishnan, F. K. Ko, D. J. Roth, and B. Lerch, *Proc. SPIE,* 2072, Fiber Optic Physical Sensors in Manufacturing and Transportation, 244 (1993).
50. W. H. Philipp, L. C. Veitch, and M. H. Jaskowiak, *NASA Tech Briefs,* Jan., 45 (1994).
51. E. K. Matthews, *Proc. SPIE,* 2072, Fiber Optic Physical Sensors in Manufacturing and Transportation, 63 (1993).
52. M. J. Finney, G. W. Tregay, and P. R. Calabrese, *Proc. SPIE,* 1799, Specialty Fiber Optic Systems for Mobile Platforms and Plastic Optical Fibers, 194 (1992).
53. M. A. Druy, L. Elandjian, and W. A. Stevenson, *Proc. SPIE,* 1170, Fiber Optic Smart Structures and Skins II, 150 (1989).

6 Polycrystalline Fibers

Leonid Butvina

CONTENTS

6.1 INTRODUCTION

Kapany[1] was perhaps the first to propose the fabrication of infrared (IR) fibers from crystalline silver chloride (AgCl) in the mid 1960s. However, systematic investigations on the fabrication of polycrystalline fibers from crystalline materials were not reported until more than a decade later by Pinnow et al.[2] These authors fabricated polycrystalline fibers made from thallium halide crystals, TlBr–TlI (KRS-5), by the extrusion method. The motivation for fabricating fibers from heavy metal halides was to realize predictions of ultralow attenuation in the IR region discussed by Pinnow et al.,[2] Bendow,[3] and Van Uitert and Wemple.[4]

Some of the earliest attempts to produce polycrystalline fibers from KRS-5,[5-8] as well as from KCl,[5] NaCl,[19] CsI,[6] KBr,[9] and AgCl[10,11] resulted in optical losses which were about two to three orders of magnitude higher than those of the original crystals. Much of the work over the years has been performed to identify the origin of this excess loss and develop techniques to fabricate low-loss crystals and fibers. Silver halides and their mutual solid solutions are the most suitable compositions for the fabrication of crystalline fibers by the extrusion method. During the last decade, these fiber compositions have received the most attention, and consequently most of this chapter is dedicated to these materials.

6.2 FIBER MATERIAL AND COMPOSITION

Besides the theoretical optical losses in crystals, there are several practical require-ments that fiber crystalline materials must satisfy. First, a crystal must be deformed plastically in a typical working range of temperature with speeds higher than 1 cm/min. This is important for obtaining a long fiber length in an acceptable time using the extrusion process. Second, the crystal must be optically isotropic due to crystallographic reorientation of grains in the fiber. Therefore, it must possess a cubic crystal structure to reduce scattering in the fiber. Third, the composition of the crystal must be a solid solution. And, finally, the recrystallization process in the fiber materials must be inhibited by composition to prevent degradation of the optical and mechanical properties of the fiber. From this point of view, the suitable crystals are solid solutions of thallium halides, solid solutions of alkali halides, and solid solutions of silver halides.

6.2.1 THALLIUM HALIDES

A number of thallium halide crystals have been used in IR fiber optics, for example, TlCl, TlBr, and solid solutions such as KRS-6 (0.5 TlCl–0.5 TlBr) and KRS-5 (0.42 TlBr–0.58 TlI). The most widely used was the composition of KRS-5. These crystals possess the cesium chloride structure and are transparent in a wide wavelength range from 0.5 to 40 μm. Thallium halide crystals have good anisotropy of mechanical properties.

It is well known that plastic deformation of single crystals is realized by dislo-cation motions in some definite planes and directions. The initial system of dislo-cation glide of CsCl-type crystals is described by the following system[12]: plane {110} and direction [001]. A secondary glide system develops with increasing deformation temperature (Table 6.1). According to Mises criterion, five independent glide systems are needed for random deformation with volume conservation.[12]

These crystals are turned on a lathe and ground in bound abrasive. The damaged surface layer is removed by etching with dilute HBr acid with H_2O_2 additive.

TABLE 6.1
Glide Systems of Crystalline Metal Halides

Metal Halides	Primary Glide System	Number of Independent Glide Systems	Secondary Glide System	T_{min} for Secondary Glide System	Total Number of Independent Systems
Tl, Cs	<001>{110}	3	Not observed until 0.6 T_m	—	3
Na, K	<110>{110}	2	{001}<110>	0.5 T_m	5
Ag	<110>{110}	2	{001}<110>	0.2 T_m	5

Although these crystals are only slightly soluble in water, they are extremely toxic. Other crystal properties are cited in Table 6.2.

Fibers made from KRS-5 crystals have one more drawback. Their polycrystalline structure changes within a year in such a way that partial separation of the intergrain boundaries takes place. This process causes significant degradation of the fiber-optical and mechanical properties within a year under ordinary conditions.[13]

6.2.2 ALKALI HALIDES

The alkali metal halides are separated into two groups: crystals having a cubic lattice of the type NaCl (e.g., NaCl, KCl, and KBr) and cesium halides (e.g., CsBr and CsI) having a lattice of the type CsCl.[12] The first group is transparent up to about 20 µm, has an order of magnitude higher thermal conductivity, and has five independent glide systems toward the direction $\langle 110 \rangle$, but is highly soluble in water. Cesium halides are transparent up to about 50 µm; they are soft at room temperature, deform similar to the thallium halides, but are very hygroscopic. They also tend to exhibit intense surface crack formation which leads to high optical scattering.[14] The major drawback of the fibers made by Harrington[15] from these materials is the fast degradation of their optical and mechanical properties under ordinary conditions.

6.2.3 SILVER HALIDES

Silver halides have been used in photography for more than 100 years, and so their properties have been well investigated.[16] The phase diagram of the solid solution system AgCl–AgBr[17] is shown in Figure 6.1. The single crystals with a cubic lattice structure can be grown from the AgCl–AgBr system at any rate with the addition of up to 1 to 10 mol% AgI.

Silver halides are characterized by easy plastic deformation to great extents without crack formation at room temperature and under normal conditions. In this case, there is a mutual small penetration of the glide lines of the dislocations through the boundary between the crystal grains.[12] This allows the materials to deform to great extents without a significant increase in optical scattering. Fibers have been fabricated from the individual halides,[18,19] AgCl and AgBr, as well as the whole range of mutual solid solutions of AgClBr,[20-22] especially KRS-13 (0.25 AgCl–0.75 AgBr),

TABLE 6.2
Properties of Several IR Transmitting Halide Crystals Compared with Some Nonhalide Crystals

Crystal	KRS-5	KCl	CsI	0.5AgCl–0.5AgBr	KRS-13 0.25AgCl–0.75AgBr	AgCl	AgBr	ZnSe	Ge	Al_2O_3
Measured absorption coefficient at 10.6 μm (dB/km)	6.5	30	40	26	26	60	40	220	—	43 at 4 μm
Refractive index at 10.6 μm	2.37	1.45	1.74	—	2.21	1.98	2.25	2.40	4.00	1.66 at 4 μm
Melting point (°C)	414	776	621	418	412	457	419	1520	940	2040
Knoop hardness (kg/mm²)	40	8	—	13	15	9.5	7	120	780	2000
Young's modulus (GN/m²)	15.85	30	5.3	20	—	—	—	67	103	335
Solubility (g/100g H_2O)	0.05	34.7	44 at 0°C	7×10^{-5} at 25°C	3×10^{-5} at 25°C	2×10^{-4} at 50°C	1.8×10^{-5} at 25°C	10^{-3} at 25°C	$<5 \times 10^{-3}$ at 25°C	10^{-4} at 25 °C
Thermal expansion coefficient (10^{-6}/°C)	60	36	48	33	32	30	35	7.8	6	—
Thermal conductivity, 10^2 cal/°C·cm·s	1.3×10^{-3}	16×10^{-3}	2.7×10^{-3}	—	—	2.6×10^{-3}	1.4×10^{-3}	4.3×10^{-2}	14×10^{-2}	6.5×10^{-2}
Density (g/cc)	7.37	1.99	4.5	—	—	5.59	6.44	5.27	5.33	3.98
Drawbacks[a]	1, 5	2, 3	3, 4	5	5	4, 5	4, 5	2	2	—

[a] 1 = toxic, 2 = fragile, 3 = hygroscopic, 4 = plastic, 5 = UV sensitive.

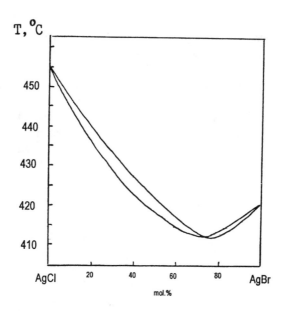

FIGURE 6.1 Phase diagram of the AgCl–AgBr solid solution.

and including AgI additive.[23] The individual AgCl and AgBr halides are very soft and possess a high rate of static recrystallization at room temperature causing the fibers to weaken with time. On the other hand, fibers made from the solid solutions exhibit higher strength due to deformation hardening, and they preserve the high strength characteristics for many years.[24]

One of the main drawbacks of silver halides is their photodecomposition upon illumination, causing free transition and the presence of photoelectrons. Another drawback is their tendency to react with most metals. The exceptions are noble metals and Ti, Ta, Zr, and Nb. These drawbacks are overcome when these fibers are used in specially designed cables. Silver halides have very low solubility in water, but are attacked readily by 5 to 10% sodium thiosulfate and 3 to 5% potassium cyanide solutions. They are nontoxic, which is very important for many applications, and are nonhygroscopic.

Unlike the chalcogenide glasses, the polycrystalline halide fibers possess a negative thermo-optic coefficient, dn/dT, defined as the change in refractive index with temperature. This prevents self-focusing during propagation of high-power laser radiation in the fibers, which may otherwise lead to thermal breakdown of the material.[3]

6.3 MOTIVATION FOR DEVELOPING THESE FIBERS

6.3.1 MINIMUM INTRINSIC LOSSES

From consideration of the minimum intrinsic optical losses, which are located in the mid-IR region (3 to 20 μm), ionic crystalline materials with heavy cations and halide ions are the most promising for IR fiber fabrication. The region of minimum intrinsic optical losses is limited by light scattering from optical phonons and by multiphonon absorption. The multiphonon absorption edge in metal halides has been theoretically calculated for alkali halides and silver halides[3] and extensively studied for thallium halides[25] and silver halides.[20,26] The level of light scattering by phonons (Mandelstam–Brillouin scattering) in these crystals was also calculated.[20,25,26] Butvina and Dianov[27] showed the need to take into consideration the absorption by free carriers in silver halides and thallium halides. It was estimated that the absorption by free polarons exceeded the previous minimal loss estimations by one to two orders of magnitude. The coefficient of optical absorption by carriers (polarons and interstitial Ag^+ ions), $\beta_{e,i}$, as a function of wavenumber, ω, is expressed as follows:

$$\beta_{e,i}(\omega) = 120\pi\sigma_{e,i}(\omega)/n \tag{6.1}$$

where the spectrum for free polarons has two regimes: $\beta_e(\omega)$ = constant for wave numbers $\omega \leq 10^3$ cm^{-1} and $\beta_e(\omega) \propto \omega^{-2.5}$ for $\omega > 10^3$ cm^{-1}, where $\sigma_e(\omega)$ is the minimal electron conductivity and n is the refractive index. The optical conductivity $\sigma_i(\omega)$ of mobile Ag^+ ions was estimated by the fluctuation–dissipation theorem.[28] The estimation showed the Lorentz character of the absorption $\beta_i(\omega) \propto \omega^{-2}$ by mobile Ag^+ ions with a maximum at the one phonon frequency.[29] The minimum fundamental optical losses estimated by these mechanisms is about 10^{-6} to 10^{-7} cm^{-1} (~0.4 to 0.04 dB/km). The real level of absorption by these mechanisms may be increased by several orders of magnitude with increased number of carriers and their frequency mobilities. The spectra of the fundamental optical losses are shown in Figures 6.2

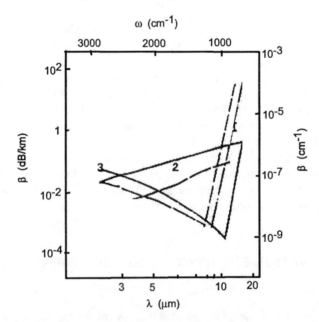

FIGURE 6.2 (1) Multiphonon absorption of AgCl (--), AgBr (–), KRS-13 (-.-); (2) Absorption by polarons; (3) Mandelstam–Brillouin scattering.

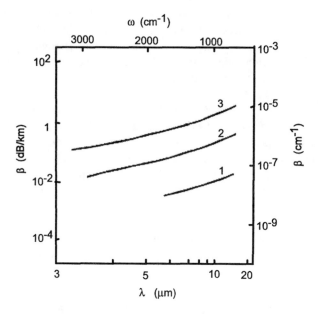

FIGURE 6.3   Calculated absorption by interstitial Ag^+ ions in (1) AgCl, (2) AgBr, and (3) $AgBr_{0.9}I_{0.1}$.

TABLE 6.3
Multiphonon Absorption Parameters

Crystal	A (cm)	ω_0 (cm^{-1})
AgCl	2.6×10^{-2}	430
AgBr	3.9×10^{-2}	306
KRS-13	2.7×10^{-2}	387

---

and 6.3. For silver halides, the coefficient of multiphonon absorption, β_{IR}, as a function of wave number, ω, is expressed as $\beta_{IR}(\omega) = \exp[A(\omega - \omega_0)]$. Values of A and ω_0 are presented in Table 6.3.

6.3.2   PRACTICAL NECESSITY OF THESE FIBERS

Only crystalline metal halides have such a wide transparency range from the visible to the far IR (0.3 to 50 μm). Fibers from these materials with an optical loss level already achieved for the cubic crystals could have widespread use in modern science.

Long lengths of fibers can be prepared from crystalline preforms by fabrication from the melt (growing of single crystalline fibers) or from the solid preform by plastic deformation. The second method of crystalline fiber preparation results in a significant increase of structure-dependent scattering and in deformational hardening. In spite of this fact, there are a number of advantages. First, the stage of purification and formation of a crystal with low optical losses is separated from the stage of fiber fabrication. Second, the deformational hardening of the material allows one to significantly improve the elastic properties and strength. The significant reduction in the extrusion-induced scattering increases the potential fiber application areas. Examples of some applications include

- The transmission of high-power IR laser radiation such as Er^{3+}:YAG at 2.91 μm, DF at 3.8 μm, CO at 5 to 6 μm, and CO_2 at 10.6 μm.
- Remote IR spectroscopic applications based on characteristic molecular vibrational modes in the fingerprint region.
- Thermovision fiber systems based on the overlap of the crystalline fiber transparency region (3 to 20 μm) with thermal radiation from an object at room and lower temperatures.

The fabrication of low-mode fibers having a core-clad structure with low optical losses is still a current problem that needs to be resolved. This will be addressed in forthcoming sections.

6.4 CRYSTAL GROWTH AND REFINING PROCESSES

The crystals of the alkali–metal halides KCl, KBr, and NaCl are available with very low optical absorption coefficients at a wavelength of 10.6 μm. The fabrication techniques of such crystals have been developed by many commercial vendors. The crystals are used as optical windows and lenses for transmission of high-power CO_2 laser energy. The lowest absorption coefficients obtained for these crystals have values of about 1×10^{-5} cm^{-1}. Plotnichenko et al.[30] have measured similar values for CsI and showed that they are about five orders of magnitude higher than the theoretical absorption coefficient.

It is known that a high absorption level is mainly due to crystal fouling by different impurity complexes, which possess absorption bands in the region of high transparency of the IR material,[31] as well as due to absorption by solid microscopic inclusions. Methods have been developed for refining thallium halides and their solid solutions.[32-34] Lichkova and Zavgorodnev[35] discussed the purification of silver halides. The main message of this work was the necessity to use the highest quality initial materials. For instance, the initial silver halide salts and their solid solutions were further refined by multiple recrystallization in acid solutions with a temperature gradient.[20] One of the most frequently used refining methods is the technique of halogenation of the melt in a pure reactive atmosphere,[32,36] in order to remove oxygen-containing inclusions. Other methods include filtering of the melt,[37] vacuum distillation,[38] zone refining,[32,38] and crystallization refining.[32] The refining is carried out in an inert gas or under vacuum. The crystal is grown using the Bridgeman–Stockbarger method in a sealed-off ampoule in an inert gas or under vacuum. The growing of solid solution single crystals is carried out at low rates of approximately 1 mm/h. The total growth cycle of a single crystal takes several weeks.

The best homogeneous solid solution crystals grow at eutectic concentrations of the components. For silver halides ultraviolet (UV) exposure should be excluded.

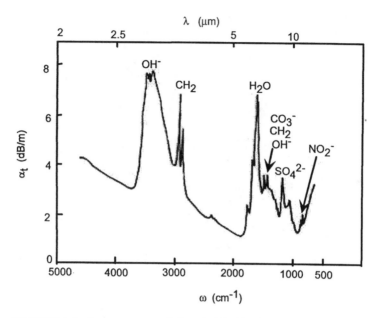

FIGURE 6.4   Loss spectrum of silver halide fiber–containing impurities.

The quality of the single crystals is determined by laser calorimetry. The silver halide crystals with absorption coefficients in the range of 1×10^{-4} to 5×10^{-5} cm^{-1} have been grown.[20,39,40] IR absorption spectra of single crystals can be used for qualitative analysis of absorption impurities down to levels of about 10^{-3} cm^{-1}. Stony and Masters[41] showed that silver halide materials available from industry were not pure enough for direct growth of high-optical-quality crystals. This is explained by the fact that the content of light elements, such as hydrogen, oxygen, nitrogen, and carbon, is prevalent in the precursors. These elements represent the main anionic impurities, such as OH^-, NO_3^-, CO_3^{2-}, SO_4^{2-} anions, along with water molecules, CO_2, and hydrocarbons, which are manifested in the loss spectra of polycrystalline fibers (Figure 6.4). In order to observe these impurities in the powders and crystals gamma- and neutron-activation methods of analysis can be used.

6.5   FIBER FABRICATION

Extrusion,[2] drawing,[19] and rolling[14] are the techniques that have been suggested for the fabrication of crystalline fibers by plastic deformation. The best results are observed using the extrusion method where the crystalline preform is extruded through a polished die. Extrusion allows the fabrication of long lengths of fiber of practically any small diameter during one process. The degree of drawing (real deformation), ε, is represented by

$$\varepsilon = 2 \ln\left(d_c/d_f\right) \tag{6.2}$$

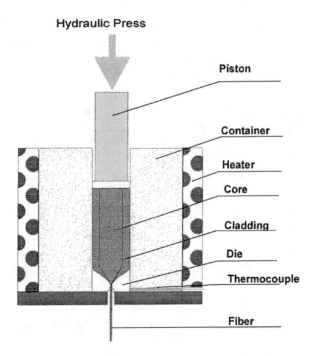

FIGURE 6.5 Direct extrusion of polycrystalline fibers.

FIGURE 6.6 Inverse extrusion of polycrystalline fibers.

and where d_c is the preform diameter and d_f is the fiber diameter. Typical values of ε lie in the range of 5 to 8. The direct extrusion (Figure 6.5) and inverse extrusion[42] (Figure 6.6) methods have both been used. The high-pressure container and die are made of high-strength materials, such as tungsten carbide (WC), boron nitride (BN),

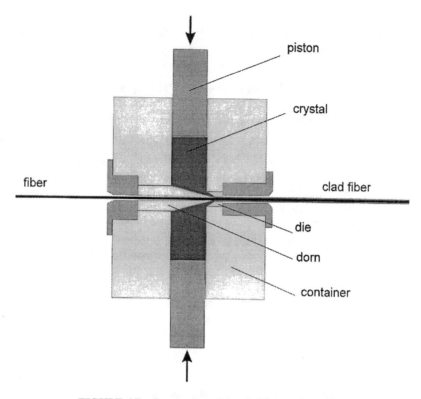

FIGURE 6.7 Coextrusion of the cladding on bare fiber.

and diamond. The inverse extrusion method has a number of advantages over direct extrusion. The most important advantage is that the preform friction on the container walls is practically eliminated, so the press pressure is practically constant during the whole extrusion. In both methods, it is possible to eliminate the near-wall layer of the crystal from being drawn into the fiber, thus reducing the fouling. Different extrusion temperatures were used in order to reduce optical losses in fibers and to improve their mechanical characteristics, as well as to increase shelf life. Typically, the temperature regime for extrusion lies in the range of 0.5 to 0.8 of the melting temperature, T_m. The fiber draw speed can range from millimeters per minute to tens of centimeters per minute. The press pressure is determined by the rheological behavior of the material, and for different materials and temperatures it ranges from units of a few tons per square centimeter to 15 to 20 ton/cm^2.

Core/clad polycrystalline fibers were fabricated by different methods. A direct preform (rod-in-tube) extrusion process[23,42,43] gave a very poor boundary between the core and cladding due to the mechanical anisotropy of single crystals. For this reason, Harrington[15] proposed a special crystallographic orientation for the rod and tube. A two-step extrusion process,[23] the inverse extrusion process[42] and the coextrusion process,[44-46] (Figure 6.7) were proposed. The cladding, along with a metal coating, has been applied by vacuum thermal evaporation.[47] Other processes include a special two-chamber container with double dies[40,48] as well as chemical[44] and sol–gel deposition of crystalline cladding materials and electrochemical metals coatings.[29] All the suggested fiber fabrication processes have their own inherent problems which must be overcome.

6.6 FIBER LOSS DATA AND LOSS MECHANISMS

6.6.1 FIBER LOSS DATA

A good earlier review of the optical properties of polycrystalline fibers as well as ideas on the loss mechanisms has been previously published.[49] Here, we present more-recent results which have not been considered earlier.

The main unresolved problem associated with polycrystalline fibers is the strong increase in scattering with respect to the initial single crystal. Optical losses have been separated into absorption and scattering using different methods, including the modified laser calorimetry method at the operating wavelengths of CO and CO_2 lasers.[39,50] It was most frequently observed that the spectral dependence of the scattering loss, α_s, on the wavenumber, ω, approximated a square relationship: $\alpha_s \propto \omega^2$.

In all of the earlier work and most of the more recent work the dependence of the scattering loss on the wavelength is represented as

$$\alpha_s = A\lambda^{-4} + B\lambda^{-2} + C \qquad (6.3)$$

Such extrapolation is valid only for defects with small refractive index difference Δn in comparison with that of the host material, that is, $\Delta n/n \ll 1$. This assumes that the fields of inhomogeneous stress,[51,52] the strain fields of dislocation,[53] the grain boundaries,[2] chemical separation,[54] as well as the surface scattering,[55,56] mainly contribute to the scattering. But detailed and more-accurate measurements[24,57] showed that the spectral dependence of the optical loss in the mid-IR region is more complex than can be represented by an additional term in a power function proportional to $\omega^{\eta(\omega)}$. Values of $\eta(\omega)$ can vary over a wide range from –0.5 to 3 depending on the wavelength, the composition, and prehistory of the crystal, as well as the extrusion and annealing parameters.

For fibers prepared by plastic deformation of a single-crystalline preform the value of the exponent, $\eta(\omega)$, decreased during the recovery, recrystallization, and annealing processes. But the level of the scattering losses can increase, as well as decrease, during the annealing of fibers extruded at high temperature.[53]

The IR absorption in extruded fibers may increase with respect to the starting crystals as a result of segregation of impurities on boundaries, pressure broadening of anionic IR absorption bands, and multiphonon absorption. Nagli et al.[39] proposed that the absorption at 10.6 μm is mainly by cation vacancies bound to dislocations.

6.6.2 LIGHT SCATTERING BY MICROPORES

Some investigations have revealed that the major cause of light scattering in polycrystalline fibers is due to microscopic pores (or microvoids) generated during extrusion.[57,58] Therefore, the effect of microscopic pores on optical scattering requires a more-detailed investigation. From scattering theory, the scattering cross section, K, as a function of x and m, where $m = n_0/n$ and $x = 2\pi Rn/\lambda$, is given by

$$K(x,m) = \alpha_s/\pi R^2 \qquad (6.4)$$

where n and n_0 are the refractive indices of the matrix and inclusion, respectively ($n_0 = 1$ for a pore), α_s is the scattering loss, and R is the radius of an inclusion or pore. The scattering cross section $K(x,m)$ is usually represented as $x^{\eta(x,m)}$ since $K(x,m)$ tends to zero as x tends to zero. The exponent $\eta(x,m)$ can be found by using the function $K(x,m)$ and performing logarithmic differentiation. Consequently, the exponent $\eta(x,m)$ can be written as

$$\eta(x,m) = \delta \ln K(x,m)/\delta \ln x \qquad (6.5)$$

The results of the numerical calculation[57] are presented in Figure 6.8. The characteristic features of the dependence of $\eta(x,m)$ are as follows:

a. $\eta(x,m) = 4$ when $x \to 0$ and at any values of m (Rayleigh scattering — λ^{-4}),
b. $\eta(x,m) \approx 2$ for a broad range of x and when $m \to 1$ (Rayleigh–Gans scattering — λ^{-2}),
c. Scattering centers with low m (e.g., pores in materials with $n > 1.5$) cause a quick and nearly linear decrease of the $\eta(x)$ in the region $4 > \eta(x) > -1$ and $x < 6$.

Therefore Equation 6.3 can not be used for a polydispersed distribution of micropores with $m < 1$, and so scattering losses must be expressed as

$$\alpha_s = \int \pi R^2 x^{\eta(x,m)} N(R) dR \qquad (6.6)$$

where $N(R)$ is the volumetric density of scattering micropores with radius R. This expression allows one to analyze the spectral variation of scattering in materials with high refractive indices. For micropores in silver and thallium halides, where $m \sim 0.5$ and 0.4, respectively, the dependence of $\eta(x)$ in the non-Rayleigh region $0.2 < \eta(x) < 2.2$ can be written in the following manner:

$$\eta(x) = 3.3 - 0.68x \quad \text{at } m = 0.5 \text{ and}$$
$$\eta(x) = 3.3 - 0.75x \quad \text{at } m = 0.4. \qquad (6.7)$$

6.6.3 SCATTERING LOSSES DUE TO COALESCENCE PROCESSES

The results of the above analysis relate the variations in the scattering (α_s) and exponent (η) with the distribution of the dimensions of the micropores. The effective average diameter (D) of pores can be evaluated in a monodispersed distribution by

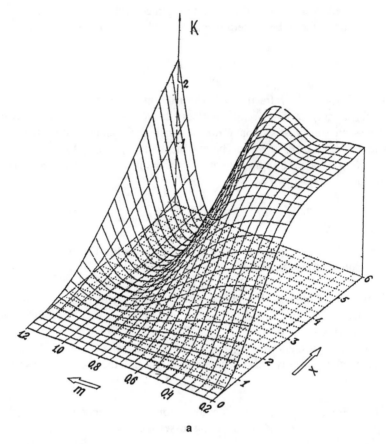

a

FIGURE 6.8 (a) Scattering cross section of a nonabsorbing particle and (b) wavelength dependence of scattering (η) as a function of m and x.

η, determined from the spectrum. From scattering loss measurement and knowing $K(D)$, we can evaluate the concentration N and the density defect P, where P is defined as

$$P = \left(\rho_0 - \rho_p\right)/\rho_0 \qquad (6.8)$$

and where ρ_0 is the initial density of the perfect crystal and ρ_p is the density of the polycrystalline material produced after extrusion. Values of P can vary quite broadly, reaching levels of 10^{-3}. An estimation from early published spectral loss data gives typical values for pore concentrations in the range of 10^5 to 10^7 cm^{-3} and dimensions D near 1 μm.

The gas associated with the vacancies, which are generated in large quantities during the extrusion process, precipitates by coagulation of vacancies on microscopic cracks and other heterogeneities with the formation of micropores. By assuming a

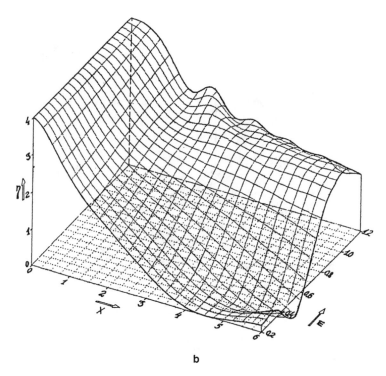

b

FIGURE 6.8 (continued)

monodispersed distribution of pores with radius R, then the bulk density defect P is related to the pore radius and concentration N by

$$N = 3P/4\pi R^3 \qquad (6.9)$$

Then the scattering losses can be represented by:

$$\alpha_s = 3P(2\pi Rn/\lambda)^{\eta(R,\lambda,n)} F(m)/4R \qquad (6.10)$$

In the stage of coagulation with a constant N, the scattering losses increase as R^6 ($\eta = 4$). Then when the pores coalescence,[28] P is constant, and the scattering losses vary as $R^{\eta(x)-1}$ and begin to decrease at shorter wavelengths when $\eta(x) < 1$ (Figure 6.9).

In the extrusion process, the vacancies are generated by jogs (steps) on moving screw dislocations and dislocation dipole annihilation. High levels of deformation and high rates of deformation produce high numbers of excess vacancies or free space generation. The dynamic recrystallization boundaries play a role as sinks for vacancies and micropores. When the boundaries move, micropores form (Figure 6.10).

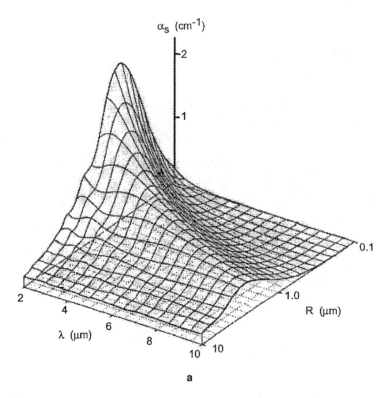

α_s (cm^{-1})

a

FIGURE 6.9 Evolution of the scattering spectra during the coalescence process of micropores: (a) $P = 10^{-4}$, $m = 0.4$; (b) $P = 10^{-4}$, $m = 0.5$.

The micropores were observed on the grain boundaries after intergranular cleavage in KRS-5 fibers and silver halide fibers under liquid nitrogen temperature.[57,58] The simplest evaluations showed that on a fiber with area A one can observe i micropores, where $i = A * D * N$ with diameter D and concentration N. The observed dimensions were in good agreement with the calculated values.

In glass materials and IR fibers, the cavities may arise during crystallization and decomposition. In crystalline fibers, the cavities are also formed by creep cavitation.

6.6.4   SCATTERING LOSS IN LOADING AND FATIGUE EXPERIMENTS

Crystalline fiber materials exhibit significant birefringence due to the stress optic effect. This results in additional scattering under the conditions of rather high dislocation densities, grain disorientation, and the presence of stress concentrators (micropores). The influence of tensile stress σ (below the yield stress) on the optical losses of KRS-5 fibers at a wavelength of 10.6 μm has been investigated by Harrington and Standlee.[52] Similar work on silver halide fibers[24] has revealed that the induced scattering losses α_s are represented by the complex relationship, $A(\sigma)\lambda^{-n(\sigma)}$. As the stress increases, the wavelength dependence of the scattering behavior changes from approximately Rayleigh (λ^{-4} dependence) to Rayleigh–Gans (λ^{-2}

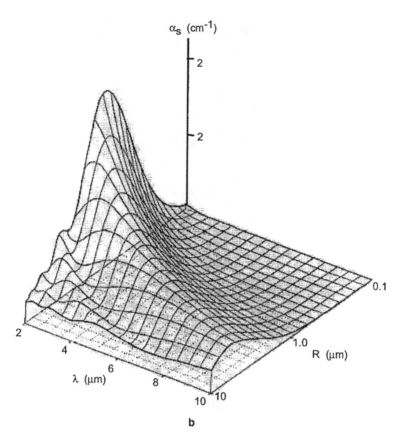

α_s (cm^{-1})

FIGURE 6.9 (continued)

dependence). At a fixed wavelength, the induced losses first increase, then decrease, and finally increase again as $\alpha \propto \sigma^2$ (Figure 6.11). Analysis of this behavior has shown that stress concentrators in the quantity of 10^7 to 10^9 cm^{-3} as well as a compressive stress of 10 to 20 MPa are present in the fibers after extrusion. The action of a tensile stress for a prolonged period of time causes irreversible creep in silver halide fibers obtained at high temperatures. The induced irreversible losses in this case exhibit a wavelength-independent character. The transmission spectra of fibers after multiple bending have also been investigated.[59,60] It was found that in harder fiber, such as AgCl$_{0.3}$Br$_{0.7}$, the irreversible losses caused by the formation and growth of the surface cracks after multiple bending with a deformation amplitude of 0.2% depended weakly on the wavelength. In softer fiber, such as AgCl$_{0.95}$Br$_{0.05}$, irreversible losses due to such fatigue tests exhibited a much higher dependence on wavelength, especially at shorter wavelengths. Shalem et al.[60] showed that for deformations smaller than one half of the static limit of elastic deformation, the losses at 10.6 μm did not grow until mechanical damage was evident after 10^6 to 10^7 bends (Figure 6.12).

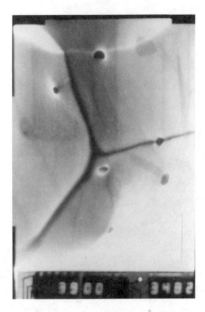

FIGURE 6.10 Electron microscope image of micropores in the KRS-13 fiber.

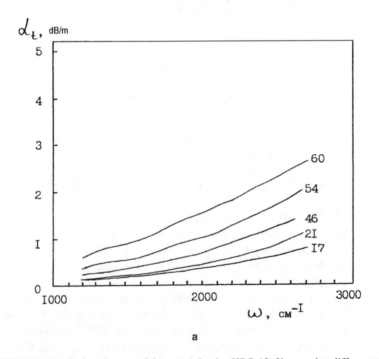

a

FIGURE 6.11 (a) Induced spectral loss (α_t) in the KRS-13 fiber under different tensile stress loading (MPa). (b) The effect of tensile loading on the wavelength dependence (η) of the induced loss (top curve) and the loss at a wavelength of 5 μm (bottom curve).

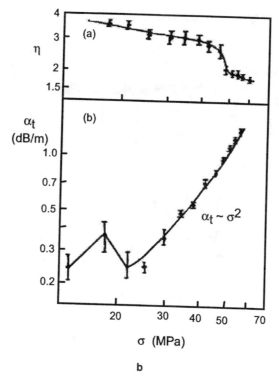

FIGURE 6.11 (continued)

6.6.5 OPTICAL LOSSES DUE TO METAL COLLOIDS

Plastic deformation of a single-crystal preform during the extrusion process leads to the generation of different point defects, such as interstitial ions and vacancies. The energy of formation of an interstitial ion is much higher than that for a vacancy formation in cubic halide crystals. The opposite correlation is valid for the activation energies of migration for these two types of defects. Therefore, metal colloid formation can be a potential problem and so its effect on the optical losses in the mid-IR region needs to be determined.

Numerical methods have been used to calculate the differential scattering cross section K by light absorption Q caused by silver colloids of radius R in silver halide materials.[23] Figure 6.13 shows the dependence of the differential cross section K of the monodispersed Ag colloids in an AgBr crystal on the colloid radius and on the wavelength. The results show that the dependence of the IR absorption by surface plasmons changes smoothly from $\lambda^{-3.7}$ to λ^0 as the colloids grow from tens of nanometers to fractions of micrometers.

The measured UV-induced losses[23] in AgClBr, AgBrI, and KRS-13 fibers initially exhibit a $\lambda^{-3.7}$ dependence. Then, the losses decrease more smoothly with increasing dosage. The rate of decrease is maximum in AgCl and AgBr fibers. In

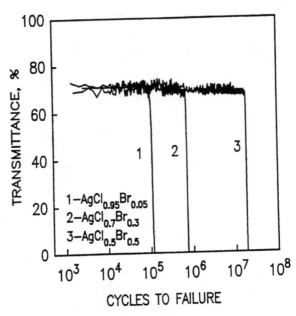

FIGURE 6.12 Transmission of CO_2 laser power through silver halide fibers after multi-bending with $R = 14$ cm. (Adapted from Nagli, A. et al., *SPIE*, 2631, 208, 1995.)

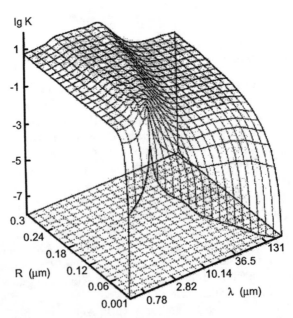

FIGURE 6.13 Calculated absorption cross section of silver colloids in $AgBr_{0.9}I_{0.1}$.

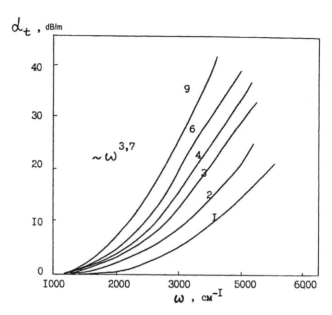

FIGURE 6.14 Effect of UV irradiation (250 W lamp at distance of 20 cm) on the absorption spectra of the $AgBr_{0.85}I_{0.15}$ fibers. (Exposure time in hours.)

high-strength $AgBr_{0.85}I_{0.15}$ fibers, the spectral dependence does not change after more than an hour of intense UV irradiation (Figure 6.14). In single crystals the induced IR absorption does not depend on the crystal composition. Thus, in the fibers, the distribution of the colloids by size turns out to be structurally dependent. In the highest-strength fibers there is a limit to the growth rate of the silver colloids. One possible explanation for this observation is that there are an increasing number of traps for interstitial silver in the fibers with higher mechanical strength.

6.6.6 CURRENT OPTICAL LOSSES

Transmission losses of silver halide fibers were measured in the IR region by the cutback method using a Bruker spectrometer (model IFS-113). The input numerical aperture (NA) in each measurement was gradually increased from 0.3 to 0.7 and fibers with lengths up to 12 m were used to increase the accuracy of the measurement. Figures 6.15 and 6.16 show the optical losses of AgClBrI unclad fibers.[40] These spectra demonstrate that it is possible to reduce the level of scattering and absorption in the near-IR region. Polycrystalline fibers have been obtained with record low optical losses of 50 dB/km at 10 μm and less than 0.5 dB/m at 3 μm. These fibers also demonstrated a weak dependence of scattering in the near-IR region on the input NA. This is due to the low surface absorption and scattering.

These fibers have ordinary waveguiding properties, similar to glass fibers with large diameters. This is in sharp contrast to typical crystalline fibers which give a significant broadening of the angular distribution of transmitted light. For example, a 1-m length of straight fiber with a diameter of 0.5 mm did not spread the angular

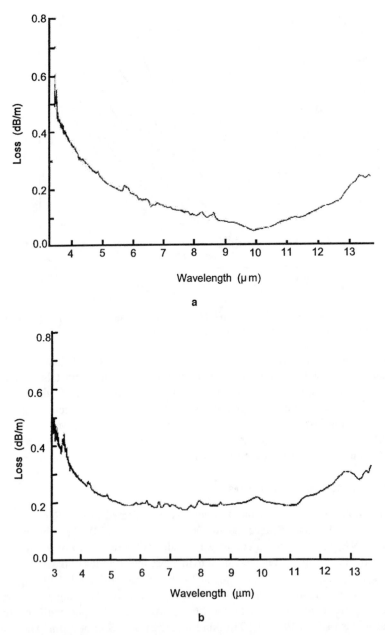

FIGURE 6.15 (a and b) Some representative optical losses for the unclad AgClBrI fibers with a diameter of 0.5 mm and using an N.A. of 0.7.

distribution of an input CO_2 laser beam. It is an important advantage not only for power delivery but for any optical application of these IR fibers.

Clad crystalline AgClBrI fibers with a core diameter of 0.7 mm and a total diameter of 0.9 mm have been fabricated in lengths up to 5 m. A typical optical loss

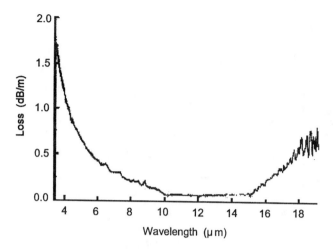

FIGURE 6.16 Optical losses in unclad $AgBr_{0.9}I_{0.1}$ fiber with a diameter 0.7 mm.

spectrum[40] for clad crystalline fibers is presented in Figure 6.17a. The NA of this fiber was 0.4. Clad fibers have been fabricated with values of NA up to 0.5 for spectroscopic applications within the broad wavelength range of 3 to 19 μm. The loss spectrum for this fiber is shown in Figure 6.17b. The low-loss AgClBrI fibers have mechanical properties similar to ordinary harder fibers from strong solid solution and these strengths are sufficient for practical applications.

Core/clad and unclad fibers are commercially available also from CeramOptec. Their optical losses have been reported elsewhere.[59] Researchers at Tel Aviv University have measured the values of NA and the output distribution of a CO_2 laser beam in core/clad fibers.[60]

Artjushenko et al.[54] have studied aging of the optical losses (increase in optical loss over a period of time) in unclad silver halide fibers with high initial scattering. In AgClBrI fibers with low initial scattering levels and low absorption, only surface reactions initiated by environmental conditions, possibly nonstoichiometry and sunlight, influenced the increase of the optical losses. Such core/clad fibers enclosed in protective coatings and with low initial losses showed negligible aging after 2 years.

6.7 MECHANICAL PROPERTIES OF POLYCRYSTALLINE FIBERS

6.7.1 STRUCTURE

Polycrystalline fibers demonstrate significantly different mechanical properties compared with brittle materials such as glass fibers. The mechanical properties of polycrystalline halide fibers are determined by their structure and composition. The structure is formed during the process of deformation and extrusion of the initial preforms and is affected by a great number of parameters. These include the composition of the initial preform, the orientation, the extrusion pressure, the deformation temperature, the deformation bulk, and speed. The process of fiber formation

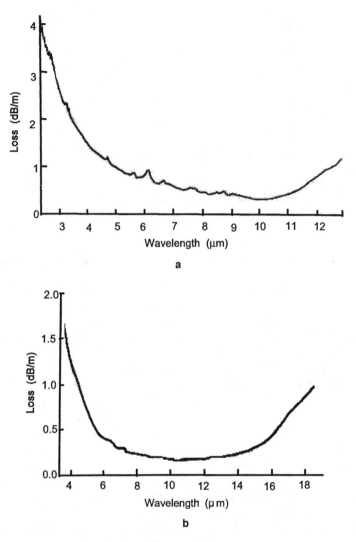

FIGURE 6.17 (a) Optical losses in the clad AgClBrI fiber with inner and outer diameters of 0.7 and 0.9 mm, respectively, and using an NA of 0.3. (b) Optical losses in the clad $AgBr_{0.9}I_{0.1}$ fiber with inner and outer diameters of 0.7 and 0.9 mm, respectively, and using an NA of 0.3.

includes the indispensable multiplication of dislocations in the deformed preform, their motion in the existing glide systems, and the formation of a "forest" of dislocations. In the case of a sufficiently high temperature, the dislocation climb processes are activated with a minimization of the energy associated with them. The dislocations unite into porous walls, forming a cell cold-work structure (Figure 6.18a).

At higher temperatures, the dislocations unite, forming a very thin topologically bounded bulk surface – boundaries separated grains. This process can occur during

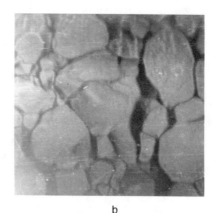

a   b

FIGURE 6.18   (a) The fractured surface of the KRS-13 fiber; (b) surface of AgCl fiber after etching.

hot deformation causing dynamic recrystallization of the fiber material with narrow distribution of the grain sizes (Figure 6.18b). The deformation process under these conditions is significantly nonlinear.[42] The speed of the extruded fiber v varies as a power function of the extrusion pressure P: $v \propto P^n$ and $n \in (3, 20)$.

It has been observed[61] that dynamic recrystallization takes place in AgCl, AgBr, and KRS-5 fibers corresponding to certain values of the degree of deformation, ε (ranging from 3 to 6), fiber flow speed (0.1 to 20 cm/min) and temperatures $T >$ 0.5 T_m. For KRS-5 fibers[42] the average grain size G was described by the following:

$$G = 2*10^7 \, v^{2/n-1} P^{-2} \exp(-U/kT) \tag{6.11}$$

where G is in µm, v is in cm/s, P is in MPa, U = 0.43 eV, and k is Boltzmann's constant. By changing the extrusion conditions, the grain size can be varied within the range of a few micrometers to hundreds of micrometers. In AgClBr and AgClBrI strong solid solutions deformed at a temperature of 100°C, a cold-work structure is formed with a high dislocation density. Very small single-crystalline cells have also been observed.[62] During annealing of these fibers in the range of 150 to 200°C, an inhomogeneous recrystallization process occurs, resulting in the formation of grains with sizes comparable to the fiber diameter. The grain growth in AgCl and AgBr takes place at room temperature, causing their plastification. The structure of the fibers made from the solid solutions does not exhibit any significant changes at room temperature after many years (>10 years). This ensures consistent mechanical properties of the fibers under ordinary conditions of temperature and humidity.

6.7.2 GRAIN BOUNDARY AND SOLID SOLUTION FIBER HARDENING

The mechanical properties of polycrystalline fibers have been evaluated using bending and tensile strength measurements. One of the main parameters determining the

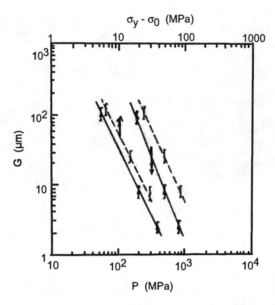

FIGURE 6.19 Dependence of the average grain size G on the extrusion pressure P for AgCl fiber (–) and AgBr fiber (--) and the influence of G on tensile hardening $(\sigma_y - \sigma_0)$ of the fibers.

elasticity limit, the yield stress and the rupture stress in polycrystalline materials is the average grain size or the dislocation density in the cold-work materials. In KRS-5,[52] AgCl, and AgBr[24] the fiber yield stress (σ_y) obeys the relationship:

$$\left(\sigma_y - \sigma_0\right) \propto G^{-1/2} \qquad (6.12)$$

where σ_0 is the yield stress of the single crystal and G is the grain size (Figure 6.19).

Nonlinear strain hardening is observed in solid solution fibers. Fibers in the compositional range of 20 to 80% of AgCl in AgBr obtained at a temperature of 100°C and with a degree of deformation ε corresponding to values in the range of 5 to 6 exhibit a room temperature elastic deformation of about 0.6% (Figure 6.20). In the AgBr–AgI solid solution system[23] even more significant hardening is observed for only 6% addition of AgI. The elastic deformation in such fibers reaches 1%. The yield stress approaches 130 to 140 MPa for a substitution of 12 to 15% of AgI (Figure 6.21). Fibers with composition KRS-13 and doped with 10% of AgI, obtained under the same conditions, exhibit greater than 1% elastic deformation and a strength of 160 MPa (Figure 6.22). Fibers fabricated at higher temperatures or at lower degrees of deformation demonstrate lower yield stress and rupture stress.[63]

The behavior of silver halide fibers after multiple bending has been investigated by Barkay et al.[63-65] The number of times the fiber is repeatedly bent (N) until mechanical failure obeys the following relationship:

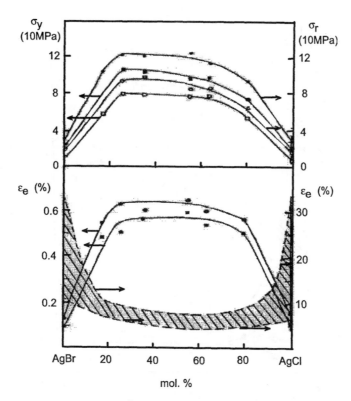

FIGURE 6.20 Composition dependence of fibers in the AgCl–AgBr solid solution of (a) σ_y and σ_r of the fibers, extruded at $\varepsilon = 6$, $T = 90°C$ (solid squares and circles), and $T = 180°C$ (open squares and circles); (b) elastic strain of fibers ($T = 90°C$ (solid squares), $T = 180°C$ (solid circles).

$$N = 0.5(\varepsilon/0.46)^{-2.4} \qquad (6.13)$$

where $\varepsilon = r/R$ and r is the fiber radius and R the bend radius. The maximum deformation is observed at the fiber surface. Practical application of a bent fiber can still take place when the static elastic deformation limit is exceeded but is lower than the destruction limit. This corresponds to values of r/R from 0.01 to 0.1. Bending with the deformation level smaller than the static elastic limit of the fiber leads to more gradual degradation. But, in this case, the number of cycles until destruction exceeds 10^6. The limit of fatigue deformation, defined as the deformation at which the number of cycles until destruction reaches 10^7 cycles of bending, depends on the composition. For AgCl–AgBr solid solution fibers,[66] this value ranges from 0.1 to 0.3%. The formation of microcracks in these experiments takes place at the intersection of the slip bends. At low temperatures the state of the fiber surface is of primary importance in these types of measurements. In silver halide polycrystalline fibers the destruction occurs after a significant one-time deformation of 10 to 50%.

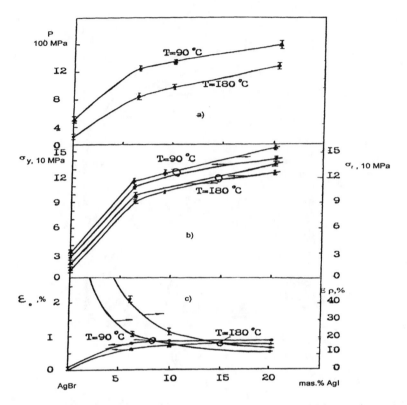

FIGURE 6.21 Composition dependence in the AgBr–AgI fibers of (a) extrusion pressure P ($\varepsilon = 6$); (b and c) mechanical parameters.

6.8 APPLICATIONS OF POLYCRYSTALLINE HALIDE FIBERS

6.8.1 DELIVERY OF CO_2 AND CO LASER POWER

One of the most important applications of polycrystalline halide fibers is the transmission of high-power mid-IR laser radiation. Results for CO_2 laser power transmission under continuous wave (CW) and pulsed operation are listed in Table 6.4. Under CW operation from 10 to 100 W, first the fiber output end and then the input end melted, more often in fibers with low absorption, thus causing the destruction of the fiber. Also, significant angular broadening of the beam (15 to 30°) was observed using a few meters of fiber. In this case, the few-mode structure of the laser radiation was transformed into a very irregular speckle pattern with significant intensity hot spots at the output end which led to damage of the output end face.

Laser-induced damage in the pulsed-periodic regime always occurred in the bulk of the fiber and within the first few centimeters from the input end face.[67] In this case, laser damage occurred at an order of magnitude lower power than in the single crystals. Typical data are listed in Table 6.4. This phenomenon can be attributed to

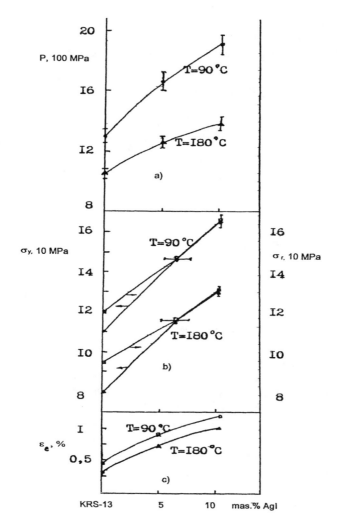

FIGURE 6.22 Composition dependence in the KRS-13–AgI fiber of (a) extrusion pressure P ($\varepsilon = 6$); (b and c) mechanical parameters.

the presence of a high concentration of micropores in the fiber. The scattering of light at the most prominent micropores causes significant broadening of the radiation, the formation of hot spots, and consequently the damage threshold can be exceeded. For the thallium halide single crystals it was found that the threshold of the pulsed laser damage at 10.6 μm was inversely proportional to the scattering level at 0.63 μm.[68] For silver halides, the seed photoelectrons as well as silver colloids already present can pose serious problems for high-power laser transmission.

For CW radiation transmission, the Fresnel reflection of 12 to 16% from each end face has a significant influence on the fiber transmission characteristics. The application of an antireflection, $As_2Se_3/KCl/As_2Se_3$, coating for KRS-5 fibers[34] increased the transmission from 67 to 97% and raised the damage threshold in the

TABLE 6.4
CO$_2$ Laser-Induced Breakdown (LIB) in Polycrystalline Fibers

Fiber Material	Operation Mode	Crystal LIB, W/mm^2	Fiber LIB, W/mm^2	Ref.
AgClBr	Long pulses (10 ms)	400	50	69
	Short pulses (100 ns)	8×10^6	2×10^5	—
	CW	4×10^3	47	—
AgClBrI	Long pulses (25 μs)	—	2×10^3	40
	CW	—	500	—
KRS-5	Short pulses (100 ns)	—	10^5	67
	CW	—	>700	34

CW regime up to 50 kW/cm^2. A BaF$_2$ antireflection coating was applied to silver halide fibers giving 98% transmission, but without increasing the damage threshold.[69] The technique used for preparing the fiber end face is important since it also affects the transmission characteristics. Mechanical polishing leads to fouling because of the softness of the material, and chemical etching leads to surface roughness. Mechanical cuts using a sharp inert edge (specially designed cutter) on silver halide fibers produces a good-quality surface.

Unlike most chalcogenide glass fibers, the silver, thallium, and alkali metal halides have negative thermo-optic coefficients, dn/dT, which inhibits thermal induced self-focusing. Therefore, the observed low destruction thresholds in the CW laser regime can be significantly increased by improving the fiber quality as well as the optical design of the input and output ends. For example, unclad silver halide fibers with losses of 50 dB/km at 10.6 μm and with specially designed fiber ends, reducing Fresnel reflection, demonstrated extremely stable CW CO$_2$ laser power delivery.[40] The total transmission in unclad fibers with diameters of 0.5 and 0.7 mm reached 86%/m and up to 80 W power was delivered without observing any nonlinear effects. Unclad fibers have demonstrated stable delivery of 40 W power for 200 h. Clad fibers have been used to deliver 20 W of CW CO$_2$ laser power over 1.5 m of fiber and with a total transmission of 82 ± 2% for 200 h. These fibers enclosed in specially designed laser cables with connectors transmit up to 18 W of CW CO laser power in the 5 to 6 μm wavelength region.

Earlier investigations of the transmission of pulsed laser radiation through polycrystalline fibers showed that the damage typically occurred in the initial section of the fibers and that the threshold was significantly lower than in the original crystals. However, the fibers with low scattering levels demonstrated a significant improvement in the delivery of pulsed laser energy. For example, damage was not observed in 0.5-mm-diameter unclad fibers using an Er:YAG laser operating at millisecond pulse widths with an energy of 0.5 J. There was no evidence of damage after an accumulation of 10^4 pulses. Preliminary experiments on power delivery of RF excited laser "Diamond" from Coherent with a pulse width 25 to 100 μs showed that this fiber is able to transmit at least an average power of 30 W.

It is well known that pulsed radiation energy 1 J/cm² with 10 μs pulse width results in very effective ablation of tissue without necrosis and carbonization. Therefore, low-loss crystalline fibers in the form of flexible delivery cables should have practical applications in medical surgery. A cable design for laser power delivery has been previously discussed.[49] Currently, several international institutes and companies are investigating and developing silver halide fibers for application in CO_2 laser power delivery using flexible cables for medical and other applications. Examples include Matsushita (Japan), Asah Medico (Denmark), Laser Industries (Israel), CeramOptec (Germany), Fiber Optic Research Center (General Physics Institute, Russian Academy of Science, Russia), and Tel Aviv University (Israel).

6.8.2 SPECTROSCOPIC CHEMICAL SENSING

Crystalline silver halide fibers are excellent optical materials for IR Fourier spectroscopy since they possess optical losses lower than 0.5 dB/m in the range of 3 to 16 μm and lower than 1 dB/m in the region of 2 to 20 μm. Furthermore, they have NAs of up to 0.7 which are well suited with the broad output beams of most commercial interferometers. Therefore, these fibers can be utilized in different, yet convenient, fiber sensor systems aimed at measurement of the transmission, reflection, and evanescent absorption spectra. The flexibility of these fibers allows significant variation in the form of the evanescent sensors, using unclad core with very tight bend radii. Typical fiber configurations for such sensors include a straight fiber with the excitation of higher modes,[76] a deformed straight fiber,[71] or an elastic bent fiber.[72] In order to reduce the solvent absorption (for example, water) which might otherwise swamp the analyte signal, special polymeric coatings have been developed for the fiber surface.[73,74] It was found that low-density polyethylene (LDPE) was the best for enrichment of chlorinated hydrocarbons (CHC) such as trichloroethylene (TCE), monochlorobenzene (MCB), and chloroform (CF). This 10-μm-thick coating was hydrophobic since it did not allow the passage of water but only the CHCs to the fiber surface. Saturation time for the CHCs in the coating was about 30 min, while removal of the CHCs took approximately 1 h. The detection limit of CHCs in water using this sensor was a few parts per million. The other type of evanescent sensor is based on the excitation of the transverse whispering gallery modes[75] in the fiber (Figure 6.23). This small-sized sensor has a very high sensitivity.

Maximization of the signal-to-noise ratio (SNR) is one of the most important parameters necessary for determining the detection limits in chemical spectral investigations. The minimum detected concentration C_{min} depends on the molecular oscillator strength b, the length of the interaction L, and SNR as follows:

$$C_{min} \propto (bL \text{ SNR})^{-1} \qquad (6.14)$$

The strength of the oscillators is maximum in the IR region (the so-called fingerprint region). The evanescent contribution of the fiber power is proportional to the wavelength λ. The maximization of the SNR is achieved by optimizing the whole design of the optical fiber system, mostly coupling fiber with an MCT

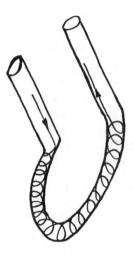

FIGURE 6.23 Fiber sensor based on whispering gallery modes.

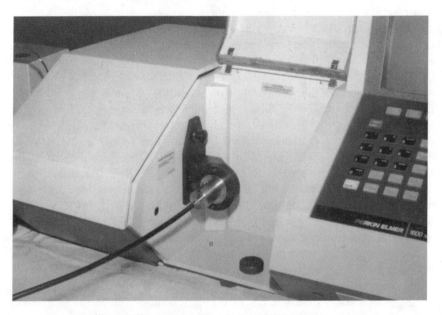

FIGURE 6.24 Input optics unit for an FTIR spectrometer.

detector.[76] Our FTIR fiber system includes a special silver halide tapered or lens input unit (Figure 6.24), the core/clad fiber cable, the replaceable sensor head from the partial unclad fiber (Figure 6.25) and the special MCT fiber detector with preamplifier cooled with liquid nitrogen (Figure 6.26).

The mechanical properties of silver halide fibers allow them to be used repeatedly in evanescent probes where the fiber touches gas, liquid, or solid objects such as polymers, papers, rubbers, and fibers. Such sensors are now beginning to be used in the promising area of medical IR diagnostics of human tissues, skin (Figure 6.27),

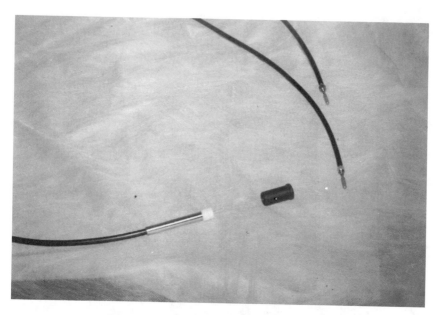

FIGURE 6.25 Replaceable bent fiber sensor.

and liquids. In particular, such sensors can be used for diagnosing cancer[77] and for control levels of glucose and other substances in liquids[78] (Figure 6.28). Fiber-optic probes provide the opportunity to make spectral measurements not only *in vitro*, but also *in vivo* with real-time processing through the channel of an endoscope.

The transmission and reflection microsensors based on low-loss silver halide fibers with spherical and hyperbolic ends demonstrate good SNR and will be very useful in remote FTIR analyses of small substances (Figure 6.29). Silver halide fibers have also been used for mid-IR tunable diode laser (TDL) evanescent spectroscopy[74] and for transmission measurements in a confocal multipass gas cell.[79]

The other promising application is Fourier IR fiber microscopy. Such a device has been fabricated using pressed fine hyperbolic lenses from silver halide fibers. A spatial resolution of 20 µm was achieved and with a SNR better than in typical IR microscopes.[80] The pressed micro-optical elements from polycrystalline fiber pieces[81] demonstrated excellent optical quality because of averaging birefringence in grains.

The low thermal conductivity and nonmagnetic character of silver halide fibers allow them to be used for launching of IR laser radiation into helium cryostats at low temperature and ultrahigh magnetic fields.[82] Silver halide fibers with low concentration of micropores exhibit stable optical properties under repeated fast cooling from room temperature to liquid nitrogen temperature. Small fiber diameters allow the measurement of the transmission or the evanescent spectra of reagents directly in a high-pressure chemical reactor.

Another promising application is the all-fiber IR spectrometer. It is possible to use the varying gap between the ends of silver halide fibers as a scanning IR Fabry–Perot interferometer.[83] Spectroscopic applications of silver halide fibers are potentially numerous, essentially for the low-loss clad fibers in appropriate cable,

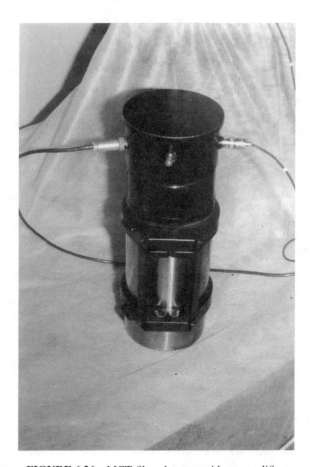

FIGURE 6.26   MCT fiber detector with preamplifier.

connectors, sensors, and fiber detector design. These are commercially available from FORC (Russia) and Infrared Fiber Sensors (Germany).

6.8.3   TEMPERATURE MONITORING AND THERMAL IMAGING

The transparency region of polycrystalline fibers coincides with the blackbody radiation of objects at room temperature. The Wine law associates the wavelength of the maximum spectral radiant photon emittance (λ_{max}) with the absolute temperature (T) according to $\lambda_{max}T = 2898$ µm K. This gives a value of 10 µm for λ_{max} at a temperature of 300 K. Fibers with an IR transmission edge at 20 µm (Figure 6.16) will transmit radiation from a body at 150 K. The temperature is determined by measuring the integral Plank radiation; in this case we need to know the operational function of the device and the grayness coefficient. The other way to measure temperature is to use spectrally selective measurements $Q(\lambda_1)$ and $Q(\lambda_2)$. The minimum object temperature measured using such noncontact techniques, as well as the temperature resolution, is determined by the SNR. However, high scattering and

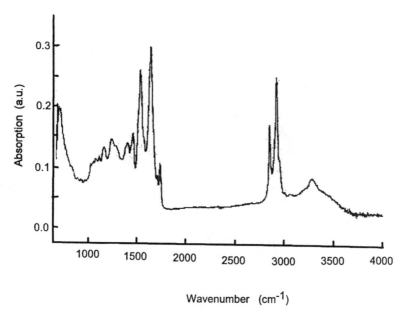

FIGURE 6.27 Evanescent spectrum of skin measured by bent fiber sensor (10 scans with an FTIR spectrometer).

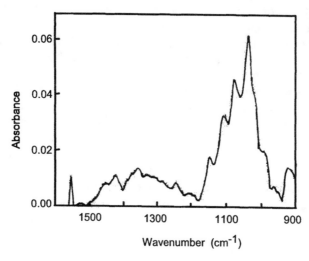

FIGURE 6.28 Absorption spectrum of a 0.5% aqueous solution of glucose (water spectrum subtracted out) using a seven loop fiber sensor.

absorption in typical polycrystalline fibers must be taken into acount. The emission of thermal radiation from the fiber itself due to IR absorption introduces perturbations into the radiometry, and this must also be taken into consideration.[84] Only polycrystalline fibers with an MCT detector allow the measurement of room temperature with a

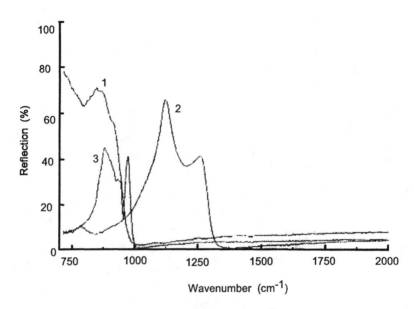

FIGURE 6.29   Reflection (45°) spectra of (1) Al$_2$O$_3$, (2) SiO$_2$ glass, and (3) ceramic GeO$_2$ measured by a reflection fiber sensor using 10 scans on an FTIR.

temperature resolution of 0.1°C.[85,86] According to Zur and Katzir,[85] the use of unclad silver halide fibers for body temperature measurement leads to a resolution of 0.1°C with a spatial resolution a fraction of the fiber diameter. One of the applications of flexible fiber systems is the noncontact temperature control of high-frequency hyperthermia.[87] Various optical detecting schemes using unclad silver halide fibers are described in the literature.[88,89]

Clad fibers with low losses allow measurements to be performed in remote and hard-to-get-to locations over distances of tens of meters. Another interesting application is surface temperature monitoring and control during the action of a CO$_2$ laser beam delivered through the same[90] or a bifurcated silver halide fiber.[91]

Flexible bundles with diameters of 1 to 3 mm consisting of 70 to 2000 silver halide fibers were fabricated by two- and three-stage extrusion techniques.[92] These bundles transmitted thermal images with a resolution less than 50 μm.

The MCT fiber detector (Figure 6.26) with a preamplifier possesses SNR = 1000 in the full frequency range from 8 Hz to 250 kHz for palm temperature measurement on ground RT chopper. The fiber detector makes it possible to scan thermal images in a remote compact optical head.

The silver halide fibers with low optical losses and anion impurity absorption less than 0.05 dB/m show no change in the mid-IR transmission under γ-irradiation from a Co^{60} source with a dose of up to 10^8 rad. This opens a perspective to remote control temperature and will provide a way to obtain thermal images from highly radioactive environments.

Due to the delicate nature of polycrystalline silver halide fibers, for any application it is very important to have full packaging, cabling, connectoring, and integrated design. The design of the application must solve the task of the end user, prevent degradation of the fiber and make handling rugged and simple.

6.9 FUTURE TRENDS AND PROSPECTS FOR FIBER DEVELOPMENT

The reduction of the optical loss level in polycrystalline silver halide fibers in the near-IR region will lead to the development of IR fiber lasers. This will be achieved in core/clad fibers with a protective coating and by using rare earth ions in the core pumped with IR diodes. Rare earth chlorides and bromides are hygroscopic and have a small solubility limit in the silver chloride and bromide matrix. As we have found, homogeneous crystals are grown at concentrations of rare earth ions lower than 0.05 wt%.[93] The measured absorption spectra of Pr^{3+}, Nd^{3+}, Sm^{3+}, Dy^{3+}, Tb^{3+}, and Ni^{2+} ions in silver bromide crystals are shown in Figure 6.30. The phonon spectra have been measured for polycrystalline fibers and single-crystal silver chloride and

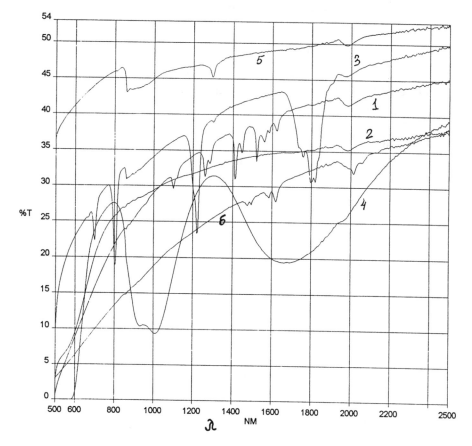

FIGURE 6.30 Transmission spectra of AgBr (5 cm length) doped with (1) 0.1% Sm^{3+}, (2) 0.05% Yb^{3+}, (3) 0.5% Tb^{3+}, (4) 0.5% Ni^{2+}, (5) 0.05% Dy^{3+}, (6) 0.1% Pr^{3+}.

bromide by Raman scattering of Nd:laser light at a wavelength of 1.32 μm.[94] It was found that the phonon spectrum was extended to 170 to 200 cm^{-1} in AgBr fibers and to 250 to 300 cm^{-1} in AgCl fibers. Consequently, the low phonon energies should cause a significant increase of the upper-level lifetimes in the known laser transitions as well the creation of new longer wavelength IR solid-state lasers. The fabrication of single- or low-mode polycrystalline fibers[95] with low losses also in the near-IR region will reduce the threshold power for such lasers.

The development of low-loss core/clad fibers will find increasingly more applications in the areas of mid-IR laser power delivery and chemical sensing. The fabrication of polycrystalline fibers from other metal halides with low scattering in the visible and UV region can enable other applications:

- Power delivery of pulsed Er^{3+} laser energy at 2.94 μm.
- Nonlinear CuCl and other Cu halide crystalline fibers for second harmonic generation.
- The CsI(Tl) scintillating fibers.
- The LiF fibers as laser host and for power delivery.

ACKNOWLEDGMENTS

The author wishes to thank E. M. Dianov for the opportunity to work on this chapter, V. G. Plotnichenko for helpful discussions, and J. Sanghera for patience.

REFERENCES

1. N. S. Kapany, *Fiber Optics: Principles and Applications,* Academic Press, New York, 1965, 272.
2. D. A. Pinnow, A. L. Gentile, A. G. Standlee, A. J. Timper, and L. M. Hobrock, *Appl. Phys. Lett.,* 33, 28 (1978).
3. B. Bendow, *Annu. Rev. Mater. Sci.,* 7, 23 (1977).
4. L. G. Van Uitert and S. H. Wemple, *Appl. Phys. Lett.,* 33, 57 (1978).
5. L. G. Gentile, M. Braunstein, D. A. Pinnow, J. A. Harrington, L. M. Hobrock, J. Mayer, R. C. Pastor, and R. R. Turk, in *Fiber Optics,* B. Bendow and S.S. Mitra, Eds., Plenum Press, New York, 1979, 105.
6. J. A. Harrington, *SPIE,* 227, 133 (1980).
7. V. G. Artjushenko, P. Bochkarev, V. F. Golovanov, T. I. Darvoid, E. M. Dianov, and A. M. Prochorov, *Sov. J. Quantum Electron.,* 8, 398 (1981).
8. S. Sakuragi, M. Saito, Y. Kubo, K. Imagava, H. Kotani, T. Morikawa, and J. Shimada, *Opt. Lett.,* 6, 629 (1981).
9. K. Takahashi, K. Murakami, and M. Yokota, in *Tech. Dig. 1981 Spring Meeting of Jap. Soc. Appl. Phys.,* 225 (1981).
10. D. Chen, R. Skogman, E. G. Bernal, and C. Butter, in *Fiber Optics,* B. Bendow and S. S. Mitra, Eds., Plenum Press, New York, 1979, 119.
11. H. J. Garfunckel, R. A. Skogman, and R. A. Walterson, *IEEE J. Quantum Electron.,* 15, 49 (1979).
12. M. T. Sprackling, *The Plastic Deformation of Simple Ionic Crystal,* Academic Press, London, 1976, 242.

13. J. A. Wysocki, R. G. Wilson, A. G. Standlee, A. C. Pastor, and R. N. Schwartz, *J. Appl. Phys.,* 63, 4365 (1988).
14. R. R. Turk, *SPIE,* 320, 128 (1980).
15. J. A. Harrington, *Appl. Opt.,* 27, 3097 (1988).
16. J. F. Hamilton, *Adv. Phys.,* 37, 359 (1988)
17. V. V. Grosnetsky, V. D. Zhuravlev, G. A. Kitaev, and L. V. Zhukova, *Sov. J. Inorg. Chem.,* 30, 1033 (1985).
18. A. Katzir, S. Simfony, R. Arieli, J. Salzman, A. Stroenberg, and E. Kapon, *SPIE,* 320, 136 (1982).
19. S. N. Avakov, V. G. Artjushenko, L. N. Butvina, and G. F. Dobrjansky, *Sov. Phys. Chem. Metal Working,* 4, 115 (1984).
20. V. G. Artjushenko, L. N. Butvina, V. V. Voitsechovsky, E. M. Dianov, L. V. Zhukova, F. N. Kozlov, S. L. Semenov, A. M. Prochorov, and N. M. Homjakova, *Sov. J. Quantum Electron.,* 13, 601 (1986).
21. K. Takahashi, N. Yoshida, and M. Yokota, *Sumitomo Electr. Tech. Rev.,* 23, 203 (1984).
22. A. Katzir and R. Arieli, *J. Non-Cryst. Solids,* 47, 149 (1982).
23. L. N. Butvina, E. M. Dianov, D. V. Drobot, Yu. G. Kolesnikov, and N. V. Lickova, *SPIE,* 1048, 17 (1989).
24. L. N. Butvina, *Structure Dependent Optical and Mechanical Properties of Crystalline Silver Halide Fibers*, Ph.D. thesis, General Physics Institute, Moscow, 1990, 137.
25. E. M. Dianov, I. S. Lisitsky, V. G. Plotnichenko, V. B. Sulimov, V. K. Sysoev, and L. N. Butvina, *Fiber Integrated Opt.,* 5, 125 (1984).
26. N. Barkay, A. Levite, F. Moser, D. Kowal, and A. Katzir, *SPIE,* 1048, 9 (1989).
27. L. N. Butvina and E. M. Dianov, *SPIE,* 484, 21 (1984).
28. L. D. Landau and E. M. Lifshitz, *Statistical Physics, Part 1,* 3rd ed., Pergamon Press, Oxford, 1976, 583.
29. L. N. Butvina, E. M. Dianov, Yu. G. Kolesnikov, and V. A. Prokashev, *SPIE,* 1228, 155 (1990).
30. V. G. Plotnichenko, V. K. Sysoev, and I. G. Firsov, *Sov. J. Opt. Technol.,* 9, 23 (1983).
31. M. Flanery and M. Sparks, *U.S. NBS Spec. Publ.,* 509, 5 (1977).
32. A. P. Belousov, E. M. Dianov, I. S. Lisitskii, T. M. Nesterova, V. G. Plotnichenko, and V. K. Sysoev, *Sov. J. Quantum Electron.,* 12, 496 (1982).
33. V. G. Artjushenko, E. P. Bochkarev, E. A. Voronina, G. G. Glavin, V. F. Golovanov, T. I. Darvoid, E. M. Dianov, and D. V. Kormilizin, *Sov. J. Quantum Electron.,* 10, 1181 (1980).
34. M. Ikedo, M. Watari, F. Tateishi, and H. Ishiwatari, *J. Appl. Phys.,* 60, 3035 (1986).
35. N. V. Lichkova and V. N. Zavgorodnev, *Sov. J. High Purity Substances,* 3, 19 (1991).
36. R. Arieli, *SPIE,* 1048, 169 (1989).
37. V. Artushenko, I. Antoniv, G. Danev, I. Garapina, L. Grigorjeva, E. Kotomin, E. Krivads, D. Millers, and L. Zhukova, *SPIE,* 1228, 140 (1990).
38. V. N. Zavgorodnev and N. V. Lichkova, *Sov. J. Inorg. Mater.,* 16, 635 (1980).
39. L. Nagli, D. Bunimovich, A. Shmilevich, N. Kristianpoler, and A.Katzir, *J. Appl. Phys.,* 74, 5737 (1993).
40. L. N. Butvina, E. M. Dianov, N. V. Lichkova, L. Kuepper, V. G. Plotnichenko, and N. Zavgorodnev, *SPIE,* 2631, 150 (1995).
41. L. H. Stony and J. I. Masters, *SPIE,* 618, 72 (1986).
42. V. G. Artjushenko, L. N. Butvina, V. V. Vojtsekhovsky, E. M. Dianov, and Yu. G. Kolesnikov, *Proc. Gen. Phys. Inst. Moscow,* 191, 12 (1986) (in Russian).
43. M. Kimura, S. Kachi, and K. Shiroama, *SPIE,* 618, 85 (1986).
44. J. A. Harrington, private communication.

45. K. Murakami and K. Takahashi, U.S. Patent 4,552,434 (1985).
46. V. G. Artjushenko, L. N. Butvina, V. V. Vojtsekhovsky, E. M. Dianov, L. V. Zukova, and Yu. G. Kolesnikov, *Sov. J. Tech. Phys. Lett.,* 14, 1667 (1988).
47. M. Saito, M. Takizawa, and M. Miyagi, *J. Lightwave Technol.,* 7, 158 (1989).
48. T. A. Fuller, V. J. Nadkarni, and J. R. Peschke, U.S. Patent 4,955,689 (1990).
49. T. Katsuyama and H. Matsumura, *Infrared Optical Fibers,* Adam Hilger, Bristol, 1989, 272.
50. V. G. Artjushenko, L. N. Butvina, and V. V. Vojtsekhovsky, *Sov. J. Opt. Technol.,* 3, 10 (1985).
51. J. A. Harrington and M. G. Sparks, *Opt. Lett.,* 8, 223 (1983).
52. J. A. Harrington and A. G. Standlee, *Appl. Opt.,* 22, 3073 (1983).
53. S. Kachi, K. Nakamura, M. Kimura, and K. Shiroyama, *SPIE,* 484, 128 (1984).
54. V. G. Artjushenko, A. A. Lerman, E. G. Litvinenko, A. O. Nabatov, V. I. Konov, R. I. Kuznetzov, V. G. Plotnichenko, I. L. Pylnov, V. A. Shtein-Margolina, A. A. Uruso-vskaja, V. V. Vojtsekhovsky, N. D. Zaharov, W. Neuberger, and K. Moran, *SPIE,* 1591, 83 (1991).
55. V. G. Artjushenko, L. N. Butvina, V. V. Vojtsekhovsky, E. M. Dianov, and Yu. G. Kolesnikov, *J. Lightwave Technol.,* 4, 461 (1986).
56. F. Moser, N. Barkay, A. Levite, E. Margalit, I. Paiss, A. Sa'ar, I. Schnitzer, A. Zur, and A. Katzir, *SPIE,* 1228, 128 (1990).
57. L. N. Butvina, E. M. Dianov, V. V. Vojtsekhovsky, and A. M. Prokhorov, *SPIE,* 843, 143 (1987).
58. L. N. Butvina, E. M. Dianov, V. V. Vojtsekhovsky, A. I. Maslakov, and A. M. Prokhorov, *Sov. Tech. Phys. Lett.,* 13, 581 (1987).
59. S. Artjushenko, V. Ionov, K. Kalaidjian, A. Kryukov, E. Kuzin, A. Lerman, A. Prokhorov, E. Stepanov, K. Bakhshpour, K. Moran, and W. Neuberger, *SPIE,* 2396, 25 (1995) .
60. S. Shalem, A. German, and A. Katzir, *SPIE,* 2631, 216 (1995).
61. V. G. Artjushenko, L. N. Butvina, V. V. Vojtsekhovsky, E. M. Dianov, and Yu. G. Kolesnikov, *SPIE,* 618, 103 (1986).
62. V. G. Artjushenko, V. M. Belous, A. A. Lerman, V. V. Metlov, A. O. Nabatov, A. A. Urusovskaja, S. A. Zhukov, and L. V. Zhukova, *SPIE,* 1228, 150 (1990).
63. N. Barkay, A. Levite, F. Moser, and A. Katzir, *J. Appl. Phys.,* 64, 5256 (1988).
64. N. Barkay and A. Katzir, *SPIE,* 1591, 50 (1991).
65. N. Barkay and A.Katzir, *J. Lightwave Technol.,* 11, 1889 (1993).
66. N. Barkay, A. German, S. Shalem, and A. Katzir, *SPIE,* 2131, 62 (1994).
67. S. S. Alimpiev, V. G. Artjushenko, L. N. Butvina, S. K. Vartapetov, E. M. Dianov, Yu. G. Kolesnikov, V. I. Konov, A. O. Nabatov, S. M. Nikiforov, and M. M. Mirakjan, *Int. J. Optoelectron.,* 3, 333 (1988).
68. E. P. Bochkarev, T. I. Darvoid, and V. N. Lebedeva, *Sov. J. Opt. Technol.,* 1, 26 (1981).
69. L. Nagli, A. German, A. Katzir, J. Tschepe, V. Prapavat, H. G. Eberle, and G. Muller, *SPIE,* 2631, 208 (1995).
70. S. Simphony, E. M. Kosower, and A. Katzir, *Appl. Phys. Lett.,* 49, 253 (1988).
71. I. Paiss, D. Bunimovich, and A. Katzir, *Appl. Opt.,* 32, 5867 (1993).
72. J. E. Walsh, B. D. MacCraith, M. Meaney, J. G. Vos, F. Regan, A. Lancia, and S. Artjushenko, *SPIE,* 2508, 233 (1995).
73. R. Krska, R. Taga, and R. Kellner, *Appl. Spectrosc.,* 47, 1484 (1993).
74. R. Kellner, R. Gobel, R. Gotz, B. Lendl, B. Edl-Mizaikoff, M. Take, and A. Katzir, *SPIE,* 2508, 212 (1995).
75. L. N. Butvina and L. Kupper, unpublished.

76. L. N. Butvina and L. Kupper, in *Program Conference High Purity Materials for IR Optics, Nizhni Novgorod,* Russia, 2–5 June, 1997, 4.
77. Natalija, I. Afanasyeva, S. F. Koljakov, V. S. Letohov, V. V. Sokolov, and G. A. Frank, *SPIE,* 2928, 16 (1996).
78. H. M. Heise, A. Bittner, L. Kupper, and L. N. Butvina, *J. Mol. Struct.,* (June 1997).
79. V. G. Artjushenko, V. G. Plotnichenko, E. V. Stepanov, A. I. Nadezchdinskii, A. I. Kuznetsov, and K. L. Moskalenko, *SPIE,* 1591, 206 (1991) .
80. L. Kupper, Ph.D. thesis, Aachen University, 1995.
81. L. Kupper and L. Butvina, PCT/EP95/01560, WO 95/29277 (1995).
82. M. Von Ortenberg, private communication.
83. M. Fendrich, M.Sc. thesis, Aachen University, 1995.
84. A. Zur and A. Katzir, *Appl. Opt.,* 31, 55 (1992).
85. A. Zur and A. Katzir, *Appl. Opt.,* 26, 1201 (1987).
86. R. J. Burger, D. A. Greenberg, and P. Kirkitelos, *SPIE,* 1228, 206 (1990).
87. A. Katzir, H. F. Bowman, Y. Asfour, A. Zur, and C. R. Valeri, *IEEE Trans. Biomed. Eng.,* 6, 634 (1989).
88. O. Eyal, A. Zur, O. Shenfeld, M. Gilo, and A. Katzir, *Opt. Eng.,* 33, 502 (1994).
89. S. Drizlikh, A. Zur, and A. Katzir, *Int. J. Optoelectron.,* 6, 451 (1991).
90. V. G. Artjushenko, V. V. Vojtsekhovsky, V. I. Masychev, I. V. Zubov, and V. K. Sysoev, *Electron. Lett.,* 20, 983 (1984).
91. O. Eyal, S. Shalem, and A. Katzir, *SPIE,* 2328, 52 (1994).
92. I. Paiss and A. Katzir, *Appl. Phys. Lett.,* 61, 1384 (1992).
93. L. N. Butvina, E. M. Dianov, V. N. Zavgorodnev, and N. V. Lichkova, in *Proc. X Conference High Purity Substance,* Nijnii Novgorod, Russia, 1995, 222.
94. S. M. Veslovskii, L. N. Butvina, and E. M. Dianov, unpublished.
95. N. Butvina, Yu. G. Kolesnikov, and V. A. Prokashev, *Sov. Lightwave Commun.,* 1, 65 (1991).

7 Hollow Waveguides

Christopher Gregory

CONTENTS

0-8493-2489-0/98/$0.00+$.50
© 1998 by CRC Press LLC

7.1   INTRODUCTION

In this chapter the history and current status of hollow waveguides for the delivery of light with a wavelength range between 2.5 and 15 μm is presented. This review will describe the motivation for developing hollow waveguides for the mid infrared (IR) and the theories that have guided their design. The optical performance, as well as the mechanical properties and manufacturing processes, of nearly all of the hollow waveguide structures that have been made and tested to date are discussed. Finally, the current and potential applications of IR hollow waveguides are reviewed. The major emphasis of research to date and, thus, of this review is the delivery of CO_2 laser radiation (10.6 μm), but applications involving noncoherent light are also presented.

Many scientists and groups have been involved in the development of mid-IR hollow waveguides over the years. Some of these will not be mentioned or will only be mentioned in passing. This is not meant to indicate that these efforts were not sound or significant to the development of hollow waveguides. To include them all, however, would be far beyond the scope of this chapter.

The concept of using hollow pipes to guide electromagnetic waves was first well described by Rayleigh[1] in 1897. Further understanding of hollow waveguides was delayed until the 1930s when microwave-generating equipment was first developed. At microwave frequencies the small length of each wave (on the order of a centimeter) and the dissipative losses in coaxial lines make hollow waveguides attractive over guides with solid metal cores. As the uses of microwave frequencies increased, the field of microwave waveguides grew tremendously.

The extension of microwave waveguide designs to the IR region was not trivial. Wavelengths in the near- and mid-IR regions are in the range of 1 to 100 μm. Microwave theory suggests that the cross section of waveguides should be on the order of one wavelength. Optical hollow waveguides of this size have excessively high losses and are prohibitively difficult to make; therefore, the field was not explored for a number of years. The use of hollow waveguides for carrying laser radiation was suggested a number of times over the years but research in the area did not become active until 1974 when Nishihara et al.[2] suggested that rectangular guides could be used for directing CO_2 laser light. Hollow waveguides for IR radiation have evolved since then from shrunken microwave guides to very carefully tuned structures with surprising flexibility and excellent reliability. They have been made of a variety of materials and in a large number of shapes. The optical performance of all of the waveguides developed to date have suffered from two problems not common with solid optical fiber. The first is a very strong dependence of attenuation on the waveguide bore size, and the second is a moderate dependence of attenuation on bending. The reliability of hollow waveguides for laser transmission due to their high damage threshold and potential high temperature capacity is, however, far superior to the solid fibers currently available for the mid-IR.

7.2 MOTIVATIONS FOR THE DEVELOPMENT OF HOLLOW WAVEGUIDES

Flexible systems for the transmission of IR radiation have been sought since detectors of IR light were developed. The introduction of IR lasers, especially the CO_2 laser at 10.6 μm, resulted in an even greater desire for a flexible transmission system. In applications, these lasers are typically rigidly mounted and the output beams are steered to a target by arrangements of mirrors. The complexity and awkwardness of these systems made the need for more-convenient ways of delivering the laser radiation to an appropriate target obvious. This has led to extensive efforts in the development of solid and hollow waveguides for the mid IR.

All of the solid waveguide materials developed for the mid IR to date suffer from the same problems. In order to be transparent in the mid-IR region, the multiphonon absorption edge of a material must lie at a longer wavelength than it does in traditional oxide glasses. The wavelength of the multiphonon edge is a result of the resonant frequencies of the atomic structure of the material. To move the multiphonon edge to a longer wavelength, the atomic masses must be increased or the bond energies between the atoms lowered. Unfortunately, this also results in the material being intrinsically weaker and having a lower melting temperature than traditional silicate glasses. Materials which are weaker and have lower melting temperatures also tend to have much lower laser damage thresholds. The low damage thresholds and melting temperatures of most IR transmitting materials have prohibited their use in many applications. Hollow waveguides do not suffer from these problems. The fact that there is no solid medium to damage or melt results in a much more reliable delivery system.

The need for a mid-IR delivery system became especially apparent with the growth of laser surgery. The traditional delivery system for mid-IR lasers in the operating room has been articulated arms. These arms consist of rigid tubes with swiveling elbows joining them at 90° angles. Each elbow has a mirror mounted at 45° to the beam path to guide the beam from the axis of each tube to the axis of the next. These systems are awkward, expensive, and difficult to align. In the past 20 years, many surgical procedures have moved to less invasive forms. Small incisions are made and viewing, manipulating, and cutting tools are passed through into the body. Articulated arms are far too large to be inserted in these small openings, and so complicated steering devices have been developed to aim the laser light externally at targets inside the body. This technique reduced the quality of the images acquired through the same opening and thus made the procedures quite difficult. Meanwhile, surgery with lasers that could use silica-based fibers was encountering none of these problems. The goal for many investigators became a flexible delivery system less than 1 mm in diameter, about 1.5 m long, with an attenuation level of less than 1 dB/m, that could deliver 80 W of continuous wave CO_2 laser energy.

More recently, the Er:YAG laser has shown promise for many surgical procedures. Its ability to cut tissue very precisely with little or no thermal damage has made it attractive for many surgeons looking for greater control and less damaged tissue. Just as with CO_2 laser radiation, Er:YAG laser radiation, at 2.94 μm, cannot be delivered from the laser to the surgical site by silica-based fiber. Great efforts have been made to produce a solid fiber delivery system for the Er:YAG laser using fluoride-based fibers, but these have run into durability issues. The high peak powers in the Er:YAG laser pulses tend to promote crystallization in the fluoride fibers during use. While fluoride fibers do have great potential for Er:YAG laser delivery, hollow waveguides are becoming an attractive alternative because of their durability and potential low cost.

7.3 THEORETICAL WAVEGUIDE LOSS MECHANISMS

There are two significant structural differences between hollow and solid waveguides. The first, and most obvious, is that the solid core is replaced by a gas or vacuum. The second, and less obvious difference, is that the cladding of a hollow waveguide is frequently opaque at the operating wavelengths. Most of the energy being guided by the hollow waveguides is confined to the hollow core of the structure. Therefore, highly opaque materials may be suitable for hollow waveguides as long as the gas within the core is transparent.

The functioning of a hollow waveguide is essentially identical to the functioning of a light pipe. The walls of the waveguide must confine the light to the core so that it can proceed down the waveguide unimpeded. The reflectivity of the wall is the most significant parameter determining the performance of a waveguide structure. A hollow waveguide reflecting wall surface may be a dielectric, a metal, or may be more complex, such as a dielectric-coated metal, to maximize the reflectivity. Hollow waveguides are frequently divided into two groups: attenuating total reflecting (ATR) and leaky waveguides. ATR guides have an inner wall surface with an index of refraction less than 1. The wall surface of a leaky guide, conversely, has an index greater than 1. The losses in both type of guides are due to less than perfect reflections of the propagating energy at the wall/core interface. Due to the large number of reflections over even a short length of waveguide, even an extremely small loss per reflection results in a sizable waveguide attenuation.

The walls of ATR guides are dielectric materials and cannot support any currents. The reflection of light is a result of the difference of the indices of refraction of the core and cladding materials. Since the index of the walls (n) is less than the core ($n_{core} = 1$ so $n < 1$) and the angle of the incident ray is greater than the critical angle, the light is totally reflected. If the angle to the wall normal is less than the critical angle, the Fresnel equations show that a percentage of the field will be transmitted into the waveguide wall and lost. If the wall is not conducting and its index is greater than the core, a large percentage of the light is refracted into the wall and thus the structure cannot act as an effective waveguide.

Losses in the $n < 1$ guides arise from the fact that even in a case of total internal reflection the light penetrates the wall material with an exponentially decreasing intensity, known as the evanescent field. This field will be strongly absorbed if the wall material is not highly transparent at the wavelength of the light. If the wall material is absorbing, as is the case in all currently available hollow waveguides, the greater the penetration of the wall, the greater the losses.

7.3.1 THEORETICAL SOLUTIONS FOR LIGHT PROPAGATION

Finding an electromagnetic solution for the attenuation of radiation in any waveguide requires the development of a mathematical description of the fields in the correct geometry. The mathematical description can then be combined with the boundary conditions for the particular waveguide finally to develop a characteristic equation which fully describes the electromagnetic fields in the waveguide structure.

The expression for the electric field propagating down a hollow waveguide including space and time dependence in rectangular coordinates is

$$E(x,y,z,t) = E_0(x,y)e^{i(kz-\omega t)}, \tag{7.1}$$

where k is real if the material is a perfect conductor. If k is real only, the entire z-dependent portion of the expression is imaginary (ikz) and represents simply a phase change as the mode propagates in the z direction. Nonsuperconducting materials with complex dielectric constants, and thus complex refractive indices, are used to make metallic waveguides. This results in an exponential decay of the electric field at the rate of $e^{-z\mathrm{Im}(k)}$. Converting this into power decay gives

$$P(x,y,t) = E(x,y,t)^2 = E_0(x,y,0)^2 e^{-2z\mathrm{Im}(k)}. \tag{7.2}$$

Therefore, the power attenuation coefficient (α) for hollow waveguides is given by

$$2\alpha = 2\,\mathrm{Im}(k). \tag{7.3}$$

This expression is identified as 2α, as most references identify α as the electric field attenuation coefficient or $\alpha = \mathrm{Im}(k)$. The effort to find the theoretical losses from the wave equation therefore centers around finding the complex propagation constant, k.

In rectangular coordinates, the solution for the shape of the electric fields is formed of sine and cosine functions. In a cylindrical waveguide the solution for the shape of the electric fields comprises Bessel functions in the core and Henkal functions in the cladding. Therefore, the solutions or approximations for the complex propagation constant in cylindrical hollow waveguides will usually depend, in some way, on Bessel functions.

7.4   RECTANGULAR HOLLOW WAVEGUIDE

7.4.1   THEORETICAL TRANSMISSION OF RECTANGULAR HOLLOW WAVEGUIDES

7.4.1.1   Wave Equation Approach to Straight Rectangular Hollow Waveguides

In 1974 Nishihara et al.[2] suggested using rectangular hollow waveguides for the transfer of IR energy. His group found expressions for the attenuation of the transverse electric (TE) and transverse magnetic (TM) modes in parallel-plate waveguides with the intent of designing a 10.6 μm transmission system. The expressions are shown in Equations 7.4a and b:

$$\alpha = \frac{n^2 \lambda^2}{16 d^3} \, \text{Re} \frac{1}{\sqrt{v^2 - 1}} \qquad \text{for TE}_n \text{ modes} \qquad (7.4a)$$

$$\alpha = \frac{n^2 \lambda^2}{16 d^3} \, \text{Re} \frac{v^2}{\sqrt{v^2 - 1}} \qquad \text{for TM}_n \text{ modes} \qquad (7.4b)$$

where λ is the wavelength, d is the distance across the guide, and v is the complex index of refraction of the wall material. These solutions neglect the extra losses due to the edges of the parallel plates, which are significant.

In 1976 Garmire[3] developed expressions as shown in Equation 7.5 for the attenuation in rectangular hollow waveguides at 10.6 μm by expanding on approximations made for microwave waveguides,

$$\alpha = \text{Re}\left(\frac{1}{\sqrt{v^2 - 1}} \right) \frac{\lambda^2}{a^3} + \text{Re}\left(\frac{v^2}{\sqrt{v^2 - 1}} \right) \frac{\lambda^2}{b^3}, \qquad (7.5)$$

where a is the gap between the waveguide walls and b is the waveguide height. This approximation is the combination of the attenuation of a TE_1 mode in an infinitely wide parallel-plate waveguide of separation a and a TM_1 mode in an infinitely tall waveguide of separation b.

Similar approximations have been found by McMahon[4] by modifying the transverse propagation constant of a perfectly conducting waveguide. McMahon added a complex term to account for the electrical resistance of the walls and developed expressions for the attenuation of TE and TM modes:

$$\alpha^{TE}(a) = m^2 \frac{\lambda^2}{a^3} \, \text{Re}\left(\frac{1}{v} \right) \qquad \alpha^{TM}(b) = n^2 \frac{\lambda^2}{b^3} \, \text{Re}(v). \qquad (7.6)$$

When n is much larger than 1, $1/\sqrt{v^2 - 1} \cong 1/v$ and $v^2/\sqrt{v^2 - 1} \cong v$, so Equation 7.5 by Garmire is equivalent to the expressions developed by McMahon (Equations 7.6) when the refractive index, v, is much larger than 1. A typical value for metals at 10.6 μm is 7.39 to 39.4i (nickel) which is a fair approximation for most bare metals.[5]

7.4.1.2 Ray Optic Approach to Straight Rectangular Hollow Waveguides

Since the modes in a rectangular hollow waveguide can be described by plane waves, it seems appropriate that a ray optic approach can be used to investigate the losses in such a waveguide. Garmire et al.[6] did such an analysis finding the reflection per unit length and the loss per reflection. The loss per unit length is the product of the number of reflections, N, and the loss per reflection, A, so $\alpha = AN$. Both of these quantities are functions of the angle of the ray of light for the mode in question. The number of reflections in a hollow waveguide per unit length is simply

$$N(\theta) = \frac{\tan \theta}{a} \quad \text{or} \quad N(\theta) = \frac{\theta}{a}, \tag{7.7}$$

since $\tan \theta \cong \theta$.

For small angles,

$$\theta = \frac{m\lambda}{2a}, \tag{7.8}$$

where m is the mode number and a is the wall separation.

Combining these equations and $\alpha = AN$ gives the loss relationship:

$$\alpha(\theta) = \frac{m\lambda}{2a^2} A(\theta). \tag{7.9}$$

For materials with $|v|^2 \gg 1$ and grazing incidence ($\theta \ll 1$), the Fresnel equations can be approximated as[6]

$$A^{\text{TE}}(\theta) = 4\theta \, \text{Re}(1/v) \quad \text{and} \quad A^{\text{TM}}(\theta) = 4\theta \, \text{Re}(v). \tag{7.10}$$

Combining Equations 7.9 and 7.10 gives,

$$\alpha^{\text{TE}}(a) = m^2 \frac{\lambda^2}{a^3} \, \text{Re}\left(\frac{1}{v}\right) \quad \alpha^{\text{TM}}(b) = n^2 \frac{\lambda^2}{b^3} \, \text{Re}(v), \tag{7.11}$$

which are equivalent to Equation 7.6 which was derived from modal analysis.

Materials used for a practical waveguide do not behave in an ideal manner. The walls may have well-defined refractive indices, but their surface quality may prevent the reflectivity from being the value predicted by the Fresnel equations. To account for this, Garmire et al.[6] used the slope of experimentally measured loss vs. angle of incidence plots instead of the theoretical results from the Fresnel equations. They showed that the relationship between the loss and the slope of this curve can be given by

$$S\theta = \frac{\partial L}{\partial \theta} \; \theta = L = 1 - R, \tag{7.12}$$

where the derivative, S, is evaluated as θ approaches 0, L is the loss, and R is the reflectivity. Just as with Equations 7.10, Fresnel equations show that the losses for the two mode classes are

$$L^{TE} = 4\theta \; \mathrm{Re}(1/v) \; \text{ and } \; L^{TM} = 4\theta \; \mathrm{Re}(v). \tag{7.13}$$

With this result, Equation 7.11 can be expressed as

$$\alpha^{TE}(a) = n^2 \frac{\lambda^2}{a^3} \left(\frac{S^{TE}}{4} \right) \qquad \alpha^{TM}(b) = n^2 \frac{\lambda^2}{b^3} \left(\frac{S^{TM}}{4} \right). \tag{7.14}$$

For a true waveguide with all four walls made of the same material the attenuation for the TE_{10} mode is

$$\alpha^{TE}(a) = \frac{\lambda^2}{a^3} \left(\frac{S^{TE}}{4} \right) + \frac{\lambda^2}{b^3} \left(\frac{S^{TM}}{4} \right). \tag{7.15}$$

7.4.1.3 Attenuation of Curved Rectangular Hollow Waveguides

In 1978, McMahon[4] developed a rigorous physical optics description of the losses in a curved rectangular hollow waveguide. His calculation assumes that the radiation does not interact with the wall on the inside of the bend, but solely bounces off of the outer wall as shown in Figure 7.1. This is known as the whispering gallery mode. The transition between a normal mode and a whispering gallery mode takes place when the bend radius falls below a critical level described by

$$R < \frac{8a^3}{\lambda^2}, \tag{7.16}$$

where a is the spacing between the two walls.[4] McMahon's description of the attenuation dependence on bending is given by

$$\alpha_b = \frac{2}{R} \; \mathrm{Re}\left(\frac{1}{v} \right), \tag{7.17}$$

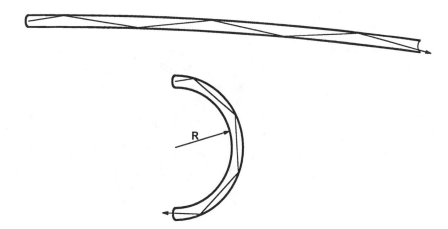

FIGURE 7.1 Normal modes (top) and whispering gallery modes (bottom) in a hollow waveguide.

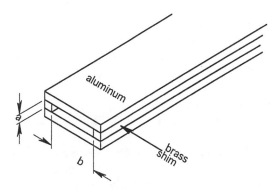

FIGURE 7.2 Ribbon metallic hollow waveguide structure.

where R is the radius of curvature and v is the complex refractive index of the wall material. The attenuation has a $1/R$ dependence on bending but does not depend upon the spacing between the walls. This is consistent with the idea of the whispering gallery mode and with observations in both rectangular and cylindrical waveguides.[7,8] A discontinuity is predicted in the attenuation due to bending at the transition between normal modes and whispering gallery modes.

7.4.2 METALLIC RECTANGULAR HOLLOW WAVEGUIDES

Although first suggested by Nishihara et al.[2] in 1974, Garmire et al.[9] were the first to publish transmission measurements of a rectangular hollow waveguide structure for the IR. Her waveguide, made of commercial aluminum strips, is illustrated in Figure 7.2. Theoretical and experimental transmission values for 1-m lengths of straight aluminum hollow waveguides compare reasonably well for a range of wall

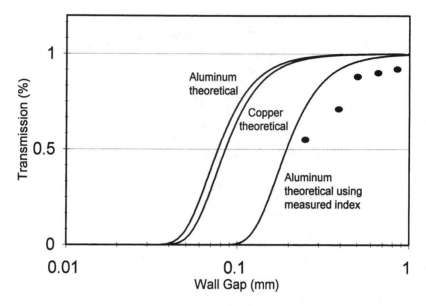

FIGURE 7.3 Normalized transmission for rectangular metallic hollow waveguides with various wall gaps. The theoretical aluminum curve uses optical constants from Table 7.4. Measured data shown by solid circles. (From Garmire, E., *Soc. Photo-Opt. Instrum. Eng.,* 320, 70, 1982. With permission.)

gaps (defined as guide height), as shown in Figure 7.3. The transmission variation due to bending of this rectangular waveguide is shown in Figure 7.4 assuming nonideal reflections.

Guides with heights greater than about 0.5 mm begin to have transmissions high enough to be practical for commercial applications. The additional attenuation due to bending radii larger than about 5 cm is also acceptable for many applications. The high power capacity of these waveguides was also demonstrated by Garmire et al.[10] She reported delivering over 200 W of continuous wave CO_2 laser energy through these waveguides and incurred no damage to the structures.

The rectangular metallic waveguides suffer from their physical geometry. The outer dimensions of the guide are approximately 10 mm by 1 mm in cross section, and this is too large for many applications. Reducing the size of the guides results in losses too high for practical use. To solve this problem, materials with orders of magnitude higher reflectivity needed to be developed.

7.4.3 DIELECTRIC COATED RECTANGULAR HOLLOW WAVEGUIDES

In 1987, Laakmann[11] patented a square dielectric-coated hollow waveguide with drastically increased surface reflectivity. This shape eliminates the tendency for the waveguide to bend in only one plane. She decreased the outside dimension of the package to under 2 mm in diameter to increase its usefulness. She had to increase the reflectivity of the inside surfaces to maintain a practical transmission level of

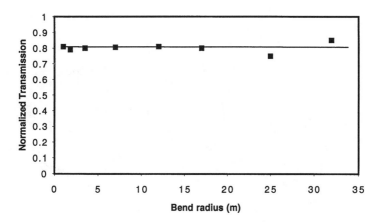

FIGURE 7.4   Attenuation coefficient due to bending for a ribbon hollow waveguide. (From Garmire, E., *Soc. Photo-Opt. Instrum. Eng.*, 320, 70, 1982. With permission.)

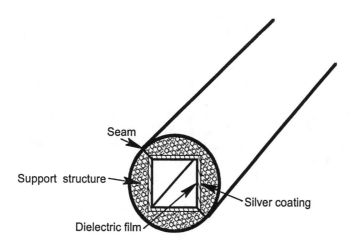

FIGURE 7.5   Rectangular hollow waveguide by Laakmann.[11]

the rectangular hollow waveguide. The form of the waveguide structure is shown in Figure 7.5. Laakmann reported using silver as the substrate metallic material and tried several dielectric coatings including ThF_4 and ZnSe. The waveguides were manufactured by forming a right angle groove in each of two silver strips. The two halves are coated with the dielectric layer, placed together to form the square guiding cavity, and then slipped inside a stainless steel tube to hold the parts in place.

The transmission of the Laakmann waveguides was adversely affected by alignment errors between the two grooves during assembly and irregularities in the surfaces. The coating of the grooves was also not ideal due to the inside geometry of the surfaces. These imperfections affected the modal quality of the output of the waveguide as well as the transmission.

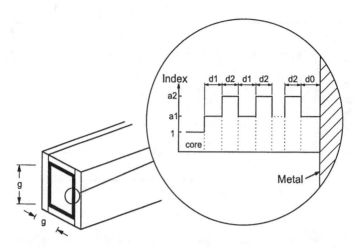

FIGURE 7.6 Structure of the coatings on the walls of a square hollow waveguide. The gap indicates any number of additional layer pairs. (From Machida, H. et al., *Appl. Opt.*, 31, 7617, 1992. With permission.)

Recently, tremendous improvements have been made in the manufacture of dielectric coated rectangular hollow waveguides. Machida et al.[12] proposed a square waveguide geometry like the previously mentioned Laakmann guide. They suggested a structure with multiple dielectric coatings of ZnS and PbTe as being capable of very low losses in normal use. This type of waveguide will have a minimum loss for polarized light, but losses due to bending would depend strongly on the direction of bending. Therefore, Machida's group described a waveguide design for circularly polarized light as shown in Figure 7.6. The thicknesses of the layers are described by the equations:

$$d_i = \frac{\pi}{2n_0 k_0 \left(a_i^2 - 1\right)^{1/2}}, \tag{7.18}$$

and

$$d_0 = \frac{1}{2n_0 k_0 \left(a_i^2 - 1\right)^{1/2}} \cdot \tan^{-1}\left(\frac{a_1}{\left(a_1^2 - 1\right)^{1/4}} \left(\frac{a_1}{a_2}\right)^m \left(\frac{a_2^2 - 1}{a_1^2 - 1}\right)^{m/2}\right), \tag{7.19}$$

where m is the number of layers and the indices are described in Figure 7.6. When these thicknesses are used, the power attenuation of the E_{nn} mode in a square hollow waveguide ($n_0 = 1$) is given by

$$2\alpha_n = \frac{16}{g^3}\left(\frac{u_n}{k_0}\right)^2 \frac{n}{n^2 + k^2}\left(\frac{a_1^2 - 1}{a_2^2 - 1}\right)^m \cdot \left(1 + \frac{a_1^2}{\left(a_1^2 - 1\right)^{1/2}}\left(\frac{a_1}{a_2}\right)^{2m}\left(\frac{a_2^2 - 1}{a_1^2 - 1}\right)^m\right)^2, \tag{7.20}$$

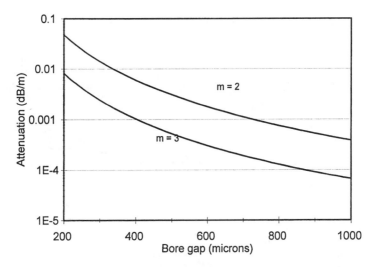

FIGURE 7.7 The theoretical transmission of a square hollow 1-mm waveguide with two and three dielectric layers.

where g is the wall-to-wall gap, n and k are the complex index of the metal, k_0 is the free-space propagation constant, and m is the number of layers. The variable u_n can be found by

$$u_n = \frac{\pi(n+1)}{2} \tag{7.21}$$

In this type of guide, all four walls of the guide are coated in a similar way. Figure 7.7 shows the theoretical attenuation of square guides with two and three layers of ZnSe and PbF$_2$ over silver. There is almost an order of magnitude decrease in attenuation by the application of the third layer. Each additional layer of dielectric produces a similar decrease in attenuation until the losses due to the dielectric materials overwhelm the reflection losses. Maximizing the difference in the index of refraction of the dielectric materials maximizes the decrease in attenuation for each additional layer.

Machida's group reported making 1-mm-bore square dielectric-coated hollow waveguides with attenuations, while straight, of about 0.1 dB/m for circularly polarized 10.6 µm laser radiation. The relative dependence of loss on bending is shown in Figure 7.8 for two 1-m-square waveguides with and without a single dielectric coating of PbF$_2$. For these measurements, light was coupled from a short piece of identical waveguide for beam conditioning. The initial behavior of both waveguides is a $1/R^2$ dependence on bending. Once the guides pass into the whispering gallery mode (at about 1 m radius), the attenuation assumes a $1/R$ dependence. The losses due to bending of this structure are moderate (80% relative transmission at a 0.5 m bend radius) and the minimum mechanical radius is large (0.3 m), but the losses are uniform and the guides could be made in long lengths without splices. The output intensity distribution of these guides exhibit several peaks, and the energy tends to

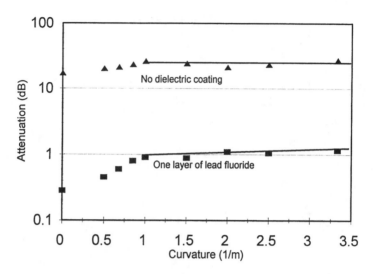

FIGURE 7.8   The measured attenuation due to bending of an uncoated and a dielectric-coated square hollow waveguide. (From Machida, H. et al., *Appl. Opt.,* 31, 7617, 1992. With permission.)

shift to the outer peak when the guide is bent.[13] This effect makes coupling the output into any other optical systems difficult.

To make these waveguides, Machida's group wrapped strips of phosphor bronze around a cylinder with the inside surface exposed. The cylinder is placed in a deposition chamber and rotated to produce a uniform coating on the phosphor bronze. The strips were coated with specific thicknesses of evaporated PbF_2 on the exposed side only. These strips were then made into the waveguide structure by soldering them together in a continuous process. Great care is required to attain a very uniform shape over the length of the guide. Other difficulties include the heat of the soldering operation damaging the coatings and solder getting into and disrupting the shape of the cavity.

7.5   CYLINDRICAL HOLLOW WAVEGUIDES

7.5.1   THEORETICAL TRANSMISSION OF CYLINDRICAL HOLLOW WAVEGUIDES

Over the last few decades, many solutions have been found describing light propagation in cylindrical step-index waveguides. Almost all of the solutions of the wave equation for solid dielectric waveguides have taken advantage of the weakly guiding approximation. This approximation assumes that there is a very small difference between the refractive index of the core and the cladding. This results in enormous simplifications in the calculations of the field components in the structure. The weakly guiding approximation, unfortunately, is not useful for describing hollow waveguides. Most hollow waveguide structures have large differences between the refractive indices of the core and cladding. Even when the refractive index of the

TABLE 7.1

Waveguide Mode Constant Dependent on the Zeros of Bessel Functions

u_{nm}	u_{n1}	u_{n2}	u_{n3}	u_{n4}
u_{0m}	3.832	7.016	10.173	13.324
u_{1m}	2.405	5.520	8.654	11.796
u_{2m}	3.832	7.016	10.173	13.324
u_{3m}	5.136	8.417	11.620	14.796
u_{4m}	6.380	9.761	13.015	16.223

cladding is near 1, the imaginary part of the cladding refractive index is large enough to make the approximation invalid. This requires more complicated solutions to be found.

7.5.1.1   Wave Equation Approach to Straight Cylindrical Hollow Waveguides

The propagation of laser light through cylindrical hollow waveguides was first thoroughly described by Marcatili and Schmeltzer[14] in 1964. Marcatili and Schmeltzer's solution for the propagation of light in a cylindrical hollow waveguide begins with the characteristic equation for a cylindrical waveguide by Stratton[15]:

$$\left(\frac{J_n'(u)}{J_n(u)} - \frac{u}{w}\frac{H_n'(w)}{H_n(w)} \right)\left(\frac{J_n'(u)}{J_n(u)} - \frac{v^2 u}{w}\frac{H_n'(w)}{H_n(w)} \right) = \left(\frac{n\lambda}{ku} \right)^2 \left(1 - \frac{u}{w} \right)^2, \qquad (7.22)$$

where a is the bore radius, $J_n(x)$ is the nth-order Bessel function of x, $H_n(x)$ is the nth-order Hankel function of x, $u = h_1 a$, and $w = h_2 a$. The expressions h_1 and h_2 are described by $h_1^2 = k^2 - h^2$, and $h_2^2 = v^2 k^2 - h^2$. k is the free space propagation constant $(2\pi/\lambda)$, h is the guided wave propagation constant, and v is the complex index of refraction of the wall material. Marcatili and Schmeltzer were able to develop an approximate solution for the characteristic equation (Equation 7.22) by making the following assumption:

$$ka \gg |v|\, u_{nm}, \qquad (7.23)$$

where u_{nm} is a characteristic of the mode being solved for and is defined by $J_{n-1}(u_{nm}) = 0$. The values for u_{nm} for a range of n and m from 0 to 4 is shown in Table 7.1. This assumption holds for glasses since they tend to have low extinction coefficients. Metals, however, can have extinction coefficients as high as 100, rendering this approximation inappropriate. Marcatili and Schmeltzer also assumed that the axial component of the guided propagation constant was nearly equal to the free-space propagation constant. The approximations that they made allowed them to replace the characteristic equation with

$$J_{n-1}(u) = iy(u/w)J_n(u),\tag{7.24}$$

where

$$y = \begin{cases} \dfrac{1}{\sqrt{(v^2-1)}} & \text{for TE}_{0m} \text{ modes,} \\[3mm] \dfrac{v^2}{\sqrt{(v^2-1)}} & \text{for TM}_{0m} \text{ modes,} \\[3mm] \dfrac{(v^2+1)/2}{\sqrt{(v^2-1)}} & \text{for HE}_{nm} \text{ modes.} \end{cases}\tag{7.25}$$

They were able to develop an approximation for the propagation constant of various modes in circular hollow waveguides, Equation 7.26, by using perturbation techniques.

$$h_{nm}(a) \cong k\left(1 - 0.5\left(\frac{u_{nm}\lambda}{2\pi a}\right)^2\left(1 - \frac{iy\lambda}{\pi a}\right)\right)\tag{7.26}$$

The attenuation constant for the nmth mode is the imaginary part of the waveguide propagation constant h. The lowest-loss mode for a hollow waveguide made of a material with an index of refraction greater than 2.02 is the TE_{01} mode. Most metallic, dielectric-coated, and high-index glasses have indices of refraction well over 2, so the low-loss mode for waveguides made of these materials is the TE_{01} mode. When the index of the cladding material is lower than 2.02, the HE_{11} mode is the lowest-loss mode. This situation arises for some glass and crystalline waveguides.[14]

7.5.1.2 Ray Optic Approach for Straight Cylindrical Hollow Waveguides

Matsuura et al.[16] have developed a very simple ray optic approximation for the attenuation of a straight hollow circular waveguide for incoherent light . In their approximation, they find the number of bounces for a ray and estimate the loss for that ray as the product of the number of bounces and the loss per bounce. The loss per bounce is found from averaging the Fresnel equations for the two polarizations and has the form[18]:

$$2\alpha(\theta) = \frac{(1 - R(\theta))\tan\theta}{2a}\tag{7.27}$$

Matsuura then integrated over the ray angles in the guide while adjusting the power for each angle according to the power distribution in the lowest-order mode.[17]

$$2\alpha = -1/z \, \log \left(\frac{\int_0^{\pi/2} P(\theta) \, \exp\left(\frac{1-R(\theta)}{2a} z \tan \theta\right) \sin \theta \, d\theta}{\int_0^{\pi/2} P(\theta) \sin \theta \, d\theta} \right) \tag{7.28}$$

The power distribution for a Gaussian input beam is described by

$$P(\theta) = A e^{-2(\theta/\theta_o)^2}, \tag{7.29}$$

where θ_o is the angle at which the $1/e^2$ intensity level is found.[18] The $1/e^2$ angle for the lowest-order mode, HE_{11}, is given by[18]

$$\sin \theta = 2.405 \lambda/2\pi a. \tag{7.30}$$

The solution to Equation 7.28 for the lowest-order mode agrees with the attenuation predicted by Marcatili and Schmeltzer. The ray optic approximation allows predictions to be made for the transmission of noncoherent light through hollow wavguides as long as the surface reflectivity and the input distribution are known.

7.5.1.3 Attenuation of Curved Cylindrical Hollow Waveguides

Marcatili and Schmeltzer also attempted to derive a solution for curved cylindrical hollow waveguides by developing corrections for the solution for straight waveguides. Their solution for the waveguide attenuation is given by

$$\alpha_{nm} = \alpha_{0nm} + \frac{a^3}{\lambda^2 R^2} \, \mathrm{Re} \left(V_{nm} \right), \tag{7.31}$$

where, α_0 is the attenuation coefficient for a straight waveguide, R is the bend radius, and $\mathrm{Re}(V_{nm})$ is the real part of V_{nm}. V_{nm} is defined by[14]

$$V_{nm} = 4/3 (y) \left(\frac{2\pi}{u_{nm}} \right)^2 \left(1 - \frac{n(n-2)}{u_{nm}^2} + 3/4 \delta \, \frac{v^2-1}{v^2+1} \cos 2\theta \right), \tag{7.32}$$

where y is defined by Equation 7.25, $\delta = 1$ when $n = \pm 1$ and 0 otherwise.

This approximation is only valid if R is large enough for σ, defined in Equation 7.33, to be much less than 1.

$$\sigma \cong 2 \left(\frac{2\pi a}{u_{nm} \lambda} \right)^2 \frac{a}{R}. \tag{7.33}$$

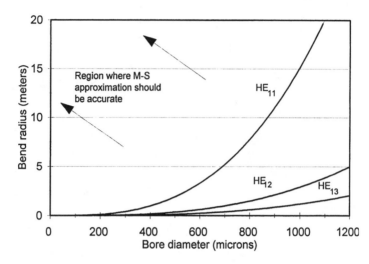

FIGURE 7.9 Region where the Marcatili–Schmeltzer (M-S) approximation is accurate for three modes in a curved cylindrical hollow waveguide.

Figure 7.9 shows the region where Marcatili and Schmeltzer's approximation for a curved waveguide is accurate. The radii where it is accurate are too large for the whispering gallery mode to exist. In these regions, the dependence of attenuation on bending goes as $1/R^2$. These bends, while not unreasonable for a laser cavity, are not appropriate for a truly flexible delivery system.

Miyagi and Karasawa[19] developed a solution for the electromagnetic fields inside of a sharply curved circular hollow waveguide in 1990. They assumed that the losses for the TE and TM components of the propagating radiation would be the same as the losses for these modes in a curved rectangular guide. They could then express the attenuation of the HE_{11} mode, polarized both perpendicular and parallel to the plane of bending, as having an inverse dependence on the bend radius. Again, this result was confirmed by experimental observations.[20]

Therefore, the attenuation of a hollow cylindrical waveguide is proportional to $1/R^2$ in the normal mode region and $1/R$ in the whispering gallery region. The transition between these two regions usually produces a discontinuity in the attenuation of the waveguide. Figure 7.10 shows the transition bend radius for a range of waveguide bore sizes.[21]

7.5.2 ENERGY COUPLING

7.5.2.1 Coupling Free Space to Waveguide Modes

The transmission of any hollow waveguide system is strongly dependent on the coupling of the energy to the waveguide modes. In 1972 Chester and Abrams[22] calculated the optimal coupling parameters by matching the energy distribution of a free-space TEM_{00} mode with the lowest-loss, the HE_{11}, waveguide mode for a circular hollow waveguide. These calculations were refined simultaneously by Hongo et al.[23] and Jenkins and Devereux,[24] and both published plots similar to the

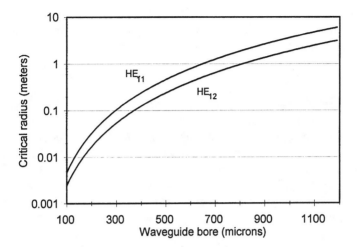

FIGURE 7.10 The transition radii of two modes from a normal mode to a whispering gallery mode at 10.6 μm.

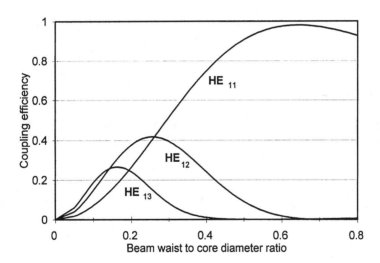

FIGURE 7.11 The power coupling coefficients for the HE_{1n} modes as a function of the ratio of the beam waist (ω) to the waveguide radius a.

one shown in Figure 7.11. The plot clearly shows that 98% coupling to the HE_{11} mode can be achieved at a ratio of beam waist diameter to bore size of 0.64. The curves also show that the coupling is not extremely sensitive to this ratio as it slowly drops off on either side of 0.64. This coupling assumes a true beam waist at the waveguide entrance with a wave front radius of infinity.

Launching incoherent light results in much higher attenuations normally due to the inability to match the field distribution of the free-space fields with the lowest-order waveguide mode. The coupling of incoherent light thus tends to produce many

higher-order modes in the waveguide. These modes are much more strongly attenuated, and thus higher overall attenuations are exhibited. The transmission of nonlaser light can still be sufficient for many applications.

7.5.2.2   Beat Length

Focusing elements are extensively used at the distal tip of products used for laser
surgery with visible and near-IR lasers such as Nd:YAG. The high power densities
and large amount of fluid present make the distal tip of a delivery system an
extremely hostile environment for any optics. At near-IR wavelengths, the water and
fluids found in the human body and likely to be on the surface of a surgical tool are
not strong absorbers. This allows focusing elements to survive even after coming in
contact with tissue. In the mid-IR region, however, these fluids absorb very strongly.
The materials used for CO_2 laser optics tend to have lower strengths and melting
temperatures than visible optical materials. This leads to almost certain destruction
of any focusing elements placed at the distal end of a mid-IR surgical system. The
ability to create a narrow beam waist at the distal end of a waveguide without any
distal optics or loss in transmission would result in an effective and durable surgical
tool.

As explained above, higher-order modes will be excited in a waveguide if the
beam waist of laser energy launched into waveguide is not reasonably close to the
ideal ratio of 0.64. Each of these modes has its own propagation constant and thus
its own attenuation and phase velocity (effective index). If several modes are coupled
into or are generated within a waveguide, the power distribution at the output can
be broadened or narrowed depending on the length of the guide due to the various
effective indices.

Roullard and Bass[25] demonstrated that the combination of the HE_{11} and the HE_{12}
modes in a waveguide results in a narrower energy distribution than the HE_{11} mode
alone at the proximal (near) end of the waveguide. These two modes propagate down
the guide at different rates and so this narrowing is reduced and then reversed when
the phase difference is π. At twice this distance, the phases of the two modes are
coincident again and the beam narrows. This is effectively a beat length, s. The beat
length can be found from Marcatili and Schmeltzer's approximation for the propagation constant of guided modes. If the wavelength is much larger than the waveguide
bore and the extinction coefficient is small,

$$\text{Im } \frac{\nu\lambda}{\pi a} \ll 1, \tag{7.34}$$

then the phase constant β can be simplified to

$$\beta_{1n} = \frac{2\pi}{\lambda} - \frac{\lambda u_{nm}^2}{4\pi a^2} . \tag{7.35}$$

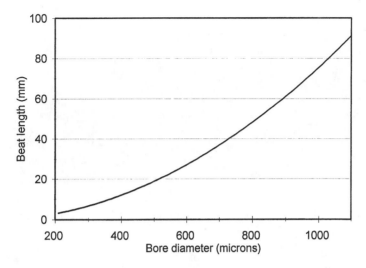

FIGURE 7.12   Beat lengths between the HE_{11} and the HE_{12} waveguide modes at 10.6 μm for a range of bore sizes.

Thus, the difference between the phase constant of the first two modes is

$$\Delta\beta = \frac{\lambda}{4\pi a^2}\left(u_{12}^2 - u_{11}^2\right).$$ (7.36)

The beat length is then found by finding the length of the phase difference for a 2π change in phase, giving

$$s = \frac{8\pi^2 a^2}{\lambda\left(u_{12}^2 - u_{11}^2\right)}.$$ (7.37)

Fortunately, in typical waveguide sizes the beat lengths are quite long because of the very small differences in the real part of the propagation constants. This distance also does not depend on the index of the material, but only on the size of the guide, the modes in question, and the wavelength. Beat lengths for a range of hollow waveguide bore sizes are shown in Figure 7.12. The large size of these beat lengths makes it easy to make waveguides with lengths that produce only small deviations from predicted intensity distributions.

Jenkins and Devereux[24] presented a description of the modal fields as they progress down a waveguide. They correlated the energy loss and temperature variations in such a waveguide with the regions of destructive interference between the two lowest-order modes. They presented evidence that the losses predominantly occur when the energy fields are concentrated near the walls of the waveguide and not uniformly down the length of the guide. They also plotted the theoretical field shapes spatially as they progress down a lossless waveguide. This was done for a

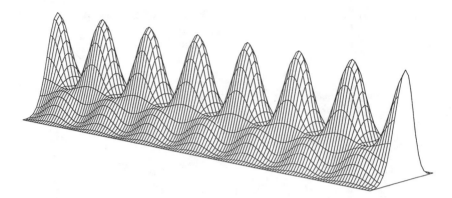

FIGURE 7.13   The energy field in a 750 µm waveguide with a launch ratio of 0.40 showing the reforming of the field each beat length.

ratio of beam waist to waveguide bore of 0.4. Figure 7.13 shows a similar simulation but includes effects of the different attenuation of the two modes and a beam waist of one half the bore.

Jenkins and Devereux do point out a potential pitfall in the design of a waveguide system using a model such as this. The phase difference between the two modes can be influenced by the location the different modes begin to interact with the waveguide walls. It is likely that the HE_{12} mode will begin to interact before the HE_{11} mode, thus offsetting the phase difference relative to the entrance of the waveguide. Care must be taken that the length chosen represents the true distance that the modes have propagated.

7.5.3   ATTENUATING TOTAL REFLECTING HOLLOW WAVEGUIDES

ATR hollow waveguides for the IR made of materials with an index less than 1 were first suggested by Hidaka et al.[26] in 1984. Total internal reflection is a label applied to situations where there is no refraction of a beam incident on an interface between two media. The situation is also usually one in which both media are highly transparent so no losses occur. Traditional optical fiber with a core index higher than the cladding index produces such a situation. A hollow waveguide with walls made of a material with an index less than 1 operates under the same principle. Although none of the energy is refracted into the wall, some of the reflected light does pass through the wall during reflection. Unfortunately, no material has been found which is essentially lossless and has an index less than 1 in the mid-IR. The light that penetrates the wall is thus strongly attenuated, resulting in the power loss of the waveguide.

The stretching, bending, or twisting of the atomic and molecular structure of a material results in internal harmonic vibrational frequencies which are purely a function of the material. If incident electromagnetic radiation has a frequency close to these material frequencies, resonant absorption will take place. The resonance interaction will cause the dielectric constant, and thus the index of refraction, of the material to fluctuate rapidly. The index of refraction of a material is indirectly

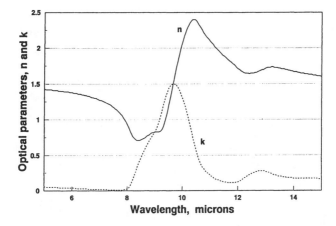

FIGURE 7.14 The index of refraction (n) and the extinction coefficient (k) of Corning 0120 glass. (From Gregory, C. and Harrington, J., *Appl. Opt.*, 32, 5302, 1993. With permission.)

dependent on the wavelength of the incident radiation. This is known as normal dispersion.[27] When the resonant frequencies are excited, the index will have regions where it increases with increasing wavelength. This effect is called anomalous dispersion. A typical form of the index of refraction in such a resonant region is shown in Figure 7.14, where n and k are the real and imaginary parts of the index of refraction, respectively. The real part of the complex index of refraction is normally identified as the index and the imaginary part as the extinction coefficient. The extinction coefficient indicates the optical density of a material at a specific wavelength. The index, n, falls to its lowest levels prior to rising in the anomalous dispersion region. In a simple oscillator model, the extinction coefficient would be quite high by the time the index drops to its lowest level. In many materials, however, the oscillations are much more complex and the extinction coefficient can still be quite low at the minimum of the index. Stronger oscillations in the material will usually produce lower values for the index, but they will also produce a narrower low-index region. Figure 7.15 shows a plot of the attenuation for a 1-mm-bore waveguide for a range of n and k found by Marcatili and Schmeltzer's electromagnetic analysis of hollow dielectric waveguides.[14]

This plot agrees with the observation of Hidaka et al.[26] from the Fresnel equations that if the index of the wall material is very much less than 1, the internal reflections will be near 100% even if the extinction coefficient is large. This is due to the low penetration of the evanescent field into the wall if the index difference is large. If little of the field penetrates the wall, the high attenuation of the wall material will not greatly increase the total waveguide loss. Thus, efficient hollow waveguides can be made from materials with nonnegligible extinction coefficients. If the index difference is not small, then the extinction coefficient must be nearly negligible for very low loss.

The behavior of the attenuation is more clearly displayed in Figure 7.16. As the index of the wall approaches 1 and the index difference approaches 0, a high loss peak develops between $k = 0.2$ and 0. When the index is 1, the peak is infinitely

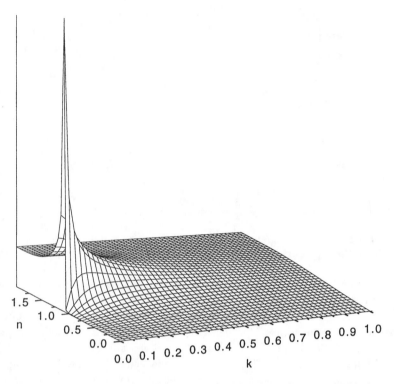

FIGURE 7.15   Attenuation of the HE_{11} mode in a perfect 1-mm-bore waveguide for ranges of the index of refraction (n) and the extinction coefficient (k).

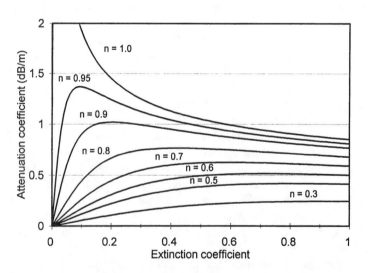

FIGURE 7.16   The attenuation of the HE_{11} mode in a 1-mm-bore hollow waveguide for a range of extinction coefficients and indices of refraction.

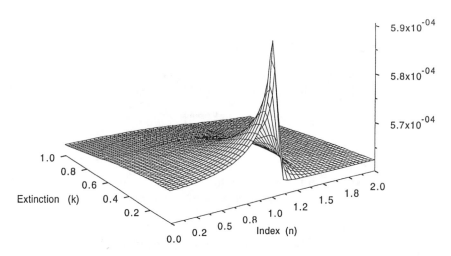

FIGURE 7.17   The relative difference in the effective indices of the HE_{11} and the HE_{12} modes in a 500-μm-core sapphire waveguide at 10.6 μm.

high and lies at $k = 0$. This is logical since, when the index difference is 0, the structure can no longer guide light; thus, the attenuation must be infinite.

The manufacture of low-loss ATR hollow waveguides requires the use of materials with indices of refraction near 0. Usually the practical use of hollow waveguides also requires the output modal pattern to be well behaved. The higher-order modes produced in many waveguides result in an output distribution that is very difficult to collimate or focus. If the energy in the waveguide remains in the lowest-order mode, the output can be easily modified by optical elements and used effectively. Different modes in the waveguide with the same effective index will intermix freely. The difference in the effective indices for the different modes should be maximized to encourage the radiation within a waveguide to stay in the lower-order modes.

The relative difference between the effective indices for the HE_{11} and the HE_{12} modes in a 500-μm-bore hollow waveguide at 10.6 μm is plotted in Figure 7.17. It should also be pointed out that the losses for higher-order modes are much greater than the loss for the HE_{11} mode. Energy that is coupled into higher-order modes is therefore much more strongly attenuated than the energy in the HE_{11} mode. This helps to maintain a low-order output, but it also increases the attenuation of the waveguide. To minimize attenuation, mode coupling needs to be minimized, so the effective index difference should be maximized.

7.5.3.1   Glass Hollow Waveguides

ATR hollow waveguides have been made from both crystalline and amorphous materials. Making hollow waveguides from glassy materials has the obvious advantage of ease of manufacture. Glass tubing is a common product, and the technologies for its manufacture are mature and widely available. High-quality surfaces and a wide variety of glass compositions for making tubing are also available. Unfortunately, while many glasses have an $n < 1$ region, very few of these regions are at

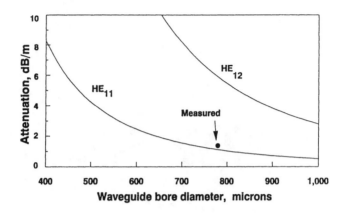

FIGURE 7.18   Theoretical and measured attenuation of a Corning 0120 glass ATR hollow waveguide. (From Gregory, C. and Harrington, J., *Appl. Opt.*, 32, 5302, 1993. With permission.)

wavelengths of interest. The regions of anomalous dispersion of most silicate glasses lie at wavelengths between 6 and 9 μm. To move the $n < 1$ region to longer wavelengths, the phonon vibrations of the glass must be moved to lower frequencies. This usually requires that the glass-forming ions be heavier or the interatomic bonds be weaker resulting in a softer and lower-melting-temperature glass. To get to the 10.6 μm region with a glass, it is necessary to avoid the high strength of the silicon–oxygen bond altogether.

Due to the amorphous nature of a glass, the resonant frequencies in glasses are broadened. This results in lower intensities in the index fluctuations and usually an index minimum close to 1. This, combined with the high extinction coefficient of most glasses, results in high-loss hollow waveguides.

Hidaka et al.[26] found that the $n < 1$ region for fused silica lies between 7.69 and 9.62 μm and that the extinction coefficient is less than 1 at wavelengths less than about 9 μm. Fused silica has an index of about 2 at 10.6 μm and so is inappropriate for a CO_2 laser waveguide. They also tested lead–silicate glasses containing 25% PbO and 71% PbO and found their $n < 1$ regions fall at 8.20 to 10.75 and 8.5 to 10.0 μm, respectively. Only the 25% PbO had potential for a hollow waveguide with transmission at 10.6 μm, but its index is too close to 1. Hidaka's group made a 1-mm-bore waveguide 1 m long and measured its attenuation as 7.7 dB/m. Lead oxide containing silicate glasses, though not useful at 10.6 μm, are appropriate for the transmission of 9.2 μm CO_2 laser radiation.[28] Gregory and Harrington[20] used Corning 0120 in experiments examining the behavior of ATR hollow waveguides. Corning 0120 glass has the composition $55SiO_2–3B_2O_3–4Na_2O–9K_2O–27PbO–2Al_2O_3$, which is similar in lead oxide content to the glasses tested by Hidaka. The complex refractive index of this glass is shown in Figure 7.14. It has an ATR region between 7.9 and 9.4 μm. Figure 7.18 shows the theoretical transmission of the lowest two symmetric modes in an 0120 hollow waveguide for a range of bore sizes. A single-size glass waveguide was drawn and its measured attenuation was quite close to theoretical. These waveguides also exhibited very little mode mixing over 1-m lengths. The output divergence of the waveguide matched the theoretical divergence

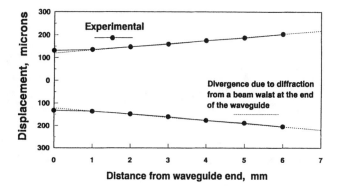

FIGURE 7.19 Measured and theoretical divergence of a Gaussian beam leaving a hollow glass waveguide. (From Gregory, C. and Harrington, J., *Appl. Opt.*, 32, 5302, 1993. With permission.)

TABLE 7.2
Complex Indices of Refraction for Some Glasses

Glass	Refractive Index	Extinction Coefficient	Attenuation of a 750-µm Waveguide, dB/m
CaAl	0.88	0.2	2.22
CaGeAl	0.81	0.58	1.80
KZnGe; U-14470	0.76	0.6	1.65
KZnGe; U-14371	1.93	1.2	3.35
27.8%PbO–SiO$_2$	1.5	0.9	2.96
TiO$_2$–SiO$_2$	1.2	1.25	2.24
SiO$_2$	2.22	0.1	4.05
GeO$_2$	0.6	0.9	1.21

Attenuations are calculated by the Marcatili and Schmeltzer's approximation.[28,29,56,60]

of a Gaussian beam with a beam waist located at the waveguide output and a diameter matching the HE$_{11}$ mode diameter of the waveguide. A comparison of the measured and theoretical divergence is shown in Figure 7.19. Gregory and Harrington[20] also showed that the polarization of an input beam is effectively maintained by these waveguides over a length of 1.25 m even with the waveguides severely bent. This is most likely a result of the minimizing of mode coupling due to the smooth surfaces and the lack of a core material. Table 7.2 shows the properties of a number of glasses at 10.6 µm and the theoretical attenuation of a 750 µm waveguide.

In 1982, Hidaka et al.[29] explored the uses of germanium-based oxide glasses for $n < 1$ hollow waveguides. The $n < 1$ region for GeO$_2$ extends from about 9.9 to about 11.5 µm, and the index is well below 1 at 10.6 µm, as can be seen in Table 7.2. Unfortunately, GeO$_2$ is not an easy glass to make and has poor environmental

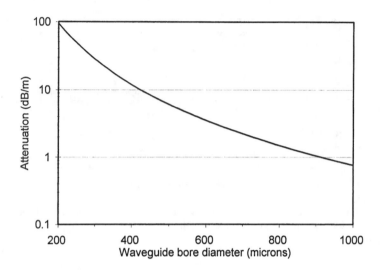

FIGURE 7.20   Calculated attenuation of the HE_{11} mode of a CaGeAl glass ATR hollow waveguide for a range of bore diameters.

stability. The 1-mm-bore waveguide they made had a theoretical loss, 2α, of 0.17 dB/m, but their experimental losses were about 2.5 dB/m.[29] Their high losses were probably a result of imperfect launching conditions and reported irregularities in the diameter of the waveguide.

Figure 7.20 is a plot of the attenuation of the CaGeAl glass calculated by Marcatili and Schmeltzer's approximation using the data in Table 7.2. As the plot shows, the attenuation drops below 1 dB/m for a bore diameter of about 1000 μm. This size and attenuation are acceptable for some applications, but the attenuation on bending is large and the glass is not chemically durable.

Worrell[30] also tested GeO_2 as a material for hollow waveguides for CO_2 laser radiation. He predicted losses could be reduced in a GeO_2 waveguide by creating a devitrified inner layer. A polycrystalline inner layer, made by postdrawing heat treatment, was predicted to reduce the attenuation by a factor of eight times.[31] The measured attenuation of a 1-mm-bore GeO_2 waveguide made by Worrell was 1.2 dB/m straight.[32] Waveguides based on attempts to devitrify germinate glasses were marketed unsuccessfully in the late 1980s.

7.5.3.2   Crystalline Hollow Waveguides

The ordered structure of crystalline materials narrows the vibrational absorption frequency bands, making resonance effects more pronounced. This results in the indices of refraction and extinction coefficients having much greater extremes in crystalline than in amorphous materials. Table 7.3 lists the complex indices for a number of common crystalline materials at 10.6 μm. These values compare well to the values of low-index glasses in Table 7.2.

In general, crystalline materials are optically superior to glasses for making ATR hollow waveguides. Many of these materials also have much higher temperature

TABLE 7.3
Complex Indices of Refraction for Some
Crystalline Materials at 10.6 μm[59-61]

Crystalline Material	Index of Refraction at 10.6 μm	Extinction Coefficient at 10.6 μm	Attenuation of a 750-μm Waveguide, dB/m
SiC	0.0593	1.21	0.11
BeO	0.0468	1.402	0.09
C–BeO	0.21	1.2	0.40
Al_2O_3	0.67	0.04	0.22
AlN	0.81	0.035	0.44

Note: Attenuations are calculated by the Marcatili and Schmeltzer approximation.

capabilities than potential glass materials. This adds to their appeal for high-power laser applications or uses in adverse environments. Crystalline materials tend to be much stiffer and more brittle than glasses, however, and so there are limitations to their applications.

Forming a hollow waveguide structure out of a crystalline material is much more difficult than making one out of a glass. Most amorphous materials have a relatively large viscosity range, while crystalline materials have clearly defined melting points. This makes forming a uniform poly- or single-crystal tube by drawing effectively impossible. Polycrystalline hollow waveguides are usually formed by extruding the ceramic material in a thick suspension. This extrusion is then sintered at high temperature producing a ridged structure. Major ceramic manufacturers have been using this technique for decades to produce refractory tubing for high-temperature applications. BeO and Al_2O_3 ceramics made by this technique have been widely used in resonators for CO_2 lasers with good results due to their optical properties. Little effort went into explaining why these materials were especially good for CO_2 lasers until the mid 1970s.

The first commercial hollow waveguide for CO_2 laser delivery was introduced in 1988 for surgical applications. The products were based on rigid alumina ceramic ATR waveguides and were designed to carry laser energy into the body through small openings. The waveguides were attached to the end of an articulated arm permitting the aiming of the output laser energy. Even though the product was designed to be ridged, Figure 7.21 shows that the transmission of CO_2 laser radiation remains essentially constant for the full bending range that the alumina was capable of withstanding. Efforts to develop a "flexible" alumina waveguide through segmentation were not successful, however. The irregular inside surface of a segmented waveguide caused excessive mode coupling and, thus, very high attenuation.

A rigid hollow waveguide made of polycrystalline beryllium oxide would have superior performance than an alumina waveguide due to the extremely low index

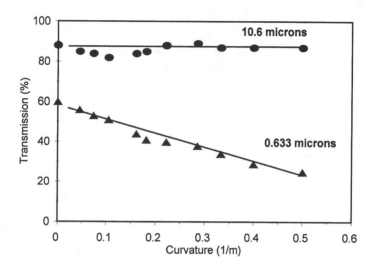

FIGURE 7.21 Transmission of bent alumina tubing for 10.6 μm and 0.633 μm laser radiation. (From Harrington, J. et al., *Soc. Photo-Opt. Instrum. Eng.,* 1048, 117, 1989. With permission.)

of BeO. Beryllia waveguides have been used for CO_2 laser cavity waveguides in a number of commercial products. The health hazards and costs of beryllia have made their uses as hollow waveguides for laser delivery unjustifiable.

To date, sapphire is the only single crystal that has been demonstrated as a hollow waveguide material.[34] Single-crystal sapphire tubing has been made for gas chromatography sample tubes for many years, but more recently its capabilities as a hollow waveguide for CO_2 laser light have been realized.

The single-crystal sapphire tubing currently manufactured is smoother, stronger, and more flexible than polycrystalline alumina tubing, but it is also much more difficult to make. Sapphire tubing is grown from a melt by the edge-defined, film-fed growth (EFG) method developed by LaBelle[35] in the late 1960s. In EFG a capillary die of molybdenum or tungsten is submerged in an Al_2O_3 melt. The capillary acts as a small, continuously replenished crucible, and long sapphire pieces are pulled from the melt by a tractor system. The aluminum oxide forms a continuous single crystal as it cools with the form of the molybdenum die.

Harrington and Gregory[34] in 1990 first reported the use of hollow sapphire as a waveguide for IR radiation. They tested bore sizes between 1060 and 330 μm, and the losses were much higher than predicted by Marcatili and Schmeltzer's theory, as shown in Figure 7.22. They demonstrated that the excess loss was predominantly due to roughness of the inside surface of the waveguides.[20] The roughness of the inside surface scatters the incident energy into higher-order modes within the waveguide. These higher-order modes are absorbed more strongly than the lowest-order HE_{11} mode, and thus the attenuation of the waveguide is increased.

The output of the hollow sapphire waveguides was found to be very well behaved. When laser energy was launched into the hollow sapphire waveguides with optimal coupling, only the HE_{11} mode in the waveguide was excited. This resulted

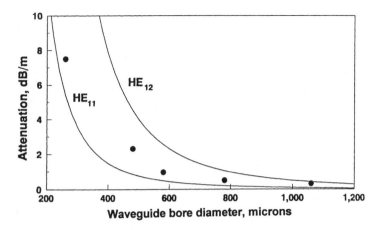

FIGURE 7.22   Attenuation vs. bore size for hollow ATR sapphire waveguides. (From Gregory, C. and Harrington, J., *Appl. Opt.,* 32, 5302, 1993. With permission.)

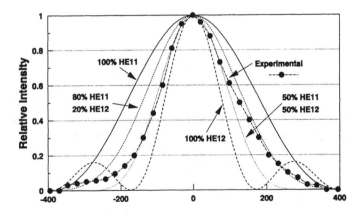

FIGURE 7.23   Output intensity cross section of a sapphire waveguide with a launch spot size less than the HE_{11} mode diameter and the theoretical distribution due to various ratios of the HE_{11} and HE_{12} modes. (From Gregory, C. and Harrington, J., *Appl. Opt.,* 32, 5302, 1993. With permission.)

in the lowest losses and nearly Gaussian output. Slightly missmatching the input beam diameter and the guided HE_{11} mode diameter resulted in the HE_{12} mode being excited as well. This produced a narrowing of the output beam, as shown in Figure 7.23, due to the destructive interference between the two modes as described in Section 7.5.2. Variations of the modes within the waveguide are most likely due to imperfections in the walls.

The lack of grain boundaries and thin walls does allow the sapphire waveguides to bend to much smaller radii than the polycrystalline alumina waveguides. The attenuation due to bending obeys the $1/R$ attenuation dependence predicted by Miyagi et al.[36] as shown in Figure 7.24. Bending the smaller size sapphire

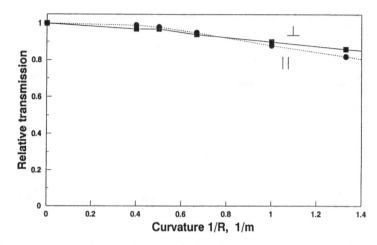

FIGURE 7.24   Attenuation due to bending for a 790-μm-bore sapphire waveguide for light polarized parallel and perpendicular to the plane of bending. (From Gregory, C. and Harrington, J., *Appl. Opt.,* 32, 5302, 1993. With permission.)

waveguides does not cause degradation to the output shape or polarization. The losses due to bending of light polarized parallel to the plane of bending are greater than the losses for light polarized perpendicular to the plane of bending. This effect is far more pronounced in metallic waveguides and has been explained by the difference in the reflection coefficient for the two polarizations.[20] Launching circularly polarized light is an effective way to eliminate variations in attenuation due to bending in orthogonal planes. Circularly polarized lights acts to average out the losses as a result of bending in each of the planes while still maintaining a good mode distribution at the output.

The high thermal stability of sapphire enhances its use as a hollow waveguide for high-power delivery. Given that these waveguides have moderate losses along their lengths, which increase with bending, low-temperature materials are limited to low-power levels. Sapphire has a melting temperature in excess of 2000°C, and, since it is a single crystal, it has no softening temperature. Sapphire waveguides have demonstrated the delivery of megawatts of pulsed transverse excitation atmospheric (TEA) CO_2 laser power with no adverse effects.[37] They have also been shown to be capable of delivering nearly 2 kw of continuous wave CO_2 laser energy with a water jacket system as shown in Figure 7.25.[38] While there is no core material to damage, at these power levels slight misalignments can cause catastrophic failure of the input end of any waveguide. The existence of a large percentage of higher-order modes in the laser output can also cause destruction of a waveguide at high power levels. The higher-order modes of the laser beam will be coupled into higher-order modes in the waveguide. This energy will be attenuated in the first few centimeters of the guide. If the power in these high-order modes is more than a few watts, the waveguide input end will not survive. Great care must be taken in the design of any waveguide system at these power levels to protect the waveguide from misalignment and non-TEM_{00} laser energy.

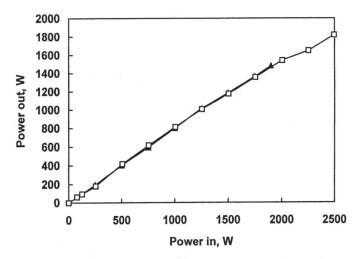

FIGURE 7.25   Power transmission by a 1-mm-bore hollow sapphire waveguide with water cooling. Solid marks are for a 2000 W laser and hollow marks are for a 3000 W laser. (From Nubling, R. and Harrington, J., *Appl. Opt.,* 1995.)

While sapphire is extremely hard and has a very high melting temperature, it is also very stiff and brittle. It is also very difficult to make with no mechanical flaws. Slight imperfections in the growth process will create defects in the grown sapphire capillary tubing. These defects in a hollow waveguide will result in excess optical losses as well as great strength reductions. Unfortunately, the stress concentrations resulting from these defects are very difficult to find without destructive testing. The high cost of the sapphire capillary makes such testing prohibitive. As of 1992, bend radii under 0.75 m for a 790 μm bore waveguide were hazardous. Sapphire is so chemically durable that great effort must be taken to find substances or environments which will etch it to remove defects.

7.5.4   DIELECTRIC-COATED CYLINDRICAL HOLLOW WAVEGUIDES

Dielectric-coated cylindrical hollow waveguides consist of three parts. They all have (1) a supporting substrate member which gives the guide its mechanical and geometric properties. The substrate supports (2) an inside metallic layer that is overcoated with (3) a dielectric to enhance the reflectivity of the surface. Creating such a waveguide has proved challenging, as traditional techniques for making high-quality reflective coatings are not compatible with this waveguide geometry. Typical waveguides are over a meter long and have bores under 1 mm in diameter. Traditional vapor deposition techniques do not produce good-quality coatings on the inside of capillary tubing this small. To overcome this difficulty, techniques for forming the waveguide shape after the coating has been deposited were developed. More recently, liquid coating techniques have been used with good success to produce low-loss waveguides.

7.5.4.1 Theoretical Attenuation of Dielectric-Coated Cylindrical Hollow Waveguides

In 1984, Miyagi and Kawakami[39] published a solution for the theoretical attenuation of dielectric-coated hollow waveguides for the IR. They developed approximations for the characteristic equation that use normalized surface impedance and admittance of the dielectric-coated surface in place of the complex indices of refraction used by Marcatili and Schmeltzer. Their approximation for the field attenuation coefficient of a waveguide with radius a has the form:

$$\alpha_{nm} = \left(\frac{u_{nm}}{2\pi n_0}\right)^2 \left(\frac{\lambda^2}{a^3}\right) \cdot \frac{1}{2} \operatorname{Re}(z_{TE} + y_{TM}), \qquad (7.38)$$

where n_0 is the index of the gas filling the bore ($n_0 = 1$) and z and y are the normalized surface impedance and admittance, respectively. Values for z and y can be found from the material properties and the thin film thickness by the equations:

$$z_{TE} = \frac{1}{\sqrt{n_s^2 - 1}} \frac{\sqrt{n_s^2 - 1} + j\sqrt{n_f^2 - 1}\ \tan\left(n_0 k_0 t\sqrt{n_s^2 - 1}\right)}{\sqrt{n_f^2 - 1} + j\sqrt{n_s^2 - 1}\ \tan\left(n_0 k_0 t\sqrt{n_s^2 - 1}\right)} \qquad (7.39)$$

and

$$y_{TM} = \frac{n_s^2}{\sqrt{n_s^2 - 1}} \frac{\dfrac{\sqrt{n_s^2 - 1}}{n_s^2} + j\dfrac{\sqrt{n_f^2 - 1}}{n_f^2}\ \tan\left(n_0 k_0 t\sqrt{n_s^2 - 1}\right)}{\dfrac{\sqrt{n_f^2 - 1}}{n_f^2} + j\dfrac{\sqrt{n_s^2 - 1}}{n_s^2}\ \tan\left(n_0 k_0 t\sqrt{n_s^2 - 1}\right)}, \qquad (7.40)$$

where n_s and n_f are the complex indices of refraction of the metal layer and dielectric film, respectively.[40] The thickness of the film is given by t, and k_0 is the free-space propagation constant. These equations reduce to Marcatili and Schmeltzer's approximation if n_s and n_f are set equal to each other.

The attenuation of these waveguides can be described by separating the effects of the metal layer and the film. The attenuation then has the form:

$$\alpha_{nm} = \left(\frac{u_{nm}}{2\pi}\right)^2 \left(\frac{\lambda^2}{a^3}\right) \left(\frac{n_s}{n_s + k_s}\right) \cdot F, \qquad (7.41)$$

where n_s and k_s are the index and extinction coefficient of the metal and F is a factor accounting for attenuation due to the dielectric film. It is quite obvious that the performance of a waveguide is still strongly dependent on the optical properties of the metal under the dielectric film. Attenuation would be minimized by using a metal

TABLE 7.4
Optical Properties of Hollow
Waveguide Metals at 10.6 μm[39]

Metal	n	k	n/(n + k)
AG	13.5	75.3	0.15
NI	9.08	34.8	0.21
Cu	14.1	64.3	0.18
Au	17.1	55.9	0.23
Al	20.5	58.6	0.26

TABLE 7.5
Optical Properties of
Dielectric Materials for
Hollow Waveguide Films[13,58]

Dielectric	n
Ge	4.0
KCl	1.47
ZnSe	2.2
ZnS	2.22
CaF$_2$	1.28
PbF$_2$	1.63
AgI	2.1
PbTe	5.7

with a low index but a very high extinction coefficient. Good choices include silver, nickel, copper, and gold as shown in Table 7.4. Other variables, such as the quality of the surface producible and the resistance to oxidation or other attack, will also influence the choice of a material for the reflecting layer.

The material and geometry of the dielectric film are obviously also critical. The optical properties of likely candidates for the dielectric film are given in Table 7.5. For a given metal coating, the dielectric film material should be selected with the maximum index of refraction to minimize the attenuation of the waveguide. Germanium has a high index relative to other dielectric materials and can be deposited by conventional techniques. This has motivated its extensive use in hollow waveguide development.

The optimal thickness for a dielectric film can be found from the equations for the attenuation of a dielectric-coated hollow waveguide above. When solved for the minimum loss, two equations are developed:

$$t_i = \frac{1}{k_0 \sqrt{n_f^2 - 1}} \left(\tan^{-1} \left(\frac{n_f}{\sqrt[4]{n_f^2 - 1}} \right) + (i - 1) \frac{\pi}{2} \right), \qquad (7.42)$$

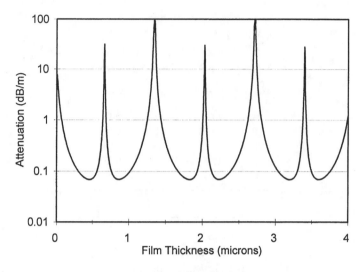

FIGURE 7.26 Theoretical attenuation of an 800-μm-bore, Ge-coated silver hollow waveguide for a range of coating thicknesses.

where $i = 1,3,5,\ldots$, and

$$t_i = \frac{1}{k_0\sqrt{n_f^2-1}}\left(\tan^{-1}\left(\frac{n_f}{\sqrt[4]{n_f^2-1}}\right)+i\,\frac{\pi}{2}\right),\qquad (7.43)$$

where $i = 2,4,6,\ldots$[41] The minimum attenuation coincides with the maximum reflection attainable for the surface at the incidence angles of the modes. As is familiar in reflective dielectric coatings, the maximum reflectance does not exist at a singular film thickness but a series of incremental thicknesses as shown in Figure 7.26. Each dip in the figure corresponds with a low-loss thickness as defined by Equations 7.42 and 7.43.

7.5.4.2 Mode Transitions in Hollow Waveguides

The behavior of electromagnetic fields inside hollow waveguides is straightforward and easy to predict as long as the bore diameter is large. As hollow waveguides get small, however, changes take place in the shape of the electric and magnetic fields within the waveguide. Kato and Miyagi[40-42] used a factor P, defined as the ratio of the H_z component to the E_z component in a propagating mode. By solving the characteristic equation numerically, they found that at large bore sizes the absolute value of P approaches 1 for the EH and HE degenerate modes. As the bore size is reduced, the P value for the EH mode approaches infinity, while the value for the HE mode moves toward 0. Thus, the EH_{11} mode becomes the TE_{12} mode and the HE_{11} mode becomes the TM_{11} mode. This transition is shown in Figure 7.27 for a metallic guide.

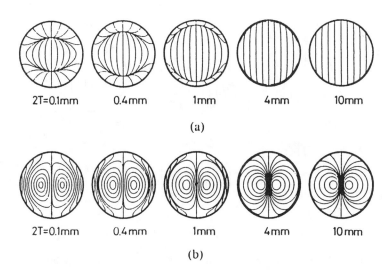

FIGURE 7.27 Mode transitions as a hollow metal waveguide size changes. (a) The electric field lines of the EH_{11} mode at $2a = 1$ mm change to a TE mode at $2a = 0.1$ mm. (b) The electric field lines of the HE_{11} mode at $2a = 1$ mm change to a TM mode at $2a = 0.1$ mm. © 1992 IEEE. (From Kator, Y. and Miyagi, M., *IEEE Trans. Microwave Theory Practice*, 40, 679, 1992. With permission.)

These transitions are strongly dependent on the waveguide materials as well as on the bore size. A metallic guide may see major field changes at bore diameters just under 1 mm. At 10.6 µm there is little change in the fields within dielectric waveguides, such as sapphire or SiC, until the bores of the guides are less than 0.2 mm in diameter. Unlike metal and ATR guides, if the index of the dielectric guide is greater than 1 and the extinction coefficient is small, the HE_{11} mode becomes like the TE_{12} mode. This is the case for SiO_2 and Ge waveguides at 10.6 µm.

The attenuation calculated by the traditional approximations is accurate for any solid dielectric waveguide, ATR or leaky, with bore sizes to well below 100 µm. The attenuation approximations for metallic waveguides, however, deviate from the exact solution at larger bore sizes. The approximation for the HE_{11} mode is reasonably accurate in most metallic guides until the bore diameter drops below 200 µm; then the solutions diverge quickly.

In a dielectric-coated waveguide, the changes in the mode shape are strongly dependent on the coating thickness.[39] As described in the previous section, a dielectric-coated metal waveguide has attenuation minimums for two series of coating thicknesses. The first series, when $i = 1, 3, 5, \ldots$ (of Equation 7.42), results in a structure where the HE_{11} mode will become like the TM_{11} mode as the bore size is reduced. The other series, $i = 2, 4, 6, \ldots$ (of Equation 7.43), results in a structure where the HE_{11} mode will become like the TE_{12} mode. The transition of the modes in a 0.1 mm CaF_2 dielectric-coated silver waveguide to the HE_{11} mode in a 1 mm waveguide for even and odd i thicknesses of the dielectric film is shown in Figure 7.28. The TE_{12} mode has lower attenuation in these waveguides, so, as the bore diameter of the waveguides is reduced, a waveguide with an even i dielectric

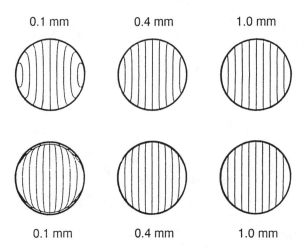

FIGURE 7.28 Mode transitions in dielectric-coated metallic waveguides: even multiples of the ideal coating thickness (top); odd multiples of the ideal coating thickness (bottom). © 1994 IEEE. (From Kato, Y. and Miyagi, M., *IEEE Trans. Microwave Theory Practice,* 42, 2336, 1994. With permission.)

thickness will have a lower loss than one with an odd i thickness. When the thicknesses of the coatings do not match the prescribed low-loss thicknesses, the fields are distorted such that much higher losses are produced. Higher-order modes, with greater electric field intensities near the wall, are even more severely attenuated. The results of attenuation calculations by the approximate theories, which ignore the effects of field distortion, lie between the results for the even and odd i thicknesses. The approximate solutions are quite good for bore sizes greater than 300 μm in most dielectric-coated metal waveguide structures at 10.6 μm.

7.5.4.3 Evaporated Coatings

In 1983, Professor Miyagi's group[7] at Tohoku University, published results of efforts to make cylindrical dielectric-coated hollow metallic waveguides. They started with a 1.2-m-long mandrel of polished aluminum tubing. This mandrel was placed in a modified sputtering chamber and coated with approximately 0.45 μm of germanium. The metallic (and substrate) layer was then created by electroplating up to 200 μm of nickel on top of the germanium. The mandrel tube was finally removed by etching the aluminum with an NaOH solution. They were able to make waveguides with bore diameters of 1.5 mm and lengths of about 1 m. The measured attenuation of these waveguides was approximately 0.4 dB/m when straight (an order of magnitude greater than theoretical) and dropped off quickly when bent. The relative transmission of light polarized parallel to the plane of bending dropped below 20% at a bend radius of 3 m. Miyagi suggested that the excess losses were probably due to the roughness of the metallic surface and variations in thickness of the dielectric layer.

These same structures were later demonstrated to have great power-transmitting capacities. Hongo et al.[23] delivered 2.6 kW of continuous wave CO_2 laser energy

TABLE 7.6
Attenuation of Three Cylindrical ZnS-Coated Ag, 1-mm-Bore
Hollow Waveguides 1 m Long at Three Laser Wavelengths

Laser Type	Wavelength, μm	Attenuation Straight, dB	Attenuation Bent, dB
Er:YAG	2.94	0.40	1.2
CO	5.3	0.50	1.1
CO_2	10.6	0.25	0.6

Note: The coatings on the surfaces of the guides were designed for the specific wavelengths (bending radius, $R = 0.5$ m).[43]

through a 1.7-mm-bore, 2-m-long waveguide. The transmission of the waveguide was 86% when straight, but dropped quickly when bent. The output of these waveguides was also multimode, making it difficult to use for materials-processing applications.

Miyagi's group continued their efforts to produce cylindrical hollow waveguides with smoother interior surfaces. They modified their technique by applying a polyimide film to the outside of the aluminum mandrel tube.[43] The aluminum tube maintained the circular shape of the waveguide, and the polyimide film helped to keep the vapor-deposited ZnS dielectric thin film smooth. Three types of ZnS-on-Ag waveguides were manufactured, each optimized for a different laser wavelength. The performance of these waveguides is summarized in Table 7.6. The attenuation for the CO laser is an average for the two polarizations, parallel and perpendicular to the plane of bending. The attenuation due to bending of the guides for CO_2 laser light was higher than the attenuation of guides made without the polyimide film (0.55 dB at a 2 m radius). These guides were able to sustain considerable power levels. They suffered no damage when about 800 mJ pulses of Er:YAG laser energy at 3 Hz was delivered to a target.

In 1990, Levy[44] patented a technique for manufacturing dielectric-coated hollow waveguides by rolling a substrate strip into a circular cross section after coating it. The coatings in these waveguides were made by traditional coating techniques on long substrates rolled lengthwise around a drum. After coating, the strips were flattened and rolled crosswise and slipped inside an outer jacket. The major difficulties encountered with the Levy design are the development of a coating which can survive the bending process and maintaining an extremely smooth and uniform inside surface. Deforming a coated metal to this extent puts the coating under a great deal of stress and can induce coating failure in the form of delaminating and flaking. A sample tested at Rutgers University performed quite well optically, but some of the coating was observed to flake off over a period of time. By using optimized coupling, the output end of the waveguide began to glow after a few seconds at a transmitted power level of 70 W. Applying an airflow to the outside of the waveguide could prevent the glowing, but without the airflow the output end would quickly heat up to a temperature that would damage the coatings.

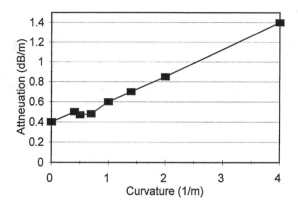

FIGURE 7.29   The effect of bending on the attenuation of a PbF$_2$-coated metallic waveguide by Levy. (From Harrington, J. and Matsuura, Y., *Soc. Photo-Opt. Instrum. Eng.*, 2396, 4, 1995. With permission.)

The Levy dielectric-coated circular waveguide was also tested for attenuation due to bending, and the results are plotted in Figure 7.29. The waveguide attenuation also has a 1/R dependence on bending with a shallow slope indicating only a moderate sensitivity to bending. These waveguides also demonstrate good durability and can be bent to radii of less than 20 cm. At smaller radii, crimping of the waveguide became likely.

The modal output of the Levy waveguide is not as well behaved as the attenuation. Figure 7.30 compares the output intensities of a hollow sapphire waveguide and a Levy waveguide made by Laser Power Optics, Inc. The launch conditions are similar, but the mode mixing in the Levy waveguide results in a highly disordered output. Efforts to recollimate the light transmitted by the Levy guide were unsuccessful for obvious reasons. Again, the mode mixing is probably due to the inside of the waveguide being noncircular and to the roughness of the coated surfaces.

7.5.4.4   Liquid Coatings on Plastic

In 1989, Croitoru et al.[45] developed a novel structure and process for dielectric-coated metallic hollow waveguides. They used a hollow plastic tubing as a substrate in which they deposited a thin and flexible metallic layer. The innermost part of this metallic coating was then converted into a dielectric thin film. One process starts with tubing that can be made of a variety of polymers. The inside of the tube is first etched with sulfochromic acid to remove any contaminants. The inside surface is then sensitized with a solution of SnCl$_2$ and HCl followed by an activation step using a solution of PdCl$_2$ and HCl. The inside wall is then coated with silver by circulating a solution of dissolved AgNO$_3$ and sodium dodecylbenzensulfonate through the tubing. The final step in the plating process requires the circulation of a reducing solution (N$_2$H$_4$:H$_2$O) through the tubing. This process, and ones similar to it, produce smooth uniform thin coatings which stick well to the inside of the plastic tubing. The final waveguide configuration is made by reacting the metallic

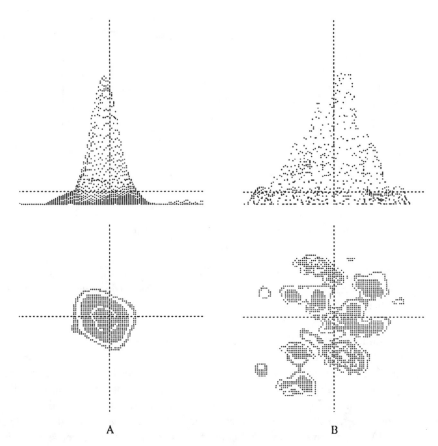

FIGURE 7.30 Side view and top contour view of the output intensity of (A) a hollow sapphire waveguide, and (B) a rolled hollow dielectric-coated metal waveguide. (From Gregory, C. and Harrington, J., *Appl. Opt.,* 32, 5302, 1993. With permission.)

layer with a second solution, such as iodine in water, to produce a thin AgI dielectric layer.

Croitoru's group has reported straight attenuations of about 0.6 dB/m for a 4-mm-bore waveguide, and, more recently, 1.8 dB/m attenuations were measured for a 1-mm-bore hollow plastic waveguide.[46,47] Both waveguides are highly flexible because of the plastic substrate, but they suffer from high attenuations due to bending. Figure 7.31 shows the transmission loss due to bending of a 1-mm hollow metal-coated plastic waveguide. The increase in attenuation is modest for bends as small as 12 cm. Unfortunately, the modest increase adds onto a relatively high attenuation when the guide is straight.

The dielectric-coated plastic waveguides also suffer from severe mode mixing. The spatial output intensity of these guides is strongly dependent on the bending of the guide.[44] The large amount of mode mixing is most likely a result of the large bore diameter of the guides and the roughness of the plastic substrate walls. Plastic tubing of this type is normally extruded through a die and therefore does not have

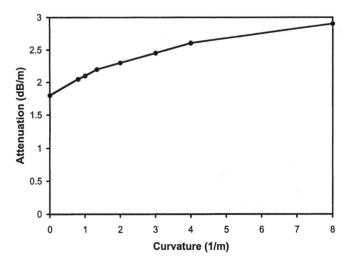

FIGURE 7.31 Attenuation due to bending of a dielectric-coated metallic waveguide on a plastic substrate. (From Harrington, J. and Matsuura, Y., *Soc. Photo-Opt. Instrum. Eng.*, 2396, 4, 1995. With permission.)

inherently smooth surfaces. The plastic normally has to flow around a spider structure supporting the core of the die. This normally leaves defect lines on the surfaces of the plastic. Surface tension does only a moderate job to reduce the defects. A smooth plastic surface typically has a root mean square (rms) roughness of 35 nm, while glass, as drawn, has a typical rms roughness of 5 nm.[48] It has been pointed out that the plastics used as substrates have greater thermal conductivity than fused silica glass used for other waveguides. This could be important at high power levels if it is necessary to cool the exterior of the waveguide to preserve the coatings. The metal tube–based waveguides would, however, have superior thermal conductivity than either plastic or fused silica. It is certainly also conceivable to use a glass with greater thermal conductivity than silica if necessary.

Mechanically, the dielectric-coated plastic waveguides are very robust. They have high flexibility and do not kink as easily as metallic waveguides. The uniform attenuation of energy along the lengths of the waveguides minimizes the hazards from high-power laser use (under 100 W), but these guides are sensitive to misalignment of the laser launch. The plastic and the coatings are easily damaged by laser radiation being focused on them. Special care must be taken to ensure that the alignment is correct before the power level is increased to even a few watts.

7.5.4.5   Liquid Coatings on Metal

In 1994, Morrow and Gu[49] reported a cylindrical waveguide in which the support structure constitutes the metallic layer. They used silver tubes as the starting material and, using a process similar to Croitoru described above, produced a silver iodide or bromide thin film on the inside surface. They have produced 1-mm-bore waveguides with attenuations below 0.1 dB/m at 10.6 μm when straight. The attenuation of the

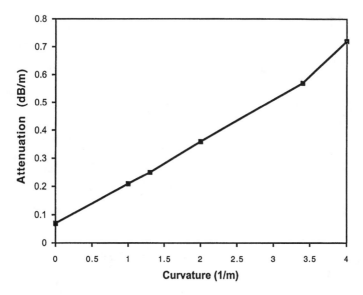

FIGURE 7.32 Attenuation of a dielectric-coated, solid silver substrate waveguide due to bending. (From Harrington, J. and Matsuura, Y., *Soc. Photo-Opt. Instrum. Eng.*, 2396, 4, 1995. With permission.)

waveguides is strongly dependent on bending, but, as Figure 7.32 shows, the attenuation is still below 0.8 dB/m when the bend radius is as small as 25 cm. Just as in all other waveguide structures, attenuations this low require the excitation of the HE_{11} mode only. This can only be done with a laser producing a nearly perfect TEM_{00} output beam and properly coupling it into the waveguide. Even with a nearly ideal beam and launching, the output of the silver tube–based waveguides is multimode. The output profile is well concentrated spatially when the guide is straight, but the beam is spread when the guide is bent. Exact calculations about the percent of energy in the lowest-order mode at the output have not been done. The multimode output of these waveguides is probably a result of roughness and distortion of the inside surface of the substrate tubing. Metallic tubing can be made by rolling and welding or drawing through a die. Neither process is very good at producing a smooth surface, and the inside of a long, small-diameter tube is very difficult to polish mechanically. Morrow and Gu describe the elaborate chemical cleaning and polishing techniques necessary to improve the inside surface of the hollow silver before the tube is treated to create the dielectric coating. These processes can improve the surface smoothness considerably, but it is questionable if it will ever be possible to achieve the finish inherent in drawn glass.

7.5.4.6 Liquid Coatings on Glass

A major step toward the commercialization of flexible hollow waveguides for the delivery of laser radiation was made with the development of dielectric-coated metallic waveguides on glass tubing substrates. This was suggested by Croitoru et al.[45] in their plastic tubing waveguide patent. The use of glass as a substrate has several

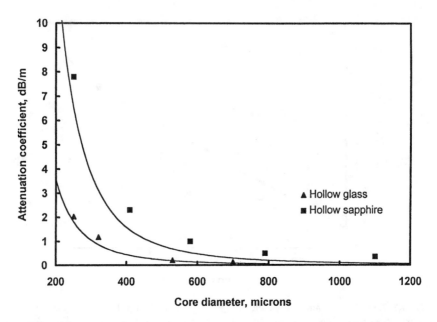

FIGURE 7.33   Attenuation of a series of hollow glass waveguides compared with theoretical (solid line) and sapphire waveguides. (From Nubling, R. and Harrington, J., *Appl. Opt.,* 1995. With permission.)

distinct advantages. Glass is one of the few materials that does not neck as it is drawn. It is therefore easy to make long, uniform tubing or fiber from most glasses. Surface tension also acts to make glass very smooth during the drawing process without extra effort. Last, glass capillary tubing of many sizes is common and inexpensive.

In 1994, Harrington's group[50] at Rutgers University presented a waveguide with a bore of 530 μm and a CO_2 laser radiation attenuation when straight of 0.3 dB/m. These guides are also made by a solution coating process of depositing a metallic layer on the inside of silica capillary tubing and then changing the surface to a dielectric thin film. The attenuation of a series of these waveguides is shown in Figure 7.33 along with the attenuation of sapphire waveguides and the theoretical attenuations for both. The outside of the tubing is polymer coated during the draw process to maximize its strength. These glass waveguides can be bent to radii as small as 5 cm and still maintain attenuation levels under 2 dB/m. Figure 7.34 plots the attenuation due to bending of a 530-μm-bore glass substrate dielectric-coated hollow waveguide. These guides have also been used to transmit over 1 kW of continuous wave CO_2 laser energy with water cooling. The power-transmitting behaviors of these waveguides when straight and bent are shown in Figures 7.35 and 7.36, respectively.

The coatings on the inside of the tubing can be modified to select the wavelength of maximum transmission. Glass tubing–based dielectric-coated waveguides have been fabricated to deliver Er:YAG laser radiation at 2.94 μm. Figure 7.37 shows the transmission of a number of hollow waveguides at 2.94 μm. The output from most

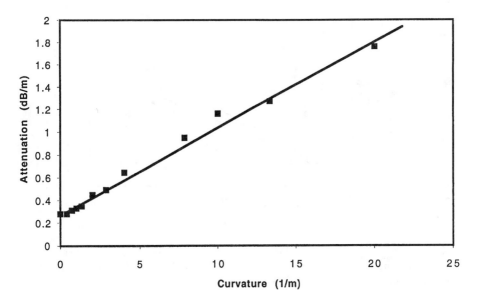

FIGURE 7.34 Attenuation due to bending of a 530-μm-bore experimental hollow dielectric-coated glass waveguide. (From Harrington, J. and Matsuura, Y., *Soc. Photo-Opt. Instrum. Eng.,* 2396, 4, 1995. With permission.)

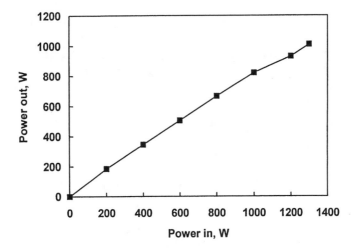

FIGURE 7.35 High-power behavior of straight 700-μm-bore dielectric-coated glass hollow waveguides at 10.6 μm. (From Nubling, R. and Harrington, J., *Appl. Opt.,* 1995. With permission.)

Er:YAG lasers is not TEM_{00}, so a portion of the attenuation may be due to the launching of higher-order modes. This excess attenuation may also be due to increased scattering at the shorter wavelength. These waveguides can also be designed to produce broadband transmission from about 3 μm to well beyond 15 μm

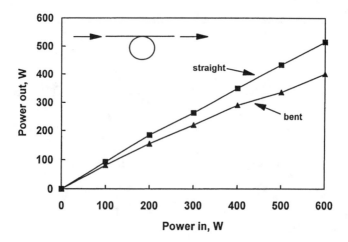

FIGURE 7.36 High-power behavior of straight and bent 700-μm dielectric-coated glass hollow waveguides at 10.6 μm. The bend was 360° at a radius of 15 cm. (From Nubling, R. and Harrington, J., *Appl. Opt.*, 1995. With permission.)

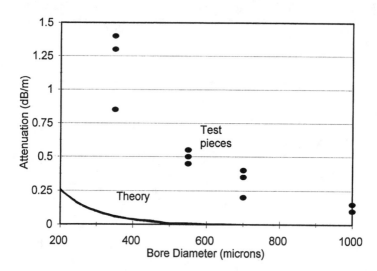

FIGURE 7.37 Transmission of several samples of hollow dielectric-coated glass waveguides at 2.94 μm. (From Nubling, R. and Harrington, J., *Appl. Opt.*, 1995. With permission.)

for spectroscopy or thermometry applications. A plot of the spectral attenuation for two waveguides, one with coatings designed for CO_2 laser energy and the other for broadband IR light transmission, is shown in Figure 7.38. The absorption peaks are actually interference bands from the dielectric coating.

The output of these waveguides is very well behaved if the guides have bore diameters less than about 1000 μm. If the input is properly coupled, the guided mode is almost entirely HE_{11} and the output is close to Gaussian. Matsuura et al.[51]

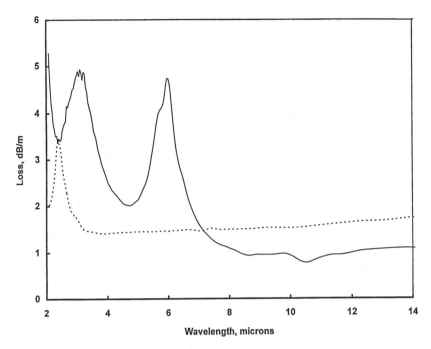

FIGURE 7.38   Spectral transmission of two dielectric-coated glass hollow waveguides. One is designed for minimum loss at 10.6 μm, the other for broadband transmission. The peaks are due to interference in the dielectric coating. (From Abel, T. et al., *Opt. Lett.,* 19, 1034, 1994. With permission.)

demonstrated that the output of a 250-μm-bore waveguide maintains 90% modal purity of the HE_{11} mode even when bent through seven loops with a radius of 25 mm. Larger-bore waveguides do not preserve the mode as well. The output of a 530-μm-bore waveguide was about 85% pure after the guide was bent to a radius of 69 mm. As in the sapphire waveguides, the polarization of an input beam is effectively maintained over the several meter length of these guides.

The good performance of these waveguides is most certainly a result of high-quality coatings being deposited on very smooth and uniform substrate surfaces. The smooth surfaces and uniform coatings allow the waveguides to be made far smaller than 1 mm and still have adequate transmission for most applications. The smaller size acts to maintain the mode quality and the polarization of the transmitted energy. The small size also allows the glass-based waveguides to be sufficiently flexible and robust for most applications.

7.5.4.7   Influence of Wall Roughness

It has been concluded by a number of researchers that surface roughness is the primary influence preventing hollow waveguides from achieving their theoretical limits.[20,47,52] Gregory and Harrington[20] demonstrated the correspondence between the roughness of sapphire waveguides and their attenuation. They also demonstrated that a glass ATR hollow waveguide, with very low rms roughness, achieved nearly theoretical attenuation levels. It has been repeatedly demonstrated that polishing the substrate surfaces before applying any coatings drastically improves waveguide transmission. Glass-based waveguides have inherently smoother surfaces because of the glass-drawing process. It has also been shown that sapphire waveguides have performance superior to alumina waveguides even though the compositions are identical. The extruded polycrystalline alumina has much rougher surfaces than the grown sapphire, but the grown sapphire is still not as smooth as glass. The smoothest waveguides have consistently produced the lowest losses and the least mode mixing.

Rough inside walls in a hollow waveguide tend to couple energy from lower-order modes to higher-order modes. In ray optic terms, the disruption of specular reflection by a rough surface explains the increased attenuation. The glancing incidence of the light makes the impact of surface defects more pronounced. Roughness of only a few nanometers (rms) can have a profound impact on the transmission of a waveguide due to the large number of reflections over the length of the guide.

Mode coupling due to roughness also has an impact on the output energy distribution of hollow waveguides. If only the HE_{11} mode is excited in a hollow waveguide and there are no perturbations in the guide, the output should be 100% HE_{11}. Small defects in the walls of hollow waveguides will couple light into higher modes distorting the Gaussian output of the waveguide. All waveguide types exhibit improved output modal quality for smaller-bore diameters. As the diameter of a waveguide is reduced, the difference in the effective indices of the different modes in the guide increases. If the effective indices of two modes are the same, they are considered degenerate and energy can couple freely between them. As the difference in the effective indices increases, the energy requires greater perturbations in the waveguide structure to transfer it to another mode. Thus, smaller waveguides maintain better mode quality. This benefit is countered, however, by the $1/r^3$ dependence of attenuation on the size of the waveguide. A high-quality guide must balance lower transmission with better mode preservation. Hollow dielectric-coated glass waveguides achieve good mode preservation because of their very smooth walls and small diameters. Their smooth walls also produce low attenuation at the small bore sizes. Increasing the bore size of the glass waveguides does reduce the modal purity of the output. If the metal or plastic substrates of the other types of liquid-deposited dielectric-coated waveguides could be made as smooth as the glass waveguides, their performance would be similar.

Wall roughness can also be described as perturbations in the shape of the waveguide on a size scale similar to the wavelength of the transmitted light. Larger-scale perturbations also adversely affect the transmission of radiation. This is most vividly seen when the substrate material is not uniformly made. Diameter variations, lack of roundness, and geometric defects from manufacture have all been seen to produce strong mode coupling and thus higher attenuation and degradation of the spatial form of the waveguide output. These perturbations are not restricted to manufacturing defects. The substrates for these waveguide structures are normally made from very thin ductile materials to maximize the flexibility of the guides and minimize manufacturing time. This results in structures that can easily be deformed during use in ways other than bending. Slight local deformations due to pressure on the outside of a guide have been seen to cause degradation to the output energy distribution. While this does cause problems in use, it can be prevented by protective, flexible packaging.

7.6 APPLICATIONS AND FUTURE POTENTIAL FOR HOLLOW WAVEGUIDES

7.6.1 SURGICAL CO_2 LASER POWER DELIVERY

Many surgeons consider the CO_2 laser the surgical tool of choice for a number of procedures. This is due to the limited penetration of the 10.6 μm light into tissue. There is enough penetration of this wavelength to allow coagulation of small blood vessels, but the light is also absorbed strongly enough to cut effectively. This has resulted in the CO_2 laser being very popular for procedures in gynecology, otorhinolaryngology, and general surgery. The CO_2 laser has, however, been very inconvenient to use. While the visible and near-IR lasers could be delivered to the patient and surgical site via silica-based optical fibers, the CO_2 laser has required the use of an articulated arm to reach the patient as mentioned earlier. Articulated arms are beam-steering devices consisting of large hollow tubes and jointed elbows. An articulated arm typically consists of several tube sections and seven elbows to allow all degrees of freedom. Each elbow requires a highly reflective mirror and a multiaxis mirror-positioning mechanism. Each elbow also requires a precision set of bearings to allow motion and to maintain alignment. In common use, articulated arms frequently lose alignment and a service call is required to repair them. All these factors result in excess expense, inconvenience, and downtime for the equipment.

Because of the large size of articulated arms, the CO_2 laser has been limited to external surgical applications or procedures which allow a straight beam path from outside the patient's body to the target. This prevented the use of the CO_2 laser in many potential minimally invasive procedures. In minimally invasive procedures, the surgery is performed through natural body orifices or very small incisions with specialized viewing optics and remote tools. Silica fiber, however, has allowed near-IR and visible lasers to be used quite conveniently in conventional as well as in minimally invasive procedures. The small diameters, high flexibility, and great durability of fused silica fibers has led to their common use in essentially all nonophthalmologic laser surgical situations.

The first commercial CO_2 laser waveguide product was a guide attached to the end of an articulated arm to direct the light to targets inside of the body. These guides, while small in diameter, were essentially rigid. Recently, CO_2 laser systems have been designed with flexible dielectric-coated glass hollow waveguides integrated into the systems to replace the articulated arm altogether. This concept has been attempted in the past but was not practical because of low transmission, drastic attenuations due to bending, and poor modal output of earlier hollow waveguides. Recent improvements in the performance optically and mechanically and reductions in their cost have made hollow waveguides a viable product for CO_2 surgical lasers.

7.6.2 SURGICAL Er:YAG LASER POWER DELIVERY

In the last 5 years, the Er:YAG laser has demonstrated potential for applications in surgery. The 2.94 μm wavelength of the Er:YAG laser falls almost exactly on a water absorption peak. Since most tissue is predominantly water, this wavelength is absorbed very strongly. This results in very clean cutting with no heating of the tissue. While this would result in serious bleeding in vascular tissue, it allows extremely precise cutting in hard and otherwise nonvascular tissue. The Er:YAG laser has already demonstrated its usefulness in hard tissue dental and ophthalmologic procedures, and it also has strong potential in arthroscopy.

Er:YAG laser radiation wavelength of 2.94 μm is not, however, transmitted by silica-based glasses. A number of other glass systems, such as fluorides, calcium aluminates, and heavier ion oxides, have been tried with limited success. The low strength, stability, or melting temperature of these systems has resulted in a low limit on transmittable power or low reliability. Sapphire solid fibers made by film-fed growth have become available, but these fibers are quite stiff and expensive. The fact that these fibers are grown in contact with a metal die results in some inherent contamination in the fibers. Sapphire fibers made by film-fed growth still have 2.94 μm attenuation levels of several decibels per meter. There are several university research programs ongoing attempting to make low-loss solid sapphire waveguides by the laser-heated pedestal growth method. In this process, a sapphire fiber is grown by laser heating the tip of a source rod of sapphire. The slow growth of the crystal and lack of contact with any other materials lead to very high purity fibers. These fibers have demonstrated very low losses, but they tend to be too short and too expensive for practical applications. A more reliable system with a higher power capacity is being sought and may be supplied by hollow waveguides.

Several hollow waveguide systems have demonstrated the ability to transmit high power levels of Er:YAG laser radiation reliably. They have larger diameters than the fluoride or sapphire fibers so they may not be useful for all applications, but their potential is especially strong in dentistry. Commercialization of an Er:YAG laser with an integrated hollow glass waveguide is expected soon.

7.6.3 INDUSTRIAL METAL PROCESSING

Industry has been using the CO_2 laser for many years for the cutting of materials. The laser is commonly used to cut steel, ceramics, plastic, leather, and paper. High peak power CO_2 lasers are used to mark many materials, such as metals and glass. All of these systems have required complicated beam-steering equipment usually consisting of many mirrors to deliver the laser light to the target. Hollow waveguides could be substituted for these beam-steering systems resulting in greater equipment flexibility, greater reliability, and lower costs. The waveguides would have to withstand up to 5 kW of continuous wave power or several megawatts of peak powers for these applications. Due to the nature of hollow waveguides, these power levels are quite possible. Glass and sapphire hollow waveguides have already demonstrated the ability to transmit kilowatt levels of continuous power and megawatts of pulsed power. Hollow waveguides have the inherent advantage of not having any solid medium to be damaged by these high power levels. The mode and polarization preservation capabilities of the hollow waveguides could allow precise laser cutting and welding without the beam-steering apparatus traditionally necessary.

Normally, CO_2 laser cutting is done on two-dimensional parts. Sheets of material are passed underneath a beam-focusing unit, which can only move in or away from the sheets. This is adequate for many processes, but, as robotics and other forms of automation become more prevalent, the ability to laser cut or weld in three dimensions will be necessary. A flexible waveguide system could be mounted on the end of a robotic arm relatively easily. The arm could be programmed for far more complicated motions than a traditional mirrored delivery system.

7.6.4 COUNTER MEASURES

The military has been interested in IR waveguides for many years. The possibility of remotely detecting IR energy incident on a vehicle has resulted in some research efforts to produce threat warnings. The military could also use such a system for countermeasures if transmitted IR laser energy were capable of disrupting the threat. It is conceivable that bursts of CO_2 laser energy could damage IR-seeking ordinance. A waveguide (solid or hollow) system could be used to create such a defensive system with a minimum of parts. Hollow waveguides have demonstrated the necessary reliability, but until recently they have not had attenuations low enough to be practical. With a little more development, they may be appropriate for a number of military applications.

7.6.5 SIGNAL DELIVERY

There are a number of nonlaser applications for IR-transmitting hollow waveguides. The solid waveguides available for the mid-IR region suffer from durability issues. Many of them cannot withstand elevated temperatures or high humidity. This makes them inappropriate for many sensing applications. Hollow waveguides can be used to collect and transmit IR light to a remote detector in a wet environment. Sapphire waveguides can even be used in extreme temperature conditions well above 1500°C.

7.6.6 THERMOMETRY

Thermometry is the most obvious use for signal transmission by mid-IR waveguides whether they are hollow or solid. Blackbody radiation near room temperature is much stronger around 10 μm than at wavelengths transmitted by silica fibers. Hollow waveguides, made for broadband transmission, have demonstrated good results in thermometry systems.[53,54] Frequently, temperature monitoring must be done in hostile chemical or very high temperature environments. The extremely high melting temperature and chemical resistance that sapphire waveguides possess have allowed them to be used *in situ*ations as adverse as combustion chambers and exhaust nozzles. There is no reason to believe that they would not also work well for thermometry in highly caustic environments.

7.6.7 IMAGING

Hollow waveguides have not been used for any imaging applications. While this is certainly conceivable, the resolution of a hollow waveguide imaging system would be extremely poor. A typical fiber-optic imaging bundle has between 5000 and 40,000 individual fibers. Each fiber constitutes a pixel in the image and typically has a size of 4 to 10 μm. Images made with fewer than several thousand fibers have extremely poor resolution. At the present time, waveguides with moderate transmission have minimum bores of greater than 250 μm. If a bundle of several thousand fibers were made of 250-μm waveguides, the final size would be quite large. Such a structure is certainly possible, but its utility has not been demonstrated. If hollow waveguides with reasonable attenuations (a few decibels per meter) can be made with cores as small as 50 μm, hollow waveguide image bundles would be a reasonable product.

7.6.8 SPECTROMETRY

Broadband hollow IR waveguides are excellent choices for spectrometry processes. Hollow waveguides have been used with FTIR spectrometers with excellent results.[53] They compete directly with chalcogenide fibers in most applications, but for high-temperature situations they are the only available choice at present. The fact that the waveguides are hollow has permitted their use as sample cells for examining gas spectra. As has been stated, hollow waveguides do not transmit higher-order modes well. This results in these guides having low effective numerical apertures. This is frequently a difficulty in spectrometry applications. The low numerical apertures result in less light being collected at the input end of the fiber. Efforts to use optics to improve the collecting ability of the waveguides have not been tremendously successful. While the utility of hollow waveguides for spectrometry has been demonstrated, there has been no commercialization of them for this purpose yet.

7.7 CONCLUSION

The search for a high-transmission, high-reliability, and low-cost mid-IR optical fiber has been long and difficult. When the extra requirement of a high laser damage threshold is added, many of the potential options disappear. Hollow waveguides cannot match the small diameters mixed with low attenuations of solid fibers in the near IR. They can, however, withstand enormous power densities. For applications where losses above 0.1 dB/m and bend radii greater than 10 cm are accessible, hollow waveguides at present offer an extremely reliable alternative. The surface quality of substrate material is currently the dominant problem in waveguide performance. The other issues preventing the widespread uses of hollow waveguides are the small effective numerical aperture, the large bending radius, the attenuation due to bending, and, most significantly, the lack of commercial sources. As material-processing and coating techniques improve, the performance of these waveguides will also improve. The influence of bending on attenuation, while not going away, will become less and less evident. As the transmission improves, the waveguide diameters will decrease allowing smaller bend radii. The performance of hollow waveguides is unlikely, however, to ever achieve the low-loss levels of ultrapure silica or fluoride optical fibers. IR hollow waveguides are not appropriate for all applications, but they do, and will continue to, play a role in IR optics.

REFERENCES

1. Lord Rayleigh, *Philos. Mag.,* 43, 125 (1897).
2. H. Nishihara, T. Inoue, and J. Koyama, *Appl. Phys. Lett.,* 25, 391 (1974).
3. E. Garmire, *Appl. Opt.,* 15, 3037 (1976).
4. T. McMahon, Ph.D. dissertation, University of Southern California, Los Angeles, 94 (1978).
5. Y. Kato and M. Miyagi, *Appl. Opt.,* 30, 3790 (1991).
6. E. Garmire, T. McMahon, and M. Bass, *IEEE J. Quantum Electron.,* QE-16, 23 (1980).
7. M. Miyagi, A. Hongo, A. Yoshizo, and S. Kawakami, *Appl. Phys. Lett.,* 43, 430 (1983).
8. C. Gregory, J. Harrington, R. Altkorn, R. Haidle, and T. Helenowski, *Soc. Photo-Opt. Instrum. Eng.,* 1420, 169 (1991).
9. E. Garmire, *Appl. Phys. Lett.,* Aug 15, 145 (1976).
10. E. Garmire, T. McMahon, and M. Bass, *Appl. Phys. Lett.,* 34, 35 (1979).
11. K. Laakmann, U.S. Patent 4,688,893 (1987).
12. H. Machida, A. Nishimura, M. Ishikawa, and M. Miyagi, *Electron. Lett.,* 27, 2068 (1991).
13. H. Machida, Y. Matsuura, H. Ishikawa, and M. Miyagi, *Appl. Opt.,* 31, 7617 (1992).
14. E. Marcatili and R. Schmeltzer, *Bell Syst. Tech. J.,* 43, 1783 (1964).
15. J. Stratton, *Electromagnetic Theory,* McGraw-Hill, New York, 1941, 524.
16. Y. Matsuura, M. Saito, M. Miyagi, and A. Hongo, *J. Opt. Soc. Am.,* A6, 423 (1989).
17. M. Satio, modified from personal correspondence.
18. E. Snitzer, *Optical Dielectric Wavguides, Advances in Quantum Electronics,* Columbia University Press, New York, 1961, 350.
19. M. Miyagi and S. Karasawa, *Appl. Opt.,* 29, 367 (1990).
20. C. Gregory and J. Harrington, *Appl. Opt.,* 32, 5302 (1993).
21. M. Miyagi, *Appl. Opt.,* 20, 1221 (1981).
22. A. Chester and R. Abrams, *Appl. Phys. Lett.,* 21, 576 (1972).
23. A. Hongo, K. Morosawa, K. Matsumoto, T. Shiota, and T. Hashimoto, *Appl. Opt.,* 31, 5114 (1992).
24. R. Jenkins and R. Devereux, *Appl. Opt.,* 31, 5086–5091 (1992).
25. F. Roullard and M. Bass, *IEEE J. Quantum Electron.,* QE-13, 813 (1977).
26. T. Hidaka, T. Morikawa, and J. Shimada, *J. Appl. Phys.,* 52, 4467 (1981).
27. O. Eshbach, *Handbook of Engineering Fundamentals,* John Wiley and Sons, New York, 1975, 1054.
28. R. Falciai, G. Gironi, and A. Scheggi, *Soc. Photo-Opt. Instrum. Eng.,* 494, 84 (1984).
29. T. Hidaka, K. Kumada, J. Shimada, and T. Morikawa, *J. Appl. Phys.,* 53, 5484 (1982).
30. C. Worrell, *Soc. Photo-Opt. Instrum. Eng.,* 843, 80 (1987).
31. C. Worrell, U.S. Patent 4,778,249 (1988).
32. C. Worrell, *Electron. Lett.,* 25, 570 (1989).
33. J. Harrington, C. Gregory, and R. Nubling, *Soc. Photo-Opt. Instrum. Eng.,* 1048, 117 (1989).
34. J. Harrington, and C. Gregory, *Opt. Lett.,* 15, 541 (1990).

35. H. LaBelle, *J. Cryst. Growth,* 50, 8 (1980).
36. M. Miyagi, K. Harada, Y. Aizawa, and S. Kawakami, *Soc. Photo-Opt. Instrum. Eng.,* 484, 117 (1984).
37. C. Gregory and J. Harrington, *Appl. Opt.,* 32, 3978 (1993).
38. R. Nubling and J. Harrington, *Appl. Opt.,* 34, 372 (1996).
39. M. Miyagi and S. Kawakami, *J. Lightwave Technol.,* LT2, 116 (1984).
40. Y. Kato and M. Miyagi, *IEEE Trans. Microwave Theory Practice,* 41, 733 (1993).
41. Y. Kato and M. Miyagi, *IEEE Trans. Microwave Theory Practice,* 42, 2336 (1994).
42. Y. Kato and M. Miyagi, *IEEE Trans. Microwave Theory Practice,* 40, 679 (1992).
43. Y. Matsuura and M. Miyagi, *Appl. Opt.,* 32, 6598 (1993).
44. M. Levy, U.S. Patent 4,913,505 (1990).
45. N. Croitoru, K. Saba, J. Dror, E. Goldenberg, D. Mendelovic, and G. Israel, U.S. Patent 4,390,863 (1990).
46. N. Croitoru, J. Dror, and I. Gannot, *Appl. Opt.,* 29, 1805 (1990).
47. J. Harrington and Y. Matsuura, *Soc. Photo-Opt. Instrum. Eng.,* 2396, 4 (1995).
48. Croitoru, *Soc. Photo-Opt. Instrum. Eng. Biomedical Optics Conference,* San Jose, CA (1995).
49. C. Morrow and G. Gu, *Soc. Photo-Opt. Instrum. Eng.,* 2131, 18 (1994).
50. T. Abel, J. Hirsch, and J. Harrington, *Opt. Lett.,* 19, 1034 (1994).
51. Y. Matsuura, T. Abel, and J. Harrington, *Electron. Lett.,* 30, 1688 (1994).
52. M. Miyagi, personal communication.
53. S. Saggese, J. Harrington, and G. Sigel, *Opt. Lett.,* 16, 27 (1991).
54. M. Saito, S. Sato, and M. Miyagi, *Opt. Eng.,* 31, 1793 (1992).
55. E. Garmire, *Soc. Photo-Opt. Instrum. Eng.,* 320, 70 (1982).
56. N. Nagano, M. Saito, M. Miyagi, N. Baba, N. Sawawobori, *Appl. Opt.,* 30, 1074 (1991).
57. T. Hidaka, T. Morikawa, J. Shimada, and K. Kumata, U.S. Patent 4,453,803. (1984).
58. E. Palik, *Handbook of Optical Constants of Solids,* Academic Press, San Diego, (1985).
59. T. Matsushima, K. Tanaka, Y. Okuda, and T. Sueta, *Jpn. J. Phys.,* 27, 1357 (1988).
60. K. Natterman, H. Hoffmann, and N. Neuroth, *Soc. Photo-Opt. Instrum. Eng.,* 929, 124 (1988).

8 Applications of IR Transmitting Fibers

Mark A. Druy

CONTENTS

8.1 INTRODUCTION

The investigation of the application of infrared (IR) transmitting optical fibers for performing remote spectroscopy started in the research laboratory in the mid 1980s. Since that time, there have been numerous other reports using, in addition to chalcogenide fibers, fibers composed of heavy metal fluoride and silver halide. Table 8.1 summarizes many of the recent reports by application area. In order not to review the literature previously cited, the focus of this chapter is on applications that involve commercially available mid-IR (MIR) cables and probes.

8.2 INFRARED TRANSMITTING OPTICAL FIBERS

IR transmitting optical fibers are a class of optical fibers that differs from conventional silica glass–based optical fibers in that they transmit the IR portion of the electromagnetic spectrum, i.e., in the 1 to 15 m region (1000 cm^{-1} to 667 cm^{-1}). IR

TABLE 8.1
Applications for IR
Fiber-Optic Sensing

Applications	Ref.
Biological determinations	1–4
Chemical identification	5–6
Reaction monitoring	7–18

---

transmitting fibers are made of halide glasses, chalcogenide glasses, as well as polycrystalline halides. These have been described in detail in Chapters 3, 4, and 6, respectively.

The development of new IR transmitting optical fibers with low optical losses, sufficient mechanical strength, and temperature range to meet the demanding conditions of many process environments and the availability of improved, ruggedized low-cost FTIR spectrometers have made *in situ* FTIR measurements possible. Until recently, the aforementioned examples were accomplished using probes and cables, that were constructed to demonstrate the potential capability of IR transmitting optical fibers for remote sensing. In 1995, the probes, cables and an interface for laboratory FTIR spectrometers were introduced to the market so that for the first time this technology is now available to the general user of IR spectrometers.

IR transmitting optical fibers can extend several meters from the spectrometer and can be used to perform conventional transmission measurements when the IR beam is shone through the analyte of interest, such as a gas, clear liquid, or transparent thin film. Optically opaque and highly absorbing materials are not amenable to transmission techniques and must be measured in reflectance. A high-index chalcogenide optical fiber makes an ideal attenuated total reflectance or MIR internal reflecting element for "fingerprint" region spectroscopic measurements. Similarly, the fibers can also be mounted in such a way that it is possible to perform specular reflectance measurements and diffuse reflectance measurements. The principles of attenuated total reflectance (ATR), specular reflectance, and diffuse reflectance are briefly described below.

8.3   PRINCIPLES OF INFRARED SPECTROSCOPY

8.3.1   Attenuated Total Reflectance

ATR spectroscopy is ideal for materials that are strong absorbers.[19,20] The phenomenon was first reported in IR spectroscopy in 1959. It was observed that if certain conditions were met, IR radiation entering a prism made of a high index of refraction IR transmitting material would be totally internally reflected. The internal reflectance creates an evanescent wave that extends beyond the surface of the crystal into the sample held in contact with the crystal. The evanescent wave will be attenuated in

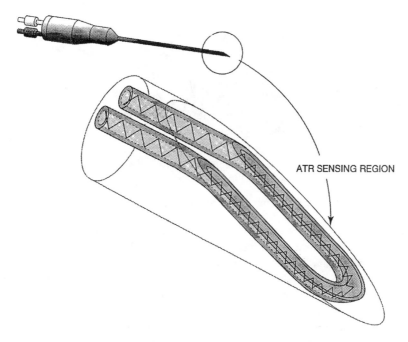

FIGURE 8.1   The ATR effect in optical fiber sensors.

regions of the IR spectrum where the sample absorbs energy. The technique is generally referred to as evanescent wave spectroscopy when applied to optical fibers, where an ensemble of reflection angles is usually present.[21-23] The spectrum that is measured through an IR transmitting fiber-optic ATR sensor in contact with an absorbing medium is dependent on the refractive indices of the fiber and sample medium, as well as the absorption coefficient of the sample. Another feature of the ATR sensor is the penetration depth of the evanescent wave into the sample. This is dependent on the angle of incidence at the interface between the ATR element and the surrounding material. The penetration depth is also dependent on the refractive index of the sample. For a chalcogenide optical fiber sensor of refractive index 2.8 and a sample of refractive index of 1.6, this typically results in penetration depths of 0.2 m at a wavelength of 3 μm (3300 cm^{-1}). This is illustrated in Figure 8.1.

8.3.2   DIFFUSE AND SPECULAR REFLECTANCE

Diffuse reflectance is widely used in the analysis of solids and powders.[24] A ray of light arrives at the surface of a particulate medium. At the interface there is a random arrangement of crystals, so that there will be a range of angles of incidence with the crystal faces. Some rays $R1$ will arrive at such an angle that reflection at the surface will occur. This gives rise to a specular or surface-reflected component and is a function of the refractive index and the absorptivity of the sample. Another ray $R2$, on the other hand, may arrive at such an angle that it is not reflected but is refracted and enters the crystal. It may then emerge or become internally reflected

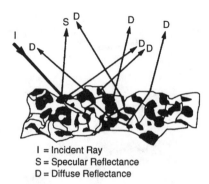

I = Incident Ray
S = Specular Reflectance
D = Diffuse Reflectance

FIGURE 8.2   Comparison of specular and diffuse reflection.

a number of times before emergence. In general, the ray may be reflected and refracted through many crystallites before emerging. The angle of emergence for this ray depends on the ray path and can take any value. This reflection mode is called diffuse specular reflectance and is a function of the complex refractive index of the sample. A third mode of interaction, true diffuse reflectance, results from the penetration of the incident radiation into one or more sample particles and subsequent scatter from the sample matrix. The resulting radiation may emerge at any angle relative to the incident radiation, and, since it has traveled through the particulates, it now contains information about the absorption characteristics of the sample material. This diffuse reflectance is optically indistinguishable from diffuse specular reflection. These interactions are summarized in Figure 8.2.

8.4   COMMERCIALLY AVAILABLE PROBES

Three commercially available probes for performing ATR spectroscopy, diffuse reflectance, and specular reflectance are shown in Figure 8.3. These probes are manufactured by Sensiv Inc. and sold as laboratory accessories for IR spectrometers by Spectra-Tech, Inc. (Shelton, CT).

The probes connect to IR transmitting optical fiber cables and a Fiberlink. The Fiberlink is a device (Figure 8.4) that fits into the sample compartment of IR spectrometers and allows the source energy from the spectrometer to be efficiently coupled into the launch cable and to couple energy from the return cable back into the detector of the spectrometer. Table 8.2 summarizes the properties of these accessories.

8.5   METHODOLOGY

The probe, cables, and Fiberlink are connected to one another using a small interconnect that enables a user to interchange probes easily. These components are linked together with the Fiberlink installed in the sample compartment of the IR spectrometer. A background spectrum of this arrangement is acquired before obtaining a sample spectrum with the probe. The ratio of the sample spectrum to the

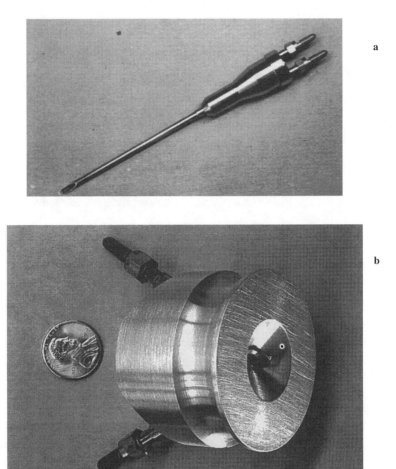

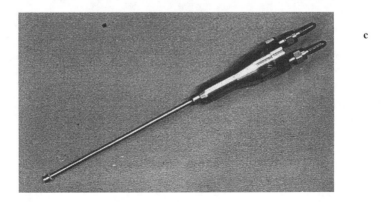

FIGURE 8.3 ATR (a), diffuse reflectance (b), specular reflectance probes (c).

FIGURE 8.4   The fiberlink.

TABLE 8.2
Properties of Probes and Cables

	ATR Needle Probe	Reflectance Needle Probe	Diffuse Reflectance Probe
Fiber material	Chalcogenide	Chalcogenide	Chalcogenide
Sampling range (cm^{-1})	4000–900	4000–900	4000–900
Probe head diameter (mm)	5	5	5
Useful temperature range (°C)	Ambient to 60	—	—
Standard cable length (m)	1.5	1.5	1.5
pH compatibility	1–9	1–9	1–9

reference spectrum is used to yield the absorbance spectrum of the analyte, where the absorbance, A, is defined as

$$A = -\log I/I_o = ecl \qquad (8.1)$$

where I is the light intensity measured by the detector after passing through the sample, I_o is the incident light intensity with no sample present, e is the absorption coefficient, c is the concentration of the sample, and l is the path length. This is known as the Beer–Lambert relationship and means that the absorption intensity is proportional to the concentration of the sample if the path length is constant. This is the fundamental basis for performing quantitative measurements with MIR spectroscopy.

8.6   EXAMPLES OF APPLICATIONS

8.6.1   Hydraulic Fluid and Lubricating Oil Analysis

A Bomem FTIR equipped with an MCT detector was used with the following accessories: a fiberlink interface for the spectrometer, two standard 1.5-m chalcogenide

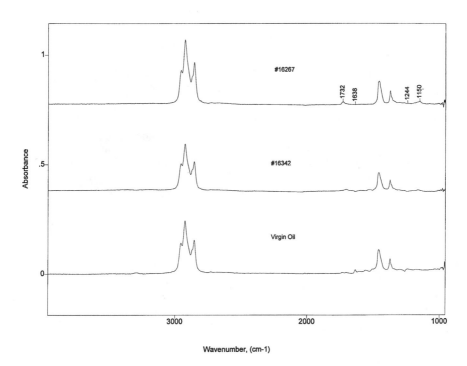

FIGURE 8.5   Spectra of fresh hydraulic oil, sample 16342, and sample 16267.

fiber-optic cables, and a standard ATR probe. The spectral data was collected at 4 cm⁻¹ resolution with coaddition of 64 scans. The data were analyzed with GRAMS Ver. 3.0 software supplied by Galactic Industries. This software runs on a PC™ compatible X86 computer under the Microsoft Windows 3.1 operating environment. Spectral subtraction of the fresh samples from the degraded samples was performed in order to examine the spectral differences between the samples. Sampling was executed by shaking each specimen and dipping the probe into the specimen. The probe was cleaned between each sampling. Nine samples were measured. Three of these were hydraulic oil, and six of these were engine oil samples. Figure 8.5 shows the spectra of the fresh hydraulic oil, used sample 16267, and used sample 16342. Figure 8.6 compares the difference spectra of the two used oils. Examination of these spectra reveals differing levels of degradation in the two hydraulic oil samples. In particular, carboxylate formation at 1733 cm⁻¹ and nitration acidic by-products at 1636 cm⁻¹ are observed. Sample 16267 also seems to contain sulfate by-products in the 1150 cm⁻¹ region.

Figure 8.7 shows the spectra of fresh engine oil and one of the degraded samples (#15322). Figure 8.8 shows the difference spectra of several degraded samples in order of apparent increasing degradation level. Increasing levels of carboxylate formation are observed at 1717 cm⁻¹, and increasing nitration acidic by-products are observed at 1633 cm⁻¹. In addition, there appears to be increasing sulfate by-products and perhaps glycol in the 1150 to 1200 cm⁻¹ region.

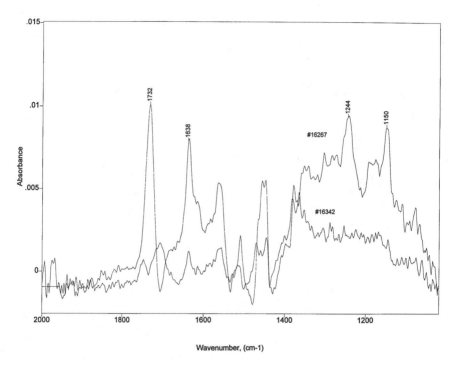

Wavenumber, (cm-1)

FIGURE 8.6   Difference spectra of samples 16342 and 16267.

8.6.2 IDENTIFICATION OF HAZARDOUS WASTE

The analysis of hazardous waste generally involves the collection of all chemical waste generated by the various facilities at a site and bringing them to a staging area. A sample from each container is then sent to an analytical laboratory for determination of its major constituents. Once the chemical composition is determined, disposal procedures are initiated. Since there is a large difference in time between sampling and analysis, volatile components may appear to be present in misleadingly lower concentrations than in the contaminated area and therefore give a false indication of low contamination. However, the use of a portable spectrometer and fiber optics enables preliminary analyses to be made on-site, which reduces the time for site characterization. To enable liquid waste to be sampled directly from the waste barrels, the ATR probe shown in Figure 8.3 is encased in a meter-long stainless steel tube, with only the sensing region of the probe exposed. The "barrel probe" is connected to the spectrometer via 6-m chalcogenide cables. Figure 8.9 shows the barrel probe inserted into a waste drum.

Representative drums containing each of the typical waste categories were sampled using the barrel probe. This includes barrels of oil and latex-based paints, lube oils, corrosion inhibitors, mixtures of oil/engine coolant, paint/organic solvents and antifreeze. Some representative spectra are shown in Figures 8.10 through 8.12.

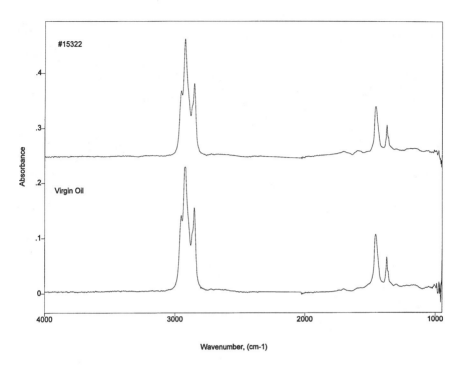

FIGURE 8.7   Spectra of fresh engine oil and degraded sample 15322.

In another environmental application, a fiber-optic-based MIR reflectance probe was used to detect contaminants in soil.[25] This detection was accomplished with the probe deployed on a cone penetrometer and tested in the field as shown in Figure 8.13. The system uses specially constructed chalcogenide fiber bundles that are up to 30 m in length. In one example, detection limits of 130 ppm of marine diesel fuel in sea sand were demonstrated using a 20-m length of cable. Figure 8.14 shows a picture of the fiber bundle that was constructed by Sensiv, Inc., from fiber that was fabricated at the Naval Research Laboratory.

8.6.3   Cure Monitoring of Urethane Cross-Linked Rubbers

Many urethane rubbers are based on the following general reactions between a hydroxyl-terminated butadiene resin and a diisocyanate as shown in Figure 8.15. In order to measure the conversion of reactants to products, specific absorption peaks in the MIR can be followed. The primary absorption peaks associated with cure are listed in Table 8.3. Figure 8.16 shows several spectra taken over a 5-min period of time after a hydroxy-terminated polybutadiene and a diisocyanate were mixed at room temperature. The peak at 2274 cm^{-1} decreases as the reaction proceeds with a concomitant increase in the peak at 1510 cm^{-1}.

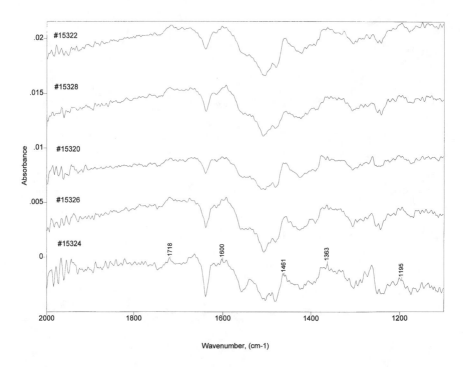

FIGURE 8.8   Difference spectra of samples 15322, 15328, 15320, 15326, and 15324.

FIGURE 8.9   ATR barrel probe inserted into hazardous waste drum.

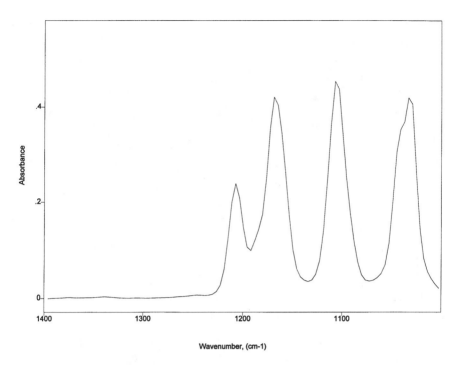

Wavenumber, (cm-1)

FIGURE 8.10   Absorbance spectrum of freon with barrel probe.

8.6.4   QUANTITATIVE GLUCOSE CONCENTRATIONS

The determination of glucose in water can be accomplished using the ATR needle probe. Figure 8.17 shows a spectra of 4000 mg/dl of glucose in water. The water peaks are at approximately 3364 and 1643 cm^{-1}, and the glucose peaks are at 1079 and 1034 cm^{-1}. The inset in Figure 8.17 shows the glucose peaks in an expanded view. Figure 8.18 shows the Beer–Lambert relationship for glucose/water solutions down to 500 mg/dl for both glucose peaks. As is evident from this data, the relationship is linear and can be used to predict the concentration of glucose.

8.6.5   SURFACE CHARACTERIZATION

The diffuse reflectance probe can be used to determine the nature of materials on surfaces. Figure 8.19 represents the absorbance spectrum of the paper that is used in "sticky yellow reminder notepaper" while Figure 8.20 shows the absorbance spectrum of the "sticky adhesive" area. Figure 8.21 shows the difference spectra of the two revealing the spectra of the sticky adhesive.

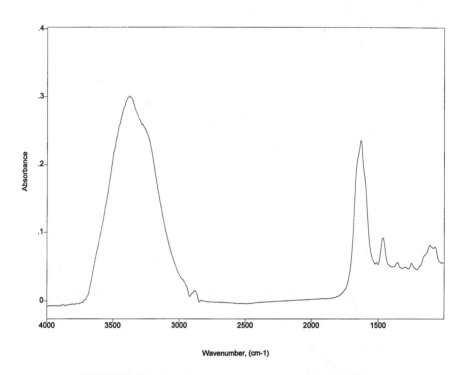

Wavenumber, (cm-1)

FIGURE 8.11   Absorbance spectrum of soap with barrel probe.

8.7   CONCLUSIONS

In this chapter we described several applications of commercially available IR transmitting optical fiber cables and probes that enable the remote analysis of materials and chemical processes. These probes can be used to monitor end points of reactions, presence or absence of reagents, degradation of materials, and identification of contaminants or functionality on surfaces. As more and more probes are developed, we can expect even more applications for this technology. It is anticipated that lower fiber optical losses will enable longer lengths of fiber sensors, thereby enhancing the remote capabilities. Furthermore, multiplexing will also enhance the capabilities of the current and future fiber sensors.

ACKNOWLEDGMENTS

The author would like to thank the following people whose many hours of effort made the work descibed in this chapter possible: Suneet Chadha, William Kyle, and Chuck Stevenson from Foster Miller; Roy Bolduc and Paul Glatkowski from Sensiv, Inc.; Aharon Bornstein and Yitzhak Weissman from Soreq Nuclear Research Center.

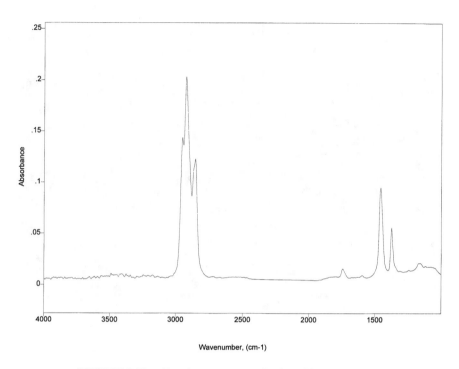

FIGURE 8.12   Absorbance spectra of paint with barrel probe.

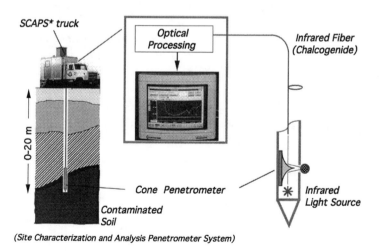

FIGURE 8.13   Detection of hazardous waste using the cone penetrometer.

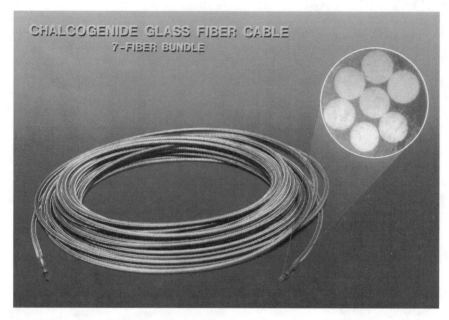

FIGURE 8.14   A 30-m-long chalcogenide fiber cable containing a seven-fiber bundle.

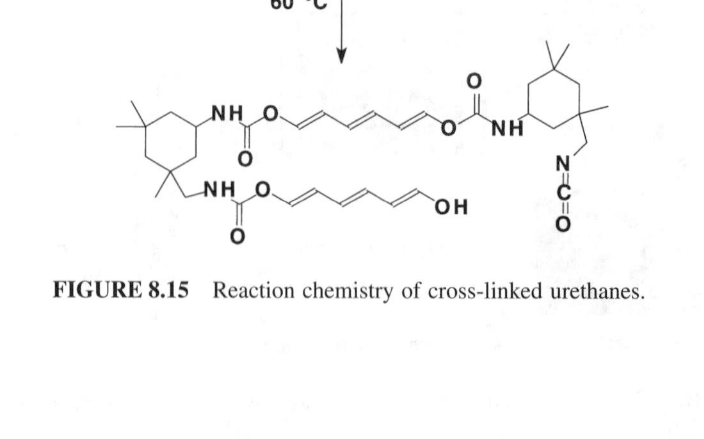

FIGURE 8.15   Reaction chemistry of cross-linked urethanes.

TABLE 8.3
MIR Peak Identification for Reactants and Products in Urethane Reactions

Compound	Functional Group	Peak (cm⁻¹)	Assignment	Comments
Polybutadiene	Primary alcohol	1080	C–O stretch	Decreases
Polybutadiene	Alkene	1640	CH=CH stretch	Unchanged
Diisocyanate	Isocyanate	2264	N=C=O stretch	Decreases
Diisocyanate	Amide	1510	CO–NH–R stretch	Increases
Diisocyanate	Ester	1725	COO stretch	Increases

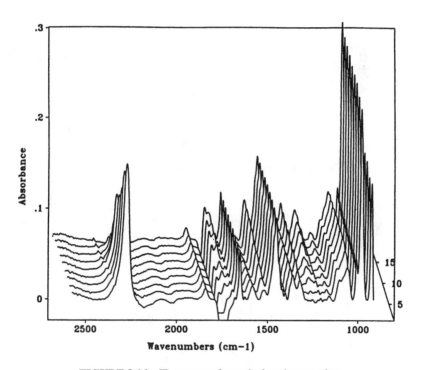

FIGURE 8.16 IR spectra of a typical curing reaction.

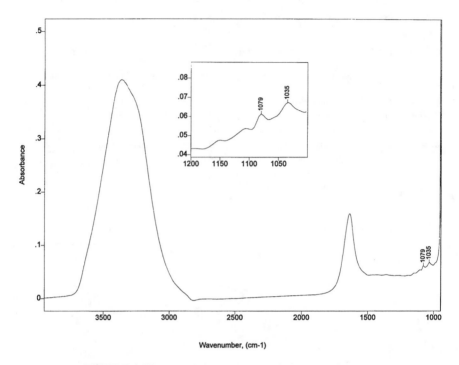

FIGURE 8.17   Absorbance spectrum of glucose and water.

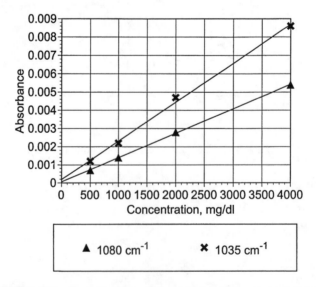

FIGURE 8.18   Beer–Lambert relationship for glucose/water solutions.

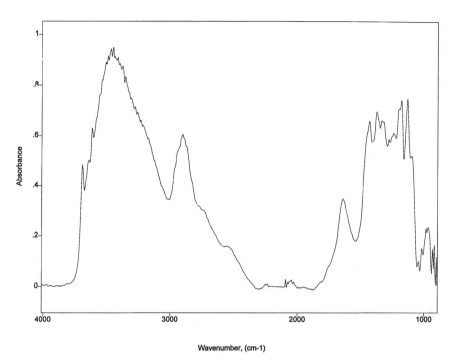

FIGURE 8.19   Absorbance spectrum of sticky yellow reminder paper.

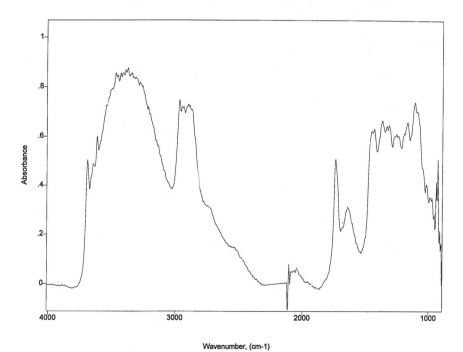

FIGURE 8.20   Absorbance spectrum of adhesive area on reminder paper.

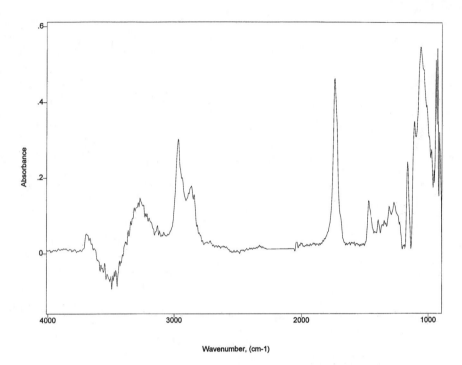

Wavenumber, (cm-1)

FIGURE 8.21   Difference spectra revealing absorbance spectra of adhesive.

REFERENCES

1. R. Simhi, D. Bunimovich, B. A. Sela, and A. Katzir, *SPIE,* 2388, 493 (1995).
2. V. G. Artjushenko, N. I. Afanasyeva, A. A. Lerman, A. P. Kryukov, E. F. Kuzin, N. Zharkova, V. G. Plotnichenko, G. A. Frank, G. I. Didenko, V. V. Sokolov, and W. Neuberger, *SPIE,* 2085, 137 (1994).
3. S. Romano, M. Balzi, L. Dei, A. A. Lerman, W. Neuberger, and A. Becciolini, *SPIE,* 2329, 303 (1995).
4. E. M. Kosower, O. Marom-Albeck, I. Pais, A. Katzir, and N. S.Kosower, *SPIE,* 2131, 71 (1994).
5. J. S. Namkung, M. L. Hoke, R. S. Rogowski, and S. Albin, *SPIE,* 2444, 447 (1995).
6. M. A. Druy, P. J. Glatkowski, R. A. Bolduc, W. A. Stevenson, and T. C. Thomas, *SPIE,* 2367, 24 (1995).
7. D. A. C. Compton., S. L. Hill, N. A. Wright, M. A. Druy, J. Piche, W. A. Stevenson and W. Vidrine, *Appl. Spectrosc.,* 42, 972 (1988).
8. P. J. Glatkowski, M. A. Druy, and W. A. Stevenson, *SPIE,* 1796, 243 (1993).
9. P. J. Glatkowski, M. A. Druy, and W. A. Stevenson, *SPIE,* 2072, 109 (1994).
10. S. Salim, C. K. Lim, K. F. Jensen, and R. D. Driver, *SPIE,* 2069, 132 (1993).
11. M. A. Serio, H. Teng, K. S. Knight, S. C. Bates, S. Farquharson, A. S. Bonanno, P. R. Solomon, W. A. Stevenson, M. A. Druy, P. J. Glatkowski, J. A. Harrington, R. K. Nubling, J. Y. Ding, and A. G. Comolli, *SPIE,* 2069, 121 (1993).
12. B. A. Sanmitra, J. A. Caughran, and J. A. De Haseth, *SPIE,* 2089, 382 (1994).
13. M. A. Druy, L. Elandjian, and W. A. Stevenson, *SPIE,* 986, 130 (1988).

14. P. R. Young, M. A. Druy, W. A. Stevenson, and D. A. C. Compton, *SAMPE J.,* 25, 11 (1989).
15. M. A. Druy, L. Elandjian, W. A. Stevenson, R. D. Driver, G. M. Leskowitz, and L. E. Curtiss, *SPIE,* 1170, 150 (1989).
16. E. Margalit, H. Dodiuk, E. M. Kosower, and A. Katzir, *Surf. Interface Anal.,* 15, 473 (1990).
17. M. A. Druy, P. J. Glatkowski, and W. A. Stevenson, *SPIE,* 1637, 174 (1992).
18. M. A. Druy, P. J. Glatkowski, and W. A. Stevenson, *SPIE,* 1591, 218 (1991).
19. N. J. Harrick, *Internal Reflection Spectroscopy,* Harrick Scientific Corporation, New York, 1967.
20. Spectra-Tech FT-IR Technical Note TN-1.
21. M.D. DeGrandpre and L. W. Burgess, *Anal. Chem.,* 60, 2582 (1988).
22. G. Stewart, J. Norris, D. Clark, M. Tribble, I. Andonvic, and B. Culshaw, *SPIE,* 990, 821 (1988).
23. M. Katz, A. Bornstein, I. Schnitzer, and A. Katzir, *SPIE,* 1591, 236 (1991).
24. Spectra-Tech FT-IR Technical Note TN-2.
25. G. Nau, F. Bucholtz, K. Ewing, S. Vohra, J. McVicker, J. Sanghera, I. Aggarwal, J. Adams, D. Eng, T. Eng, and T. King, *SPIE,* 2504, 291 (1995).

9 Summary

Jasbinder S. Sanghera and Ishwar D. Aggarwal

From the preceding chapters, it is quite apparent that IR transmitting fibers can be divided into two categories, namely those based on solid core and hollow waveguides, respectively. The solid cores can be either glassy, such as those based on silica, fluoride, and chalcogenide glasses, or crystalline. In the latter case, they can be made of single crystalline material (e.g., sapphire) or polycrystalline material (e.g., AgClBr). The hollow waveguides, however, can be made from metallic tubes or from glassy tubes with or without internal dielectric coatings to enhance transmission. Table 9.1 summarizes some of the physical, mechanical, and optical properties of several representative materials used to make optical fibers. This table clearly demonstrates that the materials possess different properties and, consequently, the particular application will dictate which specific fiber will be appropriate for that application. It is highly recommended that those interested in specific details read the respective chapters for more information on the fiber properties.

Generally speaking, the engineer is primarily interested in the optical properties of the fiber for practical applications in the infrared. Consequently, Figures 1a and 1b display the loss of the different fiber systems as a function of wavelength. The losses shown in Figure 9.1a are representative of the typical fiber quality which is obtained on a routine basis and therefore do not represent champion low losses. Figure 9.1b shows the lowest losses obtained for the fiber systems. Although these champion low losses are not routinely achieved, they indicate that lower losses are possible. Table 9.2 compares the lowest reported losses, the typical losses obtained on a routine basis, and the estimated minimum losses. As one might expect, the loss of high quality silica fibers is close to the theoretically predicted value of about 0.13 dB/km. This is attributed to the enormous worldwide effort over the years and specifically the ingenuity of the many scientists and engineers who contributed to the development of vapor phase processes (MCVD, VAD, OVD, etc.) to fabricate high quality silica preforms from which low loss telecommunications grade silica fibers are now routinely drawn. While silica fibers transmit from the UV to the near IR region (0.2 to 2 μm), other additives (e.g., Ge, F, Al, P, etc.) have been utilized to control the physical and optical properties. Standard production fibers are drawn at incredible speeds (~500 m/min), possess significant strength (proof tested to at least 100 kpsi), and exhibit very low losses (~0.15 dB/km). Once packaged in appropriate cables and implemented in telecommunication systems, they are expected to be operational for more than 20 years without failure. The implementation of silica-based fibers in telecommunications has led to the development of several other key fiber-based devices in telecommunications. Examples of the former include fiber couplers, tapered core fibers, polarization maintaining fibers, dispersion

TABLE 9.1

Some Physical, Mechanical, and Optical Properties of the Materials Used for Making Optical Fibers

	Glass				Crystalline	
	Silica	Fluoride	Chalcogenide		Single Crystal	Polycrystalline
Typical Composition	SiO_2	$53ZrF_4$-$20BaF_2$-$4LaF_3$-$3AlF_3$-$20NaF$	$As_{40}S_{60}$	$Ge_{30}As_{10}Se_{30}Te_{30}$	Al_2O_3	AgClBr
Physical Properties						
T_g or T_m (°C)[a]	~1200 (T_g)	265 (T_g)	197 (T_g)	265 (T_g)	2027 (T_m)	412 (T_m)
CTE (10^{-6}/°C)[b]	0.55	17.2	21.4	14.4	5.6 (II), 5.0 ($\perp$)	30
Thermal Conductivity (W/m-°C)	1.38	0.628	0.17	~0.2	35.1 (II), 33.0 ($\perp$)	1.1
Mechanical Properties						
Density (g/cm^3)	2.20	4.33	3.20	4.88	3.98	6.39
Knoop hardness (kg/mm^2)	600	225	109	205	1370	15.0
Fracture toughness (MPa.m$^{1/2}$)	0.7	0.3	~0.2	~0.2	2.0	—
Poisson's ratio	0.17	0.17	0.24	~0.26	0.27	—
Young's modulus (GPa)	70.0	58.3	16.0	21.9	344.5	0.14
Optical Properties						
Refractive index[c]	1.458 (0.589)	1.499 (0.589)	2.415 (3.0)	2.80 (10.6)	1.704 (3.2)	2.2 (5.0)
dn/dT (10^{-5} °C^{-1})[c,d]	+1.2 (1.064)	-1.5 (1.064)	+0.9 (5.4)	+10.0 (10.6)	+1.4 (1.064)	-1.5 (10.6)
Bulk transmission (µm)	0.16–4.0	0.2–7.0	0.6–10.0	1.0–17.0	0.17–6.5	0.5–40.0
Fiber transmission (µm)	0.2–2.0	0.25–4.0	0.8–6.5	3.0–11.0	0.5–3.5	3.0–19.0

[a] Tg is the glass transition temperature and Tm is the melting temperature.

[b] CTE is the coefficient of thermal expansion.

[c] Wavelength in µm given in parenthesis.

[d] dn/dT is the change in refractive index with temperature.

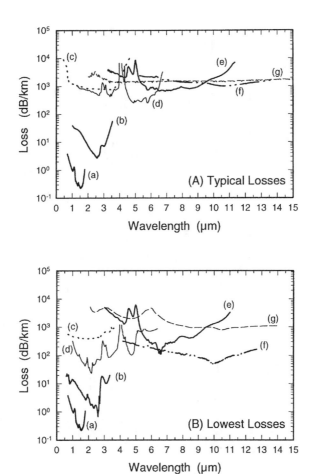

FIGURE 9.1 The (A) typical and (B) lowest reported losses as a function of wavelength for IR transmitting fibers. The examples shown are for (a) silica, (b) ZBLAN, (c) single crystalline sapphire, (d) arsenic sulphide, (e) $Ge_{30}As_{10}Se_{30}Te_{30}$, (f) polycrystalline silver halide, and (g) hollow waveguide fibers, respectively.

compensating fibers and especially erbium doped fiber amplifiers (EDFAs). The EDFAs have revolutionized telecommunications by eliminating electronic repeaters, giving rise to all optical transmission which enabled multi-wavelength, bit-rate transparent operation and thereby enhancing system capacity. More recently, the development of gratings in germano-silicate fibers has been the most enabling factor for advanced optical telecommunication systems, e.g., short cavity and single frequency fiber lasers, Raman lasers and amplifiers, dispersion compensators, ultrasharp filters, multiplexers and de-multiplexers, gain equalizers, and optical taps. While long haul telecommunications are performed at 1.55 μm, local area networks (LANs) need high bandwidth and therefore operate under optimum conditions at the wavelength of zero dispersion in silica, namely at 1.3 μm. Unlike the EDFAs which operate at 1.55 μm, the rare-earth doped silica-based fiber amplifiers at 1.3 μm

TABLE 9.2
The Lowest, Typical, and Estimated Minimum Losses for IR Fibers

| | Silica SiO$_2$ | Fluoride ZBLAN | Chalcogenide | | Single Crystalline Al$_2$O$_3$ | Polycrystalline AgClBr | Hollow Waveguide |
			As$_{40}$S$_{60}$	Ge$_{30}$As$_{10}$Se$_{30}$Te$_{30}$			
Lowest Loss (dB/km)	~0.15	0.45	23	110	~100	50	100
Wavelength (μm)	1.55	2.3	2.3	6.6	2.0	10.0	10.6
Typical Loss (dB/km)	0.2–0.35	10–30	100–500	500–1000	300–500	500–1000	500–1000
Wavelength (μm)	1.55	2.2–2.6	2.2–5.0	6.0–9.0	~2.0	10.6	10.6
Estimated minimum loss (dB/km)	0.13	0.01	~1.0	?	~0.01	0.04	~0.1
Wavelength (μm)	1.55	2.55	5.0	?	1.8	~5.0	10.6

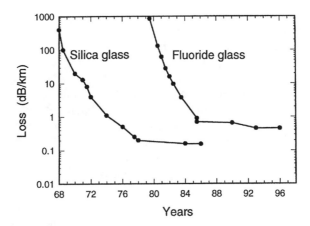

FIGURE 9.2   The reduction in the loss of silica and fluoride glass fibers with time.

are highly inefficient because of excited state absorption (e.g., for Nd^{3+}) or multiphonon quenching (e.g., for Pr^{3+}) and therefore impractical. However, a Raman fiber amplifier at 1.3 μm has been developed utilizing cascaded gratings and is commercially available. In telecommunications, more advances will be made in the areas of specialty fibers, increasing the transmission capacity, functionality, and integration with planar waveguide and semiconductor devices. Aside from telecommunications, silica-based fibers are finding more and more niche applications in the sensor area, including chemical/biomedical sensing, underwater acoustic sensing, magnetic sensing, rotation sensing (gyroscope), and temperature sensing. Recent work using Bragg grating sensors has demonstrated the capability of structural strain monitoring and spatial mapping using embedded fibers which may turn out to be a huge commercial market.

The fluoride glass fibers such as ZBLAN have received less attention than silica fibers, but significantly more than other fiber systems. A lot of emphasis has been placed on developing chemical purification and glass melting processes, as well as fiberization technology. Examples of the latter include numerous mechanical casting approaches for making preforms, as well as crucible drawing processes. As a result, losses as low as 0.45 dB/km have been achieved in fluoride fibers, while typical losses obtained on a routine basis are in the range of 10 to 30 dB/km. These values indicate that improvements need to be made in order to attain the theoretically predicted loss of around 0.01 dB/km. If we look at the improvements made to the loss of silica fiber over the years, we find that the development of vapor phase processing contributed significantly to the reduction of impurity-related losses. When we compare the reduction in the loss of fluoride fibers with silica fibers over the years, as shown in Figure 9.2, we find that the rate at which the loss has decreased (represented by the initial slope in Figure 9.2) is actually faster for the fluorides. It is evident that vapor phase processing, similar to the CVD types of processes, will be necessary to provide further improvements in the glass optical quality. Preliminary work has already demonstrated that fluoride glass films can be deposited using vapor processes. Compositions which are more resistant to crystallization will have to be developed to produce fiber with losses approaching theoretical values.

While data transmission at 2.55 μm has been demonstrated using single mode fluoride fibers, it is unlikely that long lengths of ultra-low loss fluoride fibers will be used in telecommunication systems, even though the issue of their relatively low chemical durability can be mitigated with hermetic coatings. However, there are a number of applications which require medium-loss fluoride fibers. These include chemical sensors, power delivery for mid-IR lasers and, with appropriate rare-earth dopants added to the glass composition, fiber lasers and amplifiers. Chemical sensing can be performed at longer wavelengths than for silica fibers and, furthermore, fluoride fibers have proven more resistant to radiation-induced losses in the IR than silica fibers. Fluoride fibers, due to their negative thermo-optic coefficient (dn/dT), are excellent candidates for power delivery of IR laser energy, such as Er:YAG at 2.94 μm for laser surgery. Laser oscillation in fluoride fibers has been demonstrated at many wavelengths between the visible and mid-IR (0.455 to 3.9 μm) region for numerous applications such as optical disk recording, laser printers, displays, and chemical sensing. Laser oscillation at 3.9 μm in a Ho-doped fluorozirconate fiber (11 mW output power and 1.7% slope efficiency using 1.7 W pump power at 890 nm), was demonstrated after the writing of the fluoride chapter. It represents the longest wavelength glass fiber laser, although the fiber was cooled with liquid nitrogen. It is anticipated that not only will the efficiencies be optimized, but new lasers will also be developed in the future. Single mode fiber amplifiers, operating at several discrete wavelengths between 0.8 and 2.7 μm, have been fabricated, of which the most important is probably the 1.3 μm Pr^{3+}-doped fiber amplifier (PDFA) for telecommunications. These devices are now commercially available with efficiencies of about 5% and small signal gain of about 0.1 dB/mW. While this is adequate for now, more efficient amplifiers (x10 efficiency) based on chalcogenide glasses are being developed and, if successful, will most likely replace the fluoride fiber amplifiers for LANs at 1.3 μm.

Chalcogenide glass fibers have been around since before the development of fluorozirconate fibers, but their development has only recently seen accelerated growth. Depending upon composition, these fibers transmit in the 1 to 11 μm region. For instance, sulphide-based fibers transmit to almost 7 μm, while selenide and telluride transmit to about 10 μm and beyond, respectively. Mixtures of chalcogenides and halides lead to the formation of chalcohalide glasses. Low phonon energy chalcohalide glass fibers (e.g., Te-halide based) transmit to almost 15 μm. Sulphide fibers have been fabricated with losses as low as 23 dB/km, but the losses are usually more in the range of several hundred dB/km. The theoretical losses are now considered to be around 1 to 10 dB/km and limited by the weak absorption tail (WAT). Previously, the losses were thought to be below 1 dB/km and limited by Rayleigh scattering. Such calculations need to be performed on the smaller band gap telluride glass systems. While the WAT may have a larger contribution to the loss at wavelengths near the band edge, it is expected to be negligible at relatively longer wavelengths. Nevertheless, the origin and intrinsic/extrinsic contributions to the WAT need to be identified. Although the chalcogenide fibers may not be capable of ultra-low losses, it is widely believed that relatively low fiber losses can be realized by improvements in the purification and fiberization techniques and without the need of CVD type processes.

Not only are the chalcogenide fibers being actively developed for many potential future applications, these fibers are actually being used in real applications and are commercially available, as described in Chapter 8, in chemical sensor systems. Chalcogenide fibers can and are being used in military, industrial, and medical applications, including temperature monitoring, thermal imaging using bundles, IR laser power delivery, and chemical sensing. Most of these applications arise from the fact that, depending upon composition, the fibers specifically transmit in the critical atmospheric windows of 2 to 5 μm and 8 to 11 μm, and more generally cover the 1 to 11 μm region. The ubiquitous myth that chalcogenide glasses are weak, possess high loss and, therefore, cannot be used practically has been abolished. Besides utilization as IR light conduits, these fibers can be used as flexible bright light sources and amplifiers in the IR when doped with the appropriate rare earth ions. This is attributed to their low phonon energies (<400 cm^{-1}) which reduces the nonradiative multiphonon quenching and therefore increases the radiative IR emission. This is in sharp contrast with silica (1100 cm^{-1}) and fluoride (560 cm^{-1}) glasses, which possess higher phonon energies and, therefore, higher multiphonon quenching rates. For example, a Pr^{3+}-doped selenide glass fiber has shown a broadband mid-IR emission in the 3 to 5 μm region, which is predicted to be 100 times brighter than a black-body at 900 K. Depending upon the rare-earth ion and glass host, sources could be made in 1 to 11 μm region and used for chemical sensor applications or military applications. Also, a single mode Pr^{3+}-doped $66Ga_2S_3$–$34Na_2S$ glass fiber has achieved a large maximum gain coefficient of 0.81 dB/mW at the telecommunications wavelength of 1.3 μm and with a quantum efficiency of greater than 50%. In fact, a 30 dB net gain was attained for less than 100 mW of pump power at 1.017 μm. This performance is better than that available from the commercial fluoride glass-based PDFAs and is attributed to the lower multiphonon quenching in the sulphide host compared with fluorides. This significant result will no doubt prompt a concerted effort worldwide to commercialize a PDFA device based on chalcogenide glass fibers. Other potential glass systems being investigated for this application are based on GeS_2–Ga_2S_3, GeS_2–As_2S_3–Ga_2S_3, and Ga_2S_3–La_2S_3.

While gratings in silica-based fibers have received a lot of attention over the last few years, gratings have now been written in chalcogenide glass fibers. Since the index changes are expected to be greater in chalcogenide glasses, it is predicted that the grating efficiencies will also be higher. Presumably, much more work on gratings will be done in the next few years as low loss, single mode chalcogenide glass fibers become available, especially the rare-earth doped fibers, where the gratings can be used to define the laser characteristics and performance. Also, since chalcogenide glasses possess high third-order, nonlinear susceptibilities (X^3), they may be candidates for fiber optical switching devices.

Crystalline fibers such as those based on sapphire are best characterized by their high laser damage threshold, high melting point, high tensile strength, and excellent chemical durability. Sapphire fibers transmit from the visible region to about 3.5 μm and have attained champion minimum losses of about 100 dB/km, although typical losses are in the range of several 100s dB/km. These losses are considerably higher than the predicted minimum loss of about 0.01 dB/km near 1.8 μm. This may be attributed to several factors, including the quality of the starting feed rod and the drawing conditions. Unlike glass fibers which can be pulled at high speeds (up to several 100 m/min) from a heated preform, single crystal fibers have to be drawn at a much slower rate from a melt. Nevertheless, improvements have been made in speeding up the fiber draw rate from about 3 mm/min to 20 mm/min using a modified laser heated pedestal growth technique and without sacrificing the optical quality. Current efforts are focused on developing techniques to apply a cladding onto the fibers, as well as improving the crystal fiberization technology. Single crystal sapphire fibers are prime candidates for delivery of Er:YAG laser power at 2.94 μm for medical applications due to their high threshold for damage, high chemical durability, and biocompatibility. Current fiber optical losses of 0.7 dB/m have been achieved at this wavelength, compared with the theoretical estimate of 0.13 dB/m. In addition, these fibers can be used in chemical sensor applications where high temperatures and harsh environments will be encountered. Furthermore, since bulk sapphire is a well known host for transition metal ions, there is an obvious desire to fabricate doped sapphire fibers to make visible and near IR fiber laser sources and amplifiers. For this to occur a technique would need to be developed for making single mode sapphire fibers.

Polycrystalline halide fibers (e.g., AgCl–AgBr) are made by the controlled extrusion of high purity single crystals. They are characterized by the fact that they transmit further in the infrared region than the other solid core fiber systems. For example, depending upon composition, the polycrystalline halide fibers transmit from about 3 μm to 19 μm. While theoretical losses are estimated to be about 0.4 to 0.04 dB/km, champion losses of 50 dB/km have been achieved but routine losses range in the high 100s dB/km. Most of the losses arise from scattering centers such as voids, bubbles, and grain boundaries, as well as from free-carrier absorption. The latter has the effect of shifting the wavelength of minimum loss from around 9 μm to 5 μm. One of the major concerns with polycrystalline halide fibers is their ability to undergo photo-structural aging which is manifested as an increase in the loss and a decrease in the fiber strength. The strength also deteriorates after repeated bending. Current research efforts are directed at minimizing these problems by the use of protective polymer jackets and hardening the fibers with the addition of various additives. Most polycrystalline halide fibers transmit to beyond 10.6 μm, so it is quite obvious to use them for delivery of energy from a CO_2 laser for medical applications, as well as for cutting and welding in remote locations. Owing to their wide region of transparency, these fibers can be used in numerous types of chemical sensor systems. The transparency region of polycrystalline fibers also coincides with the black-body radiation of objects at room temperature. Consequently, these fibers can be used for temperature monitoring and thermal imaging. In fact, fibers with an IR edge at 19 μm will transmit black-body radiation down to almost 150 K. Efforts are also underway to fabricate rare-earth ion doped fibers for making mid-IR and far-IR fiber sources. The development of single mode fiber fabrication technology will be critical to enable these devices.

Hollow waveguides, as the name indicates, possess a hollow core within which the light propagates. They are frequently divided into two categories, namely leaky and ATR (attenuating total reflecting) waveguides. Leaky waveguides possess opaque claddings (index >1) and require a highly conducting inner surface. Large bore (>5 mm) metal tubes give losses of a few dB/m at 10.6 μm, while small bore metal tubes (<2 mm) are typically coated with a dielectric film to maximize the reflectivity and exhibit losses of around 0.5 dB/m. Losses as low as 0.1 dB/m have been reported compared with 10^{-4} dB/m expected from theory. The surface quality of the metal tube and dielectric coating limit the losses. Also, bending leads to additional losses, proportional to 1/(bend radius). The ATR waveguides have claddings with refractive indices (n) less than unity (n < 1), and are typically made from glass or single crystal materials. In the case of glass tubes, silica fibers are inappropriate for transmission of CO_2 laser power at 10.6 μm since they exhibit n < 1 between 7.7 and 9.6 μm. Other heavy-metal oxide glasses (e.g., PbO containing, GeO_2 based) have been tried but with high losses (1 to 2 dB/m) because the glasses have nonzero absorption at these wavelengths. No material has been found to date which exhibits zero loss and has a refractive index <1 in the infrared. Therefore, light is attenuated at the interface. Crystalline materials are optically and thermally superior to glasses for making ATR hollow waveguides, but tend to be more brittle and stiffer than glasses which limits some applications. Nevertheless, both glass and crystalline hollow waveguides can withstand enormous power densities, unlike solid core fibers which undergo end face damage. Single crystal sapphire tubing has delivered megawatts of pulsed CO_2 laser power and nearly 2kW of CW CO_2 laser power with water cooling. The excess losses here are due to surface roughness, which is difficult to avoid, causing energy transfer into higher order modes which are more easily attenuated. While there is no core to damage, slight misalignment can cause damage to the input end, as can any high order modes (non-TEM_{00}) present in the laser at power levels of a few watts. The waveguides also exhibit higher attenuation when coupling to incoherent light sources. Nevertheless, the losses are tolerable for certain short length applications and the waveguides are excellent for high laser power delivery if properly aligned and only launching the lowest order mode into the fiber. Considering that metal tubes possess rough inner surfaces, and glass tubes possess smooth surfaces but high losses due to absorption, recent work has focused on dielectric coated metallic waveguides on glass tubing substrates. Bore sizes less than 500 μm can be made with losses approaching 0.1 dB/m over longer fiber lengths than was previously attainable. The coatings inside the tubing can be modified to select the wavelength of maximum transmission for laser power delivery (e.g., 2.94 μm or 10.6 μm) or produce broad band transmission from about 3 μm to beyond 15 μm for spectroscopy or thermometry applications.

While numerous infrared applications have been described in the previous chapters, there are many more applications that will be enabled depending upon the availability of low-loss fibers, single mode fibers, rare-earth doped fibers, coherent, and incoherent fiber bundles. Figure 9.3 shows a schematic representation of the IR fiber applications which can be separated into *passive* and *active*. The *passive* applications are where the fiber acts as a conduit for the IR light from one location to another and without significantly altering the optical signal, other than that based on the fiber attenuation itself. Some examples of this include optical telecommunications, laser power delivery, thermal imaging, and chemical sensing for military,

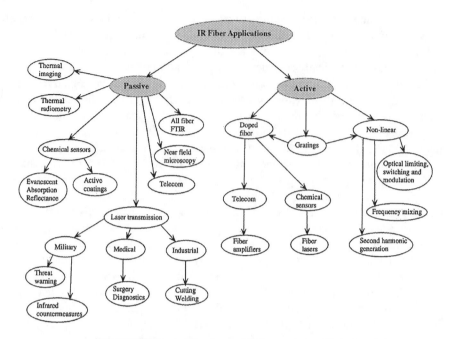

FIGURE 9.3   Applications of IR transmitting fibers.

medical, and commercial/ industrial systems. Other avenues include all-fiber FTIRs and near field IR microscopes for spectroscopy. In the case of *active* applications, the light which is incident on the fiber input face is modified in a controlled manner as it propagates through the IR fiber. This can be achieved by rare earth doping, writing gratings, and high power laser energy. Rare-earth ions can be used to capitalize upon their IR fluorescence for making IR amplifiers, bright sources, and lasers. As pointed out previously, a fiber amplifier at 1.3 µm would have a significant impact in telecommunications. Alternatively, excitation with high power light can lead to interesting nonlinear effects ranging from optical limiting, switching, frequency mixing, and second harmonic generation, to name but a few. It is not unreasonable to expect that the area of electrical poling of IR fiber materials will be investigated for enhancing the nonlinear properties for device applications. An exciting area which has enabled many device applications in silica fibers is the use of Bragg gratings. These will no doubt be investigated and exploited in the other IR fiber systems.

In summary and conclusion, the optical losses have steadily decreased and tremendous progress has been made in improving the strength of the fibers in the past several years, resulting in numerous applications and actual commercial products. We strongly believe that IR fiber optics will become increasingly more important in the future as further improvements are made to the quality of the fibers and new compositions developed. One of the most exciting developments in the future is going to be in the area of rare-earth ion doping of fibers for IR fluorescence emission. The IR light sources, lasers, and amplifiers developed using this phenomenon will be very useful in civil, medical, and military applications. Remote IR spectroscopy and imaging using flexible fibers will be realized for medical and military applications. Other future research areas which will inevitably be explored include the gratings and nonlinear optical properties and fabrication of planar waveguides, using primarily the IR glasses. The future of IR materials and fibers as described in this book looks very bright.

Finally, we would like to thank all those who participated in the preparation of this book, especially the authors for their patience and diligence.

Index